# Formal Methods for Multi-Agent Feedback Control Systems

## Cyber-Physical Systems Series

*Calin Belta, Editor*

A complete list of books published in the Cyber-Physical Systems series appears at the back of this book.

# Formal Methods for Multi-Agent Feedback Control Systems

Lars Lindemann and Dimos V. Dimarogonas

The MIT Press
Cambridge, Massachusetts
London, England

The MIT Press
Massachusetts Institute of Technology
77 Massachusetts Avenue, Cambridge, MA 02139
mitpress.mit.edu



The MIT Press would like to thank the anonymous peer reviewers who provided comments on drafts of this book. The generous work of academic experts is essential for establishing the authority and quality of our publications. We acknowledge with gratitude the contributions of these otherwise uncredited readers.

This book was set in LaTeX by Westchester Publishing Services. Printed and bound in the United States of America.

Library of Congress Cataloging-in-Publication Data

Names: Lindemann, Lars, author. | Dimarogonas, Dimos V., author.
Title: Formal methods for multi-agent feedback control systems / Lars Lindemann and Dimos V. Dimarogonas.
Description: Cambridge, Massachusetts : The MIT Press, [2025] | Series: Cyber-physical systems series | Includes bibliographical references and index.
Identifiers: LCCN 2024028681 (print) | LCCN 2024028682 (ebook) | ISBN 9780262049719 | ISBN 9780262382793 (pdf) | ISBN 9780262382809 (epub)
Subjects: LCSH: Multiagent systems. | Formal methods (Computer science)
Classification: LCC TJ215.5 .L56 2025 (print) | LCC TJ215.5 (ebook) | DDC 006.30285/436–dc23/eng/20241205
LC record available at https://lccn.loc.gov/2024028681
LC ebook record available at https://lccn.loc.gov/2024028682

10 9 8 7 6 5 4 3 2 1

EU product safety and compliance information contact is: mitp-eu-gpsr@mit.edu

To Friedrich, Inge, Dominik, and Lauren

*Lars Lindemann*
May 27, 2024

To Laura and Thomas

*Dimos V. Dimarogonas*
May 27, 2024

# Contents

# II Multi-Agent Control Barrier Functions for Spatiotemporal Constraints 89



# III Multi-Agent Funnel Control for Spatiotemporal Constraints 161

# List of Figures

# Foreword

I feel honored by the invitation to write a foreword for the research monograph "Formal Methods for Multi-Agent Feedback Control Systems," written by Lars Lindemann and Dimos V. Dimarogonas. The stated goals of this book are to provide an accessible introduction to formal methods for control, to present the authors' research results from recent years on the design of control algorithms for multi-agent systems with formal system specifications, and to highlight future research directions and inspire new generations to work at the intersection of formal methods and control. The research monograph delivers handsomely on all three.

The authors are major contributors in an active community of researchers pushing the boundaries in formal methods and control theory to build systems that satisfy complex system specifications. Formal methods allow us to reason rigorously about complex behavior, providing a way to express systematically reliability requirements, tackle system design to meet such specifications, and formally ensure them using verification techniques. However, the inherent difficulty in dealing with formal methods is computational complexity: ensuring system reliability faces scalability issues. Discretizations or abstractions quickly result in state space explosion, and this is particularly present when dealing with multi-agent systems like the ones considered in this book. This computational bottleneck limits the applicability of formal methods and sets the stage for the incorporation of new and scalable control design principles. Control theory seeks to design feedback laws for dynamical systems to achieve objectives such as stability or forward invariance, while providing robustness against various forms of uncertainty.

As this book demonstrates, embracing the continuous-time nature of physical systems, along with insights from dynamics and control, can be tremendously helpful in addressing the challenges related to the computational bottlenecks of formal methods. Besides scalability, the use of control methods also facilitates achieving robustness in the system design, another central challenge in the provably correct design of multi-agent systems operating without human intervention and with formal correctness guarantees.

The book is divided into five parts. Part I covers a diverse set of concepts and tools from dynamical systems and nonlinear control, geared toward multi-agent systems, and it also provides an introduction to formal methods and spatiotemporal logics. The latter choice of logic is rich enough to specify complex, sophisticated behaviors, and yet it is amenable to formal analysis. Parts II and III of the book are devoted to different, alternative approaches to multi-agent control (part II to control barrier functions, part III to funnel control), and both discuss centralized and distributed algorithms. Part IV is devoted to planning under spatiotemporal logic specifications. The authors explain how the feedback-control design techniques of parts II and III can only satisfy fragments of spatiotemporal logic specifications. This naturally sets the stage for part IV, where the book describes ways to decompose complex spatiotemporal logic specifications into a sequence of simpler spatiotemporal logic specifications that can be met by the sequential application of the feedback control techniques in the previous parts. Last but not least, part V provides an outlook of future research directions and outstanding problems pertaining a wide range of topics on communication, estimation, control, task decomposition, applications, and toolboxes. This is a part by which the reader can get easily inspired to jump into important questions for multi-agent systems at the intersection of formal methods and control.

It is challenging to write a book that combines disparate areas such as nonlinear control, multi-agent systems, formal methods, spatiotemporal logics, and timed automata. Beyond the obvious challenges (e.g., self-contained exposition, consistent use of the notation), one faces the dangers of omitting key context when discussing one particular area that, being standard in one community, might be completely unknown to another. It is undeniable that the authors have given careful consideration to this. Chapter 1 already sets a high bar in terms of presentation by providing a nice description of how the different parts of the book fit together, and how the material of each chapter builds on each other. The book does an excellent job at articulating what drivers are behind the use of formal methods in general and in control in particular. The presentation comes across nicely, paying careful attention to clarity, discussing the benefits and limitations of each of the proposed techniques, and sprinkling throughout the exposition key observations that help guide the reader's understanding. Myself, having worked on the multi-agent control side but, as a reader, only partially familiar with the formal methods side, found the exposition extremely helpful, to the point that sometimes it felt like the authors were anticipating what my questions were going to be and addressed them in the exposition.

We need more books like this one. This fantastic research monograph provides an excellent passage into an exciting area of research with strong theoretical foundations and significant practical applications. I hope readers enjoy reading the book as much as I did. The time invested in doing so will definitely pay off!

Jorge Cortés
University of California, San Diego
La Jolla, California

# Preface

## Motivation

This book is inspired by applications of cyber-physical and particularly multi-agent robotic systems. Cyber-physical systems integrate physical and software components and can be understood as hybrid systems (i.e., systems whose state consists of discrete and continuous states). Cyber-physical systems can be observed in almost every part of our everyday lives, ranging from spacecraft to healthcare devices. As mobile communication and computation devices become more available and affordable, such systems form interacting and dependent, often large-scale networks. We refer to these networks of systems as **multi-agent control systems**. Examples of multi-agent control systems can be found in manufacturing, agriculture, connected and smart cities, transportation, and multi-agent robotics, such as drone fleets for surveillance, automated warehouses, and mobile service robots. In fact, multi-agent control systems allow for addressing problems that cannot be addressed using single-agent systems, such as aerial surveillance of large areas by drone fleets or coordinated grasping of objects by multiple robots. Another distinct advantage of a multi-agent control system is redundancy, and thus their inherent robustness against individual agent failures. This book is particularly geared toward the provably correct design of autonomous, multi-agent control systems (i.e., systems that operate without human intervention and enjoy formal correctness guarantees).

In many cases, multi-agent control systems are safety-critical (e.g., when operating in the vicinity of humans), so system reliability becomes one of the main concerns during the design process. The increasing system complexity, however, makes it more challenging to verify the correct and safe behavior of the system. At the same time, one is interested in imposing and verifying more complex and sophisticated system specifications. The **formal methods** community, mostly using machinery from various areas of computer science, is tackling these challenges. One of their main tools are formal languages, such as spatiotemporal logics, to express system requirements that are then verified against the system under consideration using formal verification techniques, such as model checking. Formal system specifications may include nested combinations of temporal and spatial requirements; for instance, a fleet of drones may be required to repeatedly inspect particular areas of interest within certain time intervals and change its formation pattern over time, while ensuring safety at all times. In **control theory**, the goal is to explicitly design feedback control laws for dynamical systems, typically modeled using first-order principles from physics, to achieve control objectives such as stability or forward invariance of sets. The particular strength of feedback control is that feedback provides robustness against various forms of uncertainty.

Over the past two decades, a new community has formed that is interested in combining theories and tools from formal methods and control theory to build control systems that satisfy complex system specifications, and to answer questions that arise at the intersection of these two fields. In fact, formal methods–based control has attracted much attention, and a large number of invited sessions and workshops on this topic have emerged at the premier conferences over the past years. Moreover, graduate-level courses are beginning to be taught. While the developed verification and design techniques so far mainly use automata theoretical concepts (see, e.g., [43] and [287]), it is a well-known fact by now that these theories and tools are associated with a high computational cost of not scaling beyond a very small number of agents. Arguably, the biggest challenge in the field of **formal methods for multi-agent control systems** is of computational nature that requires a rethinking and the development of new and scalable control design principles. Besides scalability, another central challenge is to achieve robustness in the system design. With this in mind, the overall motivation for this book is to design control algorithms for multi-agent control systems under formal system specifications, with a focus on scalability and robustness.

# Objectives

The main objectives of this book are threefold. First, the book aims to present an accessible introduction to formal methods for control. Readers with a formal methods background can use this book to get an understanding of the control tools that are needed to solve related engineering problems, with a focus on scalability and robustness for multi-agent control systems. Readers with a control background can get familiar with the concepts of formal languages and spatiotemporal logics, as well as how these can be used to formulate system requirements. Second, the book aims to present the authors' research results from the past years. Notably, and with the motivation given here, we take a different approach than [43] and [287] based on our recent scientific findings, and we hope that this gives both the experienced and unexperienced reader some interesting new insights. Compared to [43] and [287], the main differences in our approach are:

- We especially focus on multi-agent control systems.
- We have a particular focus on integrating and using feedback control design tools to obtain scalability and robustness.
- We consider spatiotemporal logics instead of less expressive temporal logics.

As the book is based on the own work of the authors, the presentation is naturally biased. We reference, whenever needed and appropriate, relevant literature from the field and discuss differences and similarities. Third, the goal of this book is to highlight future research directions and to inspire young readers to work on topics at the intersection of formal methods and control.

# Intended Audience

This book is intended for a broad audience of scientists, engineers, and students with interests in formal methods and control for multi-agent control systems. In particular, we hope that the book will find a readership among graduate students who are pursuing their studies at the intersection of formal methods and control.

This book is neither an introductory book on formal methods nor an introductory book on systems and control theory. Rather, it is for those interested in working at the intersection of these topics and with an interest in solving multi-agent control problems. The book is more geared toward theoretical results, but it also provides algorithmic implementations. With this book, we hope to (1) provide an accessible introduction to formal methods for control of multi-agent systems, and (2) provide a basic understanding of theory, algorithms, and challenges that may arise when working in this problem space. We also hope to point the interested reader to open research questions and future research directions.

This book is mainly intended for mathematically inclined scientists, engineers, and students. General prerequisites involve some knowledge in multivariable calculus and real analysis. However, this book is self-contained, and no background in control theory or formal methods is required. Background in automata theory, dynamical systems, and convex optimization may be useful but are not required.

## Acknowledgments

This book is based on the results from Dr. Lars Lindemann's PhD studies that were conducted at KTH Royal Institute of Technology during the years 2016–2020. The writing of the book took place during the years 2021–2023. During this time, Dr. Lindemann and Dr. Dimarogonas were fortunate to be supported by the European Research Council, the Swedish Research Council, and the Knut and Alice Wallenberg Foundation.

We would like to sincerely thank Dr. Jorgé Cortes for providing the foreword for this book. We found the foreword to be thoughtful and rewarding, acknowledging our efforts and encouraging us to pursue similar projects in the future.

This book would not have been possible without interactions, support, and suggestions from a number of colleagues. In particular, we would like to thank Dr. Maria Charitidou, Adrian Wiltz, Dr. Fei Chen, Farhad Mehdifar, and Dr. Xiao Tan for their feedback on the book.

We also would like to thank Elizabeth Swayze and Matthew Valades from the MIT Press for supporting this book during the production process. In addition, we would like to thank Dr. Calin Belta for establishing the initial contact with the MIT Press and for encouraging us to pursue this project.

# I

# Foundations

# Chapter 1

# Introduction

This chapter provides a brief introduction to the general topic of this book. We will present motivating examples and introduce multi-agent systems and formal methods on an abstract level. In the end of the chapter, we provide an outline of the book and refer the interested reader to related textbooks.

## 1.1 Multi-Agent Control Systems

Cyber-physical systems integrate physical and software components and particularly deal with sensing, computation, communication, and control of physical and engineered systems. These systems form increasingly interacting and dependent networks, which we refer to as **multi-agent control systems**. Cyber-physical systems have enormous potential to transform our everyday lives, but at the same time, there are many grand challenges that we need to address [240]:

> Just as the internet transformed how humans interact with one another, cyber-physical systems will transform how we interact with the physical world around us. Many grand challenges await in the economically vital domains of transportation, health-care, manufacturing, agriculture, energy, defense, aerospace and buildings. The design, construction and verification of cyber-physical systems pose a multitude of technical challenges that must be addressed by a cross-disciplinary community of researchers and educators.

In this book, our goal is to focus on the design of provably correct control algorithms for networks of such systems following a cross-disciplinary approach. In other words, we are interested in multi-agent control systems in which we can manipulate the agents' behavior by a suitable choice of control variables such as thrust and torques of a robot to achieve a desired goal. Formally, a multi-agent control system is a collection of agents with individual actuation, computation, sensing, and communication capabilities. In this definition, an agent is hence an autonomous system (i.e., a system that operates without the intervention of a human operator) that can interact with other agents [324]. Multi-agent control systems can accomplish tasks that single-agent systems cannot address (e.g., aerial surveillance of large areas by a group of drones). In addition, the availability of multiple agents provides robustness against individual agent failures (e.g., when a single agent breaks down, another agent can account for this failure). Examples of multi-agent control systems can be found in applications such as transportation (autonomous fleets of trucks) or multi-agent robotics (mobile warehouse robots). A few examples of multi-agent robotic systems from our labs at the KTH Royal Institute of Technology and the University of Southern California (USC) are shown in figure 1.1.

In our studies, we are especially driven to achieve a high level of autonomy and safety in multi-agent control systems. In fact, for autonomous multi-agent control systems to function in practice, formal reasoning and guarantees about their performance and safety are needed, as deviations from their intended behavior may have severe consequences on human health and economics [8]. To formally reason about complex system behavior, we will introduce and use formal languages such as spatiotemporal logic that provide a convenient way of formalizing system requirements. Finally, we want to mention that algorithms solving the aforementioned challenges should ideally be robust and usable in real time.

## 1.2 Formal Methods for Multi-Agent Control Systems

In the previous section, we emphasized the need to reason formally about complex system behavior. Edmund Clarke, a famous leader and pioneer in the field of formal methods, advocated throughout his life for the use of formal methods to analyze systems. In a survey paper [79], he states:

**Figure 1.1**
Multirobot systems in the Smart Mobility Lab at KTH and the Quadrotor Swarm Lab at USC. Top pictures are by the authors, from the Smart Mobility Lab. Bottom pictures provided by Gaurav Sukhatme.

> Hardware and software systems will inevitably grow in scale and functionality. Because of this increase in complexity, the likelihood of subtle errors is much greater. Moreover, some of these errors may cause catastrophic loss of money, time, or even human life. A major goal of software engineering is to enable developers to construct systems that operate reliably despite this complexity. One way of achieving this goal is by using formal methods, which are mathematically based languages, techniques, and tools for specifying and verifying such systems.

System errors can lead to critical system failures. One example of such a tragic occurrence is the Ariane space flight V88 accident in June 1996, which was caused by software that was not correctly handling integer overflows. Other recent examples are the Boeing 737 MAX accidents in 2018 and 2019, which were caused (at least in part) by Boeing's flawed maneuvering characteristics augmentation system, a software system to control the airplane's horizontal stabilizers. Formal methods are defined as mathematical techniques for the specification, design, and verification of software and hardware systems. In particular, formal languages allow for the systematic specification of system requirements, and formal verification and synthesis techniques allow for the subsequent correct system design. These techniques are formal because system specifications are grounded in mathematical logic that allow for a formal deduction of system correctness in that logic [79, 323, 1].

The inherent difficulty with using formal methods is their computational complexity. The general problems considered in this area are NP-hard (more details are provided later in this book). To solve these problems, existing techniques explore all (or at least large parts) of the state space of the system. For complex systems, such as multi-agent systems, the state space can be extremely large and consist of millions of states. In addition, most existing techniques require a discretization, often also referred to as "abstraction," of the system at hand to lift the problem to a discrete and combinatorial one where we do not directly use continuous control design techniques, as we pursue in this book. For instance, model checking techniques represent both the system and the system specifications as finite state automata and result in the infamous state explosion problem, as an automata product between these two automata is required [27]. This **computational bottleneck** limits the applicability of existing formal methods verification and synthesis techniques and renders them infeasible for multi-agent control systems.

To address this computational challenge, this book presents a **new approach for the scalable and decentralized design of multi-agent control systems** under formal specifications. The ideas that we present in this book for accomplishing this task are thus different from existing ones. We aim to design scalable feedback control laws in parts II and III of this book, borrowing concepts from control theory, that induce a desired behavior into the closed loop of the system (i.e., the system under the designed control law) to satisfy spatiotemporal logic constraints. Spatiotemporal logic constraints impose requirements on the temporal and spatial behavior of the system, a very simple example being that three drones should establish a desired formation in no later than five seconds. The benefit of feedback control laws is their computational efficiency such that feedback control techniques scale better and are applicable to multi-agent control systems. Another benefit is the robustness of feedback control laws against system uncertainties. We also will discuss how we can decentralize the computation of these feedback control laws so each agent can calculate its control input locally. We will, however, not focus on issues related to communication in multi-agent control systems.

Naturally, we would like these feedback control laws to be as expressive as possible (i.e., to be able to satisfy a large set of spatiotemporal logic constraints). However, there are natural limitations to the expressiveness that we discover and discuss along the way in parts II and III of the book. To address these limitations, we finally present automata-based planning strategies in part IV of the book, in which we use the previously designed feedback control laws to satisfy rich classes of spatiotemporal logic constraints. This combination enables us to avoid the construction of an automata product, as is often needed in related work. The benefit is now that the complexity does not directly scale with the number of agents, as in existing approaches, but merely with the complexity of the spatiotemporal logic constraint, which only indirectly depends on the number of agents.

## 1.3 Outlook

In this book, we present an alternative approach to existing textbooks, which relies on discrete abstractions of the underlying system (see, e.g., [43] and [287]). We will design scalable feedback control laws and planning techniques for multi-agent control systems under spatiotemporal logic specifications. The spatiotemporal logic that we choose is signal temporal logic (STL)—a logic that is commonly used in the control and formal methods communities due to its simple yet expressive syntax. In addition, STL comes with a robustness score that indicates how robustly a specification is satisfied. STL specifications may be assigned locally, to individual agents, or globally, to the multi-agent control system as a whole.

Part I, including chapters 1, 2, and 3, presents the foundations for what is to come in this book. In particular, chapter 1 already has given a high-level introduction to the topic, while chapters 2 and 3 will present background on **dynamical systems, nonlinear control theory, and multi-agent control systems** and on **formal methods, spatiotemporal logics, and timed automata theory**, respectively.

Part II, including chapters 4 and 5, presents **time-varying control barrier functions (CBFs) for spatiotemporal constraints**. As mentioned earlier, we focus on specific fragments of spatiotemporal logic specifications, which impose requirements on the transient and steady-state behavior of the agents. Chapter 4 presents centralized CBFs for global, collaborative multi-agent system specifications, while chapter 5 presents decentralized CBFs for global, collaborative or local and potentially adversarial specifications (adversarial in the sense that the set of local specifications may not be satisfiable, as local specifications are assigned independently to each agent).

Part III, including chapters 6 and 7, presents **funnel control for spatiotemporal constraints**. While part II focused on time-varying CBFs, this part presents funnel control as an alternative way of enforcing transient constraints for multi-agent control systems. Chapter 6 presents centralized funnel control for global, collaborative multi-agent system specifications, while chapter 7 presents decentralized funnel control for local and adversarial specifications.

The feedback control techniques presented in parts II and III are based on time-varying CBFs and funnel control. These techniques follow the main idea of inducing a desired temporal behavior, according to the STL specification at hand, into the closed loop of the system. Alternative techniques are based on mixed-integer linear programming [241, 243] and nonconvex optimization [225, 226]. These techniques, however, do not apply to continuous-time control systems, as is the focus of this book, while they also may face computational challenges, as no closed-form control law can be obtained. The proposed control laws have the desirable property of being robust against uncertainties in the system dynamics, as well as couplings among agents. All the theoretical results are illustrated here in simulations and experiments involving mobile robots.

Part IV of the book, including chapter 8, deals with **planning under spatiotemporal logic specifications**. As we remarked before, the feedback control laws from parts II and III are limited in the sense that they can satisfy only fragments of spatiotemporal logic specifications. This is not surprising, as the general control problem that we consider is NP-hard and caused by natural limitations of feedback control that are well known and not only limited to our setting. Therefore, we decompose an STL specification into a sequence of simpler spatiotemporal logic specifications that can be realized by sequentially applying the previously designed feedback control laws. Chapter 8 shows how such a sequence can be obtained from a timed automaton that encodes the STL specification. The combination of timed automata-based planning and feedback control is efficient in the sense that the computationally expensive planning algorithm can be run offline, while only the efficient feedback control laws are run online.

The last part of the book, which includes chapter 9, concludes by presenting an outlook and listing open problems.

## 1.4 Notes and References

We now briefly review textbooks that are related to the theme of this work. In the main parts of the book, we detail peer-reviewed papers at the end of each chapter.

Closest to our work is the textbook [43] in which automata-based verification and control synthesis techniques are deployed to find fixed-point solutions to reachability/invariance automata games. Therefore, system abstraction techniques are employed using notions of simulation and bisimulation relation to abstract (infinite) state systems into finite state automata. Similar to [43], the textbook [287] focuses on system abstractions and uses automata games for verification and control synthesis, but without explicitly considering temporal logics. In this book, we will directly consider continuous time dynamical systems and not abstract the system.

Also related to our book are [165, 11, 211, 233], as they present detailed introductions to cyber-physical systems and verification of these systems. However, the focus of these works is fundamentally different, and the authors only briefly touch upon verification and control under temporal logic specifications, and without discussing multi-agent feedback control. The textbooks [27, 77] provide a detailed introduction to temporal logics, model checking, and system verification, but again with no focus on multi-agent systems. In [84], the verification of multi-agent systems under formal specifications is considered, while control design in such a framework remains challenging. All the aforementioned books consider temporal logics, while we focus on the more expressive class of spatiotemporal logics, allowing us to express quantitative temporal requirements such as deadlines.

A general introduction to multi-agent systems is provided in [324]. The textbook in [293] discusses computation and communication constraints on control and has a chapter dedicated to multi-agent control under temporal logic specifications. The textbook in [94] focuses on long-duration autonomy and is related to our work, as it considers Boolean composition of specifications. Also related are textbooks on multi-agent control systems, such as [209] and [57]. The focus in these books is on designing distributed feedback control laws for rather simple system specifications, such as consensus and formation control, but with a special focus on communication. Our book differs in the sense that the focus is not on communication, but instead on scalable and robust control algorithms for more complex system specifications formulated in spatiotemporal logics that go beyond consensus and formation control problems.

## 1.5 Notation

For later convenience, we now introduce a common notation system that is used throughout the book. Additional, less common notation will be introduced when needed.

Logical true and false are denoted by the symbols $\top$ and $\bot$, respectively. The Boolean domain, hence, is $\mathbb{B} := \{\top, \bot\}$. Let $\mathbb{R}$ denote the set of real numbers, while $\mathbb{R}^n$ is the $n$-dimensional real vector space. The sets of nonnegative and positive real numbers are $\mathbb{R}_{\geq 0}$ and $\mathbb{R}_{>0}$, respectively. Similarly, the set of rational numbers is denoted by $\mathbb{Q}$, while the sets of nonnegative and positive rational numbers are $\mathbb{Q}_{\geq 0}$ and $\mathbb{Q}_{>0}$, respectively. The sets of natural numbers and integers (both including 0 as an element) are denoted by $\mathbb{N}$ and $\mathbb{Z}$, respectively.

Let $\text{bd}(\mathcal{X})$, $\text{int}(\mathcal{X})$, and $\text{cl}(\mathcal{X})$ denote the boundary, interior, and closure of a set $\mathcal{X} \subseteq \mathbb{R}^n$. For two sets $\mathcal{X}_1, \mathcal{X}_2 \subseteq \mathbb{R}^n$, we denote the Minkowski sum by $\mathcal{X}_1 \oplus \mathcal{X}_2$ and the Minkowski difference by $\mathcal{X}_1 \ominus \mathcal{X}_2$. Formally, we have

$$\mathcal{X}_1 \oplus \mathcal{X}_2 := \{x_1 + x_2 \in \mathbb{R}^n | x_1 \in \mathcal{X}_1, x_2 \in \mathcal{X}_2\},$$
$$\mathcal{X}_1 \ominus \mathcal{X}_2 := \{x_1 - x_2 \in \mathbb{R}^n | x_1 \in \mathcal{X}_1, x_2 \in \mathcal{X}_2\}.$$

The Euclidean norm of a vector $x \in \mathbb{R}^n$ is denoted by $\|x\| := \sqrt{x^T x}$, where $x^T$ is the transpose of $x$. The infinity norm of a vector $x \in \mathbb{R}^n$ is $\|x\|_\infty := \max(|x_1|, \ldots, |x_n|)$, where $|x_i|$ denotes the absolute value of the $i$th element $x_i$ of $x$. The sum norm of a vector $x \in \mathbb{R}^n$ is $\|x\|_1 := |x_1| + \ldots + |x_n|$. More generally, and for $p \geq 1$, the $p$-norm is defined as $\|x\|_p := \big(|x_1|^p + \ldots + |x_n|^p\big)^{\frac{1}{p}}$.

Let $f : \mathbb{R}^n \to \mathbb{R}$ be a real-valued and differentiable function. Then $\frac{\partial f(x')}{\partial x_i} := \frac{\partial f(x)}{\partial x_i}|_{x=x'}$ denotes the partial derivative of $f$ with respect to the $i$th component $x_i$ of $x$, evaluated at $x'$. Similarly, let $\frac{\partial f(x')}{\partial x} := \frac{\partial f(x)}{\partial x}_{|x=x'}$ be the gradient of $f$ evaluated at $x'$. By convention, we assume that the gradient is a row vector (unless stated otherwise). For a differentiable function $x : \mathbb{R}_{\geq 0} \to \mathbb{R}^n$, we denote the derivative of $x(t)$ with respect to $t$ as $\dot{x}(t) := \frac{\mathrm{d}x(t)}{\mathrm{d}t}$. In this book, we will refer to $x$ as a signal or trajectory and think of $x$ as the solution to an initial value problem. With this in mind, we will interpret parameter $t$ as time. Accordingly, we can compute the time derivative of function $V : \mathbb{R}^n \times \mathbb{R}_{\geq 0} \to \mathbb{R}$, evaluated at $(x(t), t) \in \mathbb{R}^n \times \mathbb{R}_{\geq 0}$, by using the chain rule as follows:

$$\dot{V}(x(t), t) := \frac{\mathrm{d}V(x(t), t)}{\mathrm{d}t} = \frac{\partial V(x(t), t)}{\partial x} \frac{\mathrm{d}x(t)}{\mathrm{d}t} + \frac{\partial V(x(t), t)}{\partial t}.$$

Finally, we define the inverse image of a function $f : \mathcal{X} \to \mathcal{Y}$ (for some arbitrary input and output domains $\mathcal{X}$ and $\mathcal{Y}$, respectively) under the set $\mathcal{M} \subseteq \mathcal{Y}$ as

$$\mathrm{inv}\big(f(\mathcal{M})\big) := \{x \in \mathbb{R}^n \,|\, f(x) \in \mathcal{M}\}.$$

# Chapter 2

# Multi-Agent Systems and Control Theory

In the first two sections of this chapter, we provide background on nonlinear dynamical systems and selected topics from nonlinear control theory that will be relevant in the study of this book. Our first goal is to make the reader familiar with ordinary differential equations (ODEs) as our modeling formalism, and to illustrate how such a model, together with an initial value of a nonlinear system, can describe the motion of the system. We then introduce the concepts of forward invariance and stability to capture distinct desirable system behaviors. Our second goal is to provide an understanding of how we can design feedback control laws so that a nonlinear dynamical system satisfies a priori given state constraints (e.g., to avoid unsafe regions in the state space). For this purpose, we present a general introduction to nonlinear control theory, but specifically focus on control barrier functions (CBFs), control Lyapunov functions (CLFs), and funnel control as our main control design tools. While we do not require the reader to be familiar with linear dynamical systems, we refer the interested reader to the textbooks [129, 23, 68] or the excellent online lectures by Stephen Boyd [131] for a detailed exposition.

In the third section of this chapter, we provide background on multi-agent control systems, which are defined as a collection of coupled dynamical systems. Multi-agent control systems research has sparked interest in both industry and academia in areas such as robotics (e.g., drone fleets, warehouse automation), transportation (e.g., truck platooning, autonomous driving), and manufacturing (e.g., collaborative grasping of objects), among others. Multi-agent control systems are essentially high-dimensional dynamical systems (i.e., systems with a large state space). The high-dimensional nature of these systems creates a computational bottleneck, which makes the study and design of said systems challenging. Our goal in this section is to introduce multi-agent control systems and give an overview of multi-agent objectives to set the stage for the main theme of this book.

## 2.1 Dynamical Systems

In this section, we first show how we can model dynamical systems, such as mobile ground robots or drones, in continous time by using ODEs. This book is solely focused on the continuous-time case, and we will not consider the discrete-time case, in which a dynamical system would be modeled by a difference equation. Together with an initial condition $x_0$ of the system, this ODE results in an initial value problem (IVP). A solution to the IVP describes the motion of the system, starting from this initial condition. We therefore study the existence, uniqueness, and completeness of solutions to IVPs. Finally, we introduce the notions of stability and safety and show how the stability and safety of a dynamical system can be certified using Lyapunov and barrier functions, respectively.

### 2.1.1 Ordinary Differential Equations as System Models

At time $t \in \mathbb{R}_{\geq 0}$, let $x(t) \in \mathbb{R}^n$ be the state of the nonlinear dynamical system that is described by the ODE:

$$\dot{x}(t) = f(x(t), t) + g(x(t), t)u(x(t), t), \tag{2.1}$$

where the functions $f : \mathcal{D} \times \mathbb{R}_{\geq 0} \to \mathbb{R}^n$, $g : \mathcal{D} \times \mathbb{R}_{\geq 0} \to \mathbb{R}^{n \times m}$, and $u : \mathcal{D} \times \mathbb{R}_{\geq 0} \to \mathbb{R}^m$ denote the internal dynamics, the input dynamics, and the feedback control law by which the system is controlled, respectively. The set $\mathcal{D} \subseteq \mathbb{R}^n$ is nonempty and open, and describes the domain in which the system is supposed to operate. When $\mathcal{D}$ is not explicitly specified, we will assume that $\mathcal{D} := \mathbb{R}^n$ for convenience. The system in equation (2.1) is input-affine (i.e., the control input $u$ enters the system dynamics linearly).[1] Input-affine dynamics allow for modeling a broad class of nonlinear systems.

For readers unfamiliar with ODEs, an intuitive way to think of how an ODE describes the motion of a system is as follows: Assume that we know the value of $x(t)$, and thus also the value of $\dot{x}(t)$ at time $t$. For an infinitesimally small time step $\delta$, the state at the next time is obtained as follows:

$$x(t+\delta) = x(t) + \delta \dot{x}(t).$$

The system in equation (2.1) is called linear (see, e.g., [68]) if it can be written as

$$\dot{x}(t) = A(t)x(t) + g(t)u(t), \tag{2.2}$$

where the input dynamics $g$ and the control law $u$ do not depend on the state $x$, and the internal dynamics $f$ are linear in the state $x$ and written as $f(x(t),t) := A(t)x(t)$ for a matrix-valued function $A : \mathbb{R}_{\geq 0} \to \mathbb{R}^{n \times n}$. Let us briefly look at an example of a linear system to get some insight into this.

---

**Example 2.1** Consider a double integrator, which is a model for a wide variety of physical systems (e.g., a mass pulled frictionless by a force in one direction), and hence often used as a canonical example in systems and control theory. A double integrator consists of two states, position and velocity, and is described by the following dynamics:

$$\dot{x}(t) = \begin{bmatrix} v(t) \\ u(t) \end{bmatrix} = \underbrace{\begin{bmatrix} 0 & 1 \\ 0 & 0 \end{bmatrix}}_{=:A} x(t) + \underbrace{\begin{bmatrix} 0 \\ 1 \end{bmatrix}}_{=:g} u(t),$$

where $x(t) := \begin{bmatrix} p(t) & v(t) \end{bmatrix}^T \in \mathbb{R}^2$ is the state at time $t$ consisting of position $p(t)$ and velocity $v(t)$, and the control input $u(t) \in \mathbb{R}$ directly controls the time derivative $\dot{v}(t)$ of the velocity $v(t)$ (i.e., the acceleration of $p(t)$). Note specifically that the double integrator is described by linear dynamics as in equation (2.2). By integrating the system dynamics, it is easy to see that the system trajectory $x$ starting at time $t = 0$ has to satisfy the following two equations:

$$p(t) = p(0) + \int_0^t v(s)\mathrm{d}s$$
$$v(t) = v(0) + \int_0^t u(s)\mathrm{d}s,$$

---

[1] We naturally refer to the ODE that describes the dynamics of a system as "system dynamics."

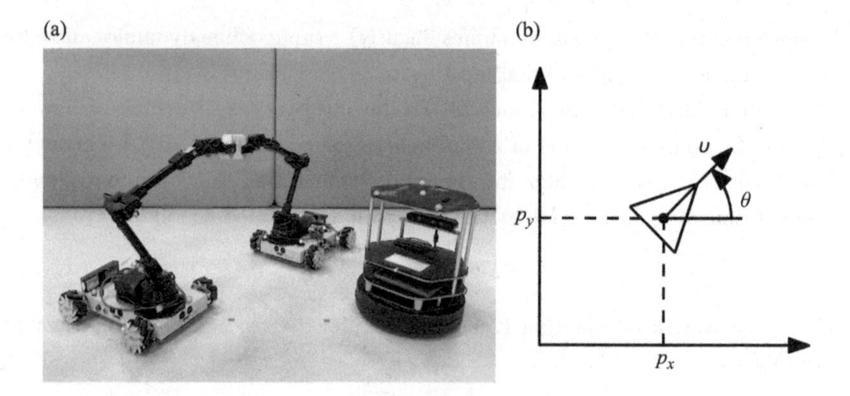


**Figure 2.1**
Mobile robots, particularly a Turtlebot (robot on the right in (a)) that can be modeled by unicycle dynamics; see schematic on (b).

highlighting the origin of the name "double integrator." A double integrator can be used as an abstract model for the planar motion of a mobile robot; for instance, see the two omnidirectional robots shown in figure 2.1(a). To do so, let us consider a two-dimensional double integrator system described by the dynamics

$$\dot{x}(t) = \begin{bmatrix} v_x(t) \\ u_x(t) \\ v_y(t) \\ u_y(t) \end{bmatrix} = \underbrace{\begin{bmatrix} 0 & 1 & 0 & 0 \\ 0 & 0 & 0 & 0 \\ 0 & 0 & 0 & 1 \\ 0 & 0 & 0 & 0 \end{bmatrix}}_{=:A} x(t) + \underbrace{\begin{bmatrix} 0 & 0 \\ 1 & 0 \\ 0 & 0 \\ 0 & 1 \end{bmatrix}}_{=:g} u(t), \tag{2.3}$$

where the state $x(t) := \begin{bmatrix} p_x(t) & v_x(t) & p_y(t) & v_y(t) \end{bmatrix}^T \in \mathbb{R}^4$ consists of the two-dimensional positions $p_x(t)$ and $p_y(t)$ and the two-dimensional velocities $v_x(t)$ and $v_y(t)$. The control input $u(t) := \begin{bmatrix} u_x(t) & u_y(t) \end{bmatrix}^T \in \mathbb{R}^2$ again directly controls the acceleration of the system.

---

Linear systems are an important class of systems that are well understood, and for which powerful analysis and control design tools are available [129, 23, 68]. In this book, however, we focus on nonlinear systems as in equation (2.1), as it allows to model a broader class of systems. In addition, it will become apparent later in this book that the control input $u$ has to be a nonlinear function of the state $x$ if we want to enforce system-relevant safety constraints or spatiotemporal logic constraints.

The nonlinear system in equation (2.1) is time-invariant if $f$ (or $A$ in the linear case) and $g$ do not depend on time $t$; that is, the function values of $f$ and $g$ at time $t$ are given by $f(x(t))$ and $g(x(t))$. Note that the linear double integrator discussed in example 2.1 is time-invariant. While the nonlinear system in equation (2.1) is time-varying (i.e., not time-invariant), we assume for simplicity that $f$ and $g$ do not have any time dependency for most parts of this book. We stress, however, that time dependence of the control law $u$ will be pivotal. Indeed, one of the main goals in this book is to design time-varying feedback control laws $u$ such that the system satisfies complex system requirements, which will be given as spatiotemporal logic constraints. In other words, we will design time-varying feedback control laws to satisfy transient system constraints.

We next want to highlight an important distinction that can be made with respect to the control law $u$. If the control law $u$ has no dependency on the state $x$ (i.e., if the value of $u$ at time $t$ is determined by $u(t)$ only), then we say that the control law is open-loop, as it does not use feedback in the form of $x(t)$. Such control laws are also refered to as "feedforward control laws." It is even possible that the control law has no dependency on time $t$, so the control input is a constant. However, if the value of $u$ at time $t$ is determined by $u(x(t), t)$, then we say that the control law is closed-loop and does use feedback. We will elaborate more on the concept of feedback later in this chapter.

To get an intuition for the proposed system modeling formalism, we next present two examples of nonlinear dynamical systems.

---

**Example 2.2** Consider again the mobile robots shown in figure 2.1(a). In particular, consider the Turtlebot on the right, which we can model by the nonlinear unicycle dynamics

$$\dot{x} = \begin{bmatrix} v\cos(\theta) \\ v\sin(\theta) \\ \omega \end{bmatrix} = \underbrace{\begin{bmatrix} 0 \\ 0 \\ 0 \end{bmatrix}}_{=:f(x)} + \underbrace{\begin{bmatrix} \cos(\theta) & 0 \\ \sin(\theta) & 0 \\ 0 & 1 \end{bmatrix}}_{=:g(x)} u, \tag{2.4}$$

where $x := \begin{bmatrix} p_x & p_y & \theta \end{bmatrix}^T$ is the state of the system consisting of its two-dimensional positions $p_x$ and $p_y$, as well as its orientation $\theta$, and where $u := \begin{bmatrix} v & \omega \end{bmatrix}^T$ is the control input consisting of translational and rotational velocities $v$ and $\omega$, respectively. Note that for convenience, we have dropped the time dependency of $x(t)$, as well as the time and state dependency of $u(x, t)$ in equation (2.4).[2] A schematic of this system is also shown in figure 2.1(b).

---

[2]We will continue dropping the time dependency in $x$ and the time and state dependency in $u$ throughout this book when it is appropriate and clear from the context.

**Example 2.3** Based on the previously presented unicycle dynamics, let us next present a model for the lateral control of a vehicle with body length $L$ and fixed velocity $v$. In this case, we can use the bicycle dynamics

$$\dot{x} = \begin{bmatrix} v\cos(\theta) \\ v\sin(\theta) \\ \frac{v}{L}\tan(\delta) \end{bmatrix} = \underbrace{\begin{bmatrix} v\cos(\theta) \\ v\sin(\theta) \\ 0 \end{bmatrix}}_{=:f(x)} + \underbrace{\begin{bmatrix} 0 \\ 0 \\ \frac{v}{L} \end{bmatrix}}_{=:g} u,$$

where $x := \begin{bmatrix} p_x & p_y & \theta \end{bmatrix}^T$ is again the state of the system consisting of its two-dimensional position, as well as its orientation, and where $u := \tan(\delta)$ is the control input with steering angle $\delta$, see also [230].

## 2.1.2 Initial Value Problems

The ODE in equation (2.1) describes the dynamical system under consideration, as discussed in the previous section. However, to be able to talk about specific realizations of this system, we have to take its initial condition $x_0 \in \mathbb{R}^n$ into account. Together, the ODE and the initial condition result in the following IVP:

$$\dot{x}(t) = f(x(t),t) + g(x(t),t)u(x(t),t), \quad x(0) = x_0 \in \mathcal{D}. \tag{2.5}$$

In this section, we discuss conditions under which a solution to an IVP exists, and under which conditions such a solution is unique and complete (i.e., defined at all times). Let us assume throughout this section and the next one that we are given a control law $u$ (i.e., the function $u$ is fixed). For such a given control law $u$, the IVP in equation (2.5) can more compactly be written as

$$\dot{x}(t) = H(x(t),t), \; x(0) = x_0 \in \mathcal{D}, \tag{2.6}$$

where the function $H : \mathcal{D} \times \mathbb{R}_{\geq 0} \to \mathbb{R}^n$ is defined as

$$H(x,t) := f(x,t) + g(x,t)u(x,t).$$

We next discuss what conditions a function $x(t)$ has to satisfy in order to be a solution to the IVP in equation (2.6). While this appears to be a formal definition, it is an important one, as a solution to the IVP in equation (2.6) factually describes actual system behavior that can be used to analyze a system (e.g., in terms of safety of the system). In fact, one can think of a solution to the IVP in equation (2.6) as a trajectory of the dynamical system.

**Definition 2.1 (Solution to the IVP in Equation (2.6))** *A solution to the IVP in equation (2.6) over a time interval $\mathcal{I} \subseteq \mathbb{R}_{\geq 0}$ is an absolutely continuous function $x : \mathcal{I} \to \mathcal{D}$ that satisfies the following two conditions:*

- $x(0) = x_0$ *(initial condition), and*
- $x(t) = x_0 + \int_0^t H(x(s), s)\mathrm{d}s$ *for all* $t \in \mathcal{I}$ *(integral condition).*

In this definition, we assumed that the function $x$ is absolutely continuous, and then we considered the integral condition based on the integral of the equation $\dot{x}(t) = H(x(t), t)$ from equation (2.6). A solution to the IVP in equation (2.6) hence satisfies $\dot{x}(t) = H(x(t), t)$ for almost all $t \in \mathcal{I}$. The choice of absolutely continuous functions $x$ as the solution is motivated to allow for systems described by functions $H$ that are discontinuous in their second argument (e.g., when a discontinuous control input is used that may be caused by switching system objectives). If the function $H$ is continuous, then the solution $x$ will be continously differentiable, and when the solution $x$ is continuously differentiable, we note that the integral conditions are equivalent to the derivative condition $\dot{x}(t) = H(x(t), t)$ for all $t \in \mathcal{I}$.

Let us now apply this solution concept to a one-dimensional linear system, as well as to the nonlinear unicycle dynamics from example 2.2.

---

**Example 2.4** Consider the linear system $\dot{x} = -5x + 1$, with initial condition $x_0 = 3$. For such a linear system, we can easily find the solution analytically as $x(t) = a\exp(-5t) + b$, where $b := 1/5$ and $a := x(0) - b = 14/5$. Note that $x(t)$ converges to the steady-state value $b$ as $t$ converges to $\infty$.

---

**Example 2.5** For the unicycle dynamics in equation (2.4) with constant control inputs $v(t) := 1$ and $\omega(t) := 0$ and initial condition $x_0 := \begin{bmatrix} p_{x,0} & p_{y,0} & \theta_0 \end{bmatrix}^T$, the solution is given by $x(t) = \begin{bmatrix} p_{x,0} + \cos(\theta_0)t & p_{y,0} + \sin(\theta_0)t & \theta_0 \end{bmatrix}^T$.

---

Note that we were able to analytically derive the solution $x$ in these two examples. For general nonlinear systems as in equation (2.6), we can rarely (indeed, almost never) obtain analytical solutions. Only for the special case of linear time-invariant systems,

$$\dot{x}(t) = Ax(t) + gu(t), \quad x(0) = x_0 \in \mathcal{D},$$

can we always find an analytical solution as

$$x(t) := e^{At}x_0 + \int_0^t e^{A(t-s)}gu(s)\mathrm{d}s,$$

where $e^X := \sum_{k=0}^{\infty} \frac{1}{k!}X^k$ denotes the matrix exponential of a matrix $X$. The fact that the solution to the IVP in equation (2.6) is not known in general makes the study of nonlinear systems challenging. As a consequence, nonlinear analysis and control design tools such as the ones that we discuss in this book reason about system properties without using explicit knowledge of the system solution.

Another important question in the study of nonlinear dynamical systems is when a solution $x$ is maximal, which is when the time domain $\mathcal{I}$ of the solution cannot be extended to the right so that the solution $x$ is not defined beyond the right end point of $\mathcal{I}$.

**Definition 2.2 (Maximal Solutions)** *A solution $x : \mathcal{I} \to \mathcal{D}$ to the IVP in equation (2.6) over the time interval $\mathcal{I} := [0, \tau_{max}) \subseteq \mathbb{R}_{\geq 0}$ with maximal time $\tau_{max} \in \mathbb{R}_{\geq 0}$ is a maximal solution to equation (2.6) if there is no other solution $\hat{x} : \hat{\mathcal{I}} \to \mathcal{D}$ to equation (2.6) over a time interval $\hat{\mathcal{J}} := [0, \hat{\tau})$ with maximal time $\hat{\tau} \in \mathbb{R}_{\geq 0}$, such that $\hat{\tau} > \tau_{max}$ and $x(t) = \hat{x}(t)$ for all $t \in [0, \tau_{max})$.*

Depending on the value of the maximal time $\tau_{\text{max}}$ in definition 2.2, a solution $x$ is either defined for a bounded time or for all times. Closely related to the concept of a maximal solution hence is the concept of a complete solution, which is a solution where the time interval $\mathcal{I}$ is unbounded and the solution is defined for all times (as is often required).

**Definition 2.3 (Complete Solutions)** *A solution $x : \mathcal{I} \to \mathcal{D}$ to the IVP in equation (2.6) over the time interval $\mathcal{I} := [0, \tau_{max}) \subseteq \mathbb{R}_{\geq 0}$ with maximal time $\tau_{max}$ is complete if $\tau_{max} = \infty$.*

While maximal solutions of linear systems are guaranteed to be complete, this is not necessarily the case for nonlinear systems. We next present an example that illustrates this point.

---

**Example 2.6** Consider the nonlinear system $\dot{x} = x^2$, with initial condition $x_0 := 1$. The solution to this IVP is $x(t) = -1/(t-1)$ and hence is only defined on $\mathcal{I} := [0, 1)$. Therefore, the solution is maximal but not complete. We also say that the solution has finite escape time. On the other hand, when the initial condition is set to $x_0 := 0$, the solution to this IVP is $x(t) = 0$ and defined on $\mathcal{I} := [0, \infty)$ (i.e., maximal and complete).

---

So far, we have defined the concept of a solution to the IVP in equation (2.6) and discussed when a solution is maximal and complete. However, there are two fundamental questions that are still open. Is it always guaranteed that a solution to the IVP in equation (2.6) exists? And if such a solution exists, is the solution unique or are there multiple solutions? To answer these questions, we next state a standard result regarding the existence and uniqueness of solutions to the IVP in equation (2.6) according to [281, theorem 54], which will be used extensively in this book.

**Lemma 2.1 (Existence of Unique Solutions)** *Assume that the function* $H:\mathcal{D}\times\mathbb{R}_{\geq 0}\rightarrow\mathbb{R}^n$ *is locally Lipschitz continuous in its first argument,*[3] *and piecewise continuous in its second arguments.*[4] *Then, there is a unique and maximal solution* $x:\mathcal{I}\rightarrow\mathcal{D}$ *to the IVP in equation (2.6) over the time interval* $\mathcal{I}:=[0,\tau_{max})\subseteq\mathbb{R}_{\geq 0}$ *for some maximal time* $\tau_{max}$.

When the first condition in lemma 2.1 is replaced by the weaker condition that the function $H$ is only continuous in its first argument (i.e., continuous in $x$ for each fixed $t\in\mathbb{R}_{\geq 0}$), solutions are guaranteed to exist, but they are not necessarily unique. We will illustrate this point in another example.

---

**Example 2.7** Consider the nonlinear system $\dot{x}=\sqrt{|x|}$ with initial condition $x_0:=0$. It turns out that the IVP has infinitely many solutions. In particular, for each time $t'\in[0,\infty)$, we have that

$$x(t)=\begin{cases}0 & \text{if } t\leq t'\\ (t-a)^2/4 & \text{if } t>t'\end{cases}$$

is a solution to the IVP under consideration. Note that the function $\sqrt{|x|}$ is not locally Lipschitz continuous at $x=0$ so that solutions are not necessarily unique, as discussed previously. However, the function $\sqrt{|x|}$ is continuous at $x=0$, which guarantees the existence of solutions. In contrast, and as studied in example 2.6, the IVP $\dot{x}=x^2$ with initial condition $x_0:=0$ has the unique solution $x(t)=0$, as the function $x^2$ is locally Lipschitz continuous (in fact, continuously differentiable) everywhere.

---

We now know under which conditions solutions to the IVP in equation (2.6) exist, which is important, as it give us information about the existence of meaningful system behavior described by an absolutely continuous function.

[3]This means locally Lipschitz continuous in $x$ for each fixed $t\in\mathbb{R}_{\geq 0}$.
[4]This means piecewise continuous in $t$ for each fixed $x\in\mathcal{D}$.

However, as of now, it is not clear under which conditions solutions are also complete (i.e., defined for all times). Completeness of solutions is important, as we expect a system solution to be defined for all times. How could we otherwise reason over system properties, such as safety? The next result, according to [281, proposition C.3.6], states a property that all noncomplete solutions must satisfy, which we will use later to design control laws for dynamical systems that are in fact complete.

**Lemma 2.2 (Property of Noncomplete Solutions)** *Let the conditions from lemma 2.1 hold. For a maximal solution $x:\mathcal{I}\rightarrow\mathcal{D}$ to the IVP in equation (2.6) with $\tau_{max}<\infty$ (i.e., $x$ is not complete), and for any compact set $\mathcal{D}'\subset\mathcal{D}$, there is $t'\in\mathcal{I}$ such that $x(t')\notin\mathcal{D}'$.*

An intuitive explanation of lemma 2.2 is that every noncomplete solution $x$ has to escape any compact set $\mathcal{D}'$ within the domain $\mathcal{D}$ in finite time. It can even be shown that lemma 2.2 still applies when the first condition in lemma 2.1 is not met (i.e., when the function $H$ is discontinuous in its first argument).

So far, we have discussed properties of solutions to an IVP. As mentioned before, the solution to an IVP is unfortunately rarely known and can only be approximately computed numerically. In the next two subsections, we will instead focus on system properties such as stability and safety, which we will analyze by looking at the associated ODE and for a range of initial conditions (i.e, locally and globally). In order to analyze stability and safety, we will resort to tools that analyze the instantaneous flow of a system instead of its system solutions.

### 2.1.3 Forward Invariance and Safety

Let us now discuss how we can define and certify safety for systems. We now define safety as the ability of a system to stay within a set of states that are labeled safe (e.g., states that satisfy a minimum safety distance). As the focus in this book is on designing systems that satisfy spatio-temporal logic constraints, we consider time-varying safe sets $\mathcal{C}(t)$. We use the concept of forward invariance to formally capture the ability of a system to stay within the safe set.

**Definition 2.4 (Forward Invariance)** *A set $\mathcal{C}(t)$ is forward invariant with respect to the IVP in equation (2.6) if, for each $x_0\in\mathcal{C}(0)$, each solution $x:\mathcal{I}\rightarrow$*

$\mathcal{D}$ *to equation (2.6) with* $x(0)=x_0$ *is complete and such that* $x(t)\in\mathcal{C}(t)$ *for all* $t\in\mathcal{I}$.

Stated in words, forward invariance of the set $\mathcal{C}(t)$ guarantees that if the system starts in $\mathcal{C}(0)$, the system will stay in $\mathcal{C}(t)$ for all times and never leave it. In order to show how we can certify forward invariance, and hence safety, we state two additional lemmas up front that we will invoke for the task at hand. We first introduce the Comparison Lemma (taken from [163, theorem 1.10.2]), which is extensively used in the proofs of this book and for the purpose of establishing forward invariance.

**Lemma 2.3 (Comparison Lemma)** *Let* $W:\mathbb{R}\times\mathbb{R}_{\geq 0}\to\mathbb{R}$ *be a function that is locally Lipschitz continuous in its first argument and piecewise continuous in its second argument, and describe the following IVP:*

$$\dot{w}(t)=W(w(t),t),\ w(0)=w_0\in\mathbb{R}.$$

*Let* $w:\mathcal{I}\to\mathbb{R}$ *be the unique and maximal solution to this IVP. Now let* $v:\mathbb{R}_{\geq 0}\to\mathbb{R}$ *be an absolutely continuous function, such that*

- $v(0)\geq w(0)$ *(initial condition), and*
- $\dot{v}(t)\geq W(v(t),t)$ *for almost all* $t\in\mathcal{I}$ *(derivative condition).*

*Then it holds that*

$$v(t)\geq w(t)\ \textit{for all}\ t\in\mathcal{I}.$$

Intuitively, the Comparison Lemma gives us a lower bound for the potentially unknown function $v$ in case we know that the derivative constraint $\dot{v}(t)\geq W(v(t),t)$ is satisfied for almost all $t$. The lower bound of $v$ is in fact given by the function $w$ and obtained by comparison of the derivatives of $v$ and $w$.

The Comparison Lemma assumes that we know the solution $w$ to the IVP $\dot{w}(t)=W(w(t),t)$ with initial condition $w_0\in\mathbb{R}$. Lemma 2.4 provides a solution $w$ for a specific choice of $W$ that will be used in various parts of this book.

**Lemma 2.4** *Let* $\alpha:\mathbb{R}_{\geq 0}\to\mathbb{R}_{\geq 0}$ *be a locally Lipschitz continuous class* $\mathcal{K}$ *function.*[5] *Then the IVP*

$$\dot{w}(t)=-\alpha(w(t)),\quad w(0)=w_0\geq 0$$

[5] A class $\mathcal{K}$ function $\alpha:\mathbb{R}_{\geq 0}\to\mathbb{R}_{\geq 0}$ is a continuous and strictly increasing function with $\alpha(0)=0$.

*has a unique solution $w(t)$ that satisfies*

$$w(t) = \beta(w(0), t) \geq 0$$

*for all $t \in \mathbb{R}_{\geq 0}$ where $\beta : \mathbb{R}_{\geq 0} \times \mathbb{R}_{\geq 0} \to \mathbb{R}_{\geq 0}$ is a class $\mathcal{KL}$ function.*[6] *For $\epsilon \in \mathbb{R}_{\geq 0}$ if $\alpha \in \mathbb{R}_{>0}$ is a constant, the IVP*

$$\dot{w}(t) = -\alpha w(t) - \epsilon, \quad w(0) = x_0 \geq -\epsilon/\alpha$$

*has a unique solution $w(t)$ that satisfies*

$$w(t) \geq \beta(w(0) + \epsilon/\alpha, t) - \epsilon/\alpha \geq -\epsilon/\alpha$$

*for all $t \in \mathbb{R}_{\geq 0}$.*

*Proof: The first part follows from [151, lemma 4.4]. The second part can easily be verified, as $\alpha$ is a constant making $\dot{w}(t) = -\alpha w(t) - \epsilon$ a linear ODE for which we can construct the solution explicitly, as done in example 2.4.*

---

**Remark 2.1** Using lemma 2.4, and assuming that $\alpha : \mathbb{R}_{\geq 0} \to \mathbb{R}_{\geq 0}$ is a locally Lipschitz continuous extended class $\mathcal{K}$ function,[7] it can be shown that the IVP $\dot{w}(t) = -\alpha(w(t))$, with $w(0) \leq 0$ (i.e., with a negative initial condition), has the solution $w(t) = -\beta(|w(0)|, t)$ for all $t \in \mathbb{R}_{\geq 0}$.

---

Let us now get back to the problem of certifying forward invariance. We assume for simplicity that the set $\mathcal{C}$ is static (i.e., not depending on time), and return to the case of time-varying sets $\mathcal{C}(t)$ later in this book. To get a mathematical handle on the problem, assume that $\mathcal{C}$ is defined by a continuously differentiable function $\mathfrak{b} : \mathbb{R}^n \to \mathbb{R}$ as the zero superlevel set of the function $\mathfrak{b}$; that is, as

$$\mathcal{C} := \{x \in \mathbb{R}^n \mid \mathfrak{b}(x) \geq 0\}.$$

We will assume in the remainder that $\mathcal{C}$ has a nonempty interior (i.e., that the set $\mathcal{C}$ has some volume), and is contained within the domain $\mathcal{D}$ (i.e., such that $\mathcal{C} \subseteq \mathcal{D}$). As it will be important later in this discussion, we remark that the set $\mathcal{C}$ is closed, as the function $\mathfrak{b}(x)$ is continuous. We will now set up the function $\mathfrak{b}$ such that it serves as the function $v$ used in the Comparison Lemma 2.3.

[6] A class $\mathcal{KL}$ function $\beta : \mathbb{R}_{\geq 0} \times \mathbb{R}_{\geq 0} \to \mathbb{R}_{\geq 0}$ is a continuous function that is such that (1) for a fixed $s \in \mathbb{R}_{\geq 0}$, the function $\beta(\cdot, s)$ is a class $\mathcal{K}$ function; and (2) for fixed $r \in \mathbb{R}_{\geq 0}$, the function $\beta(r, \cdot)$ is decreasing such that $\lim_{s \to \infty} \beta(r, s) = 0$.

[7] An extended class $\mathcal{K}$ function is a function $\alpha : \mathbb{R} \to \mathbb{R}$; that is, like a class $\mathcal{K}$ function, a continuous and strictly increasing function with $\alpha(0) = 0$, but with $\mathbb{R}$ as the input domain.

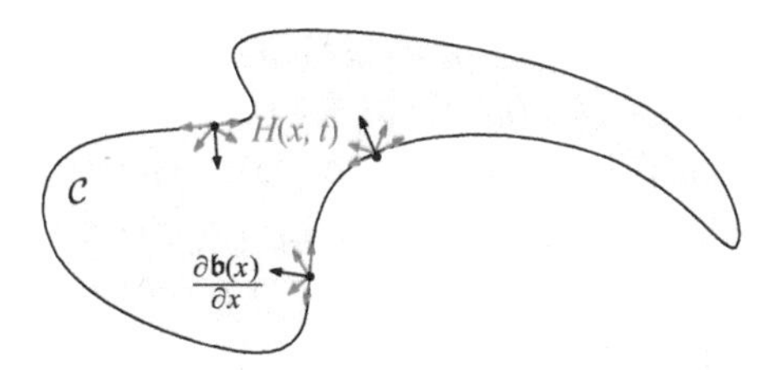


**Figure 2.2**
Geometric interpretation of barrier functions.

**Definition 2.5 (Valid Barrier Function)** *A continuously differentiable function* $\mathfrak{b}:\mathbb{R}^n\to\mathbb{R}$ *is said to be a* valid barrier function *for equation (2.6) on* $\mathcal{D}$ *if* $\mathcal{C}\subseteq\mathcal{D}$*, and if there exists a locally Lipschitz continuous extended class* $\mathcal{K}$ *function* $\alpha:\mathbb{R}\to\mathbb{R}$ *such that, for all* $(x,t)\in\mathcal{D}\times\mathbb{R}_{\geq 0}$*, it holds that*

$$\frac{\partial \mathfrak{b}(x)}{\partial x}H(x,t)\geq -\alpha(\mathfrak{b}(x)). \tag{2.7}$$

In equation (2.7), note that we have considered the inner product of the gradient of the function $\mathfrak{b}$ with the system dynamics $H$. Using the chain rule, we note that it holds that $\dot{\mathfrak{b}}(x(t)):=\frac{\partial \mathfrak{b}(x(t))}{\partial x}H(x(t),t)$ for almost all $t\in\mathcal{I}$ for a solution $x:\mathcal{I}\to\mathbb{R}^n$ to the IVP in equation (2.6). In this way, the inner product in equation (2.7) captures the time derivative of the function $\mathfrak{b}$.

We can give the definition of a barrier function a geometric interpretation by focusing on the boundary of the safe set (see figure 2.2). For $\mathfrak{b}$ to be a barrier function, the angle between the vectors $\frac{\partial \mathfrak{b}(x)}{\partial x}$ and $H(x,t)$ has to be acute at all times. In other words, since the normal vector $\frac{\partial \mathfrak{b}(x)}{\partial x}$ is pointing inside the safe set, the vector field $H(x,t)$ (which describes the instantaneous motion of the system) is not allowed to point outside the safe set to guarantee that the solution of the system starting from point $x$ stays inside the safe set. While the proof that we provide next is based on the Comparison Lemma, an alternative proof technique that directly follows this geometric intuition is based on Nagumo's theorem. However, by using Nagumo's theorem, one also has to assume that the function $\mathfrak{b}$ is regular in the sense that $\frac{\partial \mathfrak{b}(x)}{\partial x}\neq 0$ has to hold for all states on the boundary $x\in\text{bd}(\mathcal{C}):=\{x\in\mathbb{R}^n|h(x)=0\}$. This is because directional information is lost when the normal vector $\frac{\partial \mathfrak{b}(x)}{\partial x}$ is zero.

Finally, we show that the set $\mathcal{C}$ is forward invariant with respect to the IVP in equation (2.6) if $\mathfrak{b}$ is a valid barrier function.

**Theorem 2.1 (Forward Invariance via Valid Barrier Functions)** *Let the function* $H$ *that describes the system in equation (2.6) be continuous in* *its first and piecewise continuous in its second argument. Assume that* $\mathfrak{b}$ *is a valid barrier function for equation (2.6) on* $\mathcal{D}$. *If the initial condition* $x_0$ *is such that* $x_0 \in \mathcal{C}$, *then each solution* $x : \mathcal{I} \to \mathcal{D}$ *to the IVP in equation (2.6) is such that* $x(t) \in \mathcal{C}$ *for all* $t \in \mathcal{I}$. *If solutions* $x$ *are complete, it holds that* $\mathcal{C}$ *is forward invariant with respect to the IVP in equation (2.6); that is,* $x_0 \in \mathcal{C}$ *implies* $x(t) \in \mathcal{C}$ *for all* $t \geq 0$.

*Proof: First, note that* $\dot{w}(t) = -\alpha(w(t))$ *with initial condition* $w_0 \geq 0$, *where* $\alpha$ *satisfies equation (2.7), admits a unique and complete solution* $w : \mathbb{R}_{\geq 0} \to \mathbb{R}$ *such that* $w(t) \geq 0$ *for all* $t \geq 0$ *due to lemma 2.4 (it is easy to see that lemma 2.4 still holds despite* $\alpha$ *being an extended class* $\mathcal{K}$ *function). Since* $\mathfrak{b}$ *is a valid barrier function, each solution* $x : \mathcal{I} \to \mathbb{R}^n$ *to equation (2.5), which are guaranteed to exist due to the assumption made on* $H$, *is now (using the chain rule) such that*

$$\dot{\mathfrak{b}}(x(t)) = \frac{\partial \mathfrak{b}(x(t))}{\partial x} H(x(t), t) \geq -\alpha(\mathfrak{b}(x(t)))$$

*for almost all* $t \in \mathcal{I}$. *Assume that* $x_0 \in \mathcal{C}$; *that is,* $\mathfrak{b}(x(0)) \geq 0$. *Using the Comparison Lemma and picking* $w(0)$ *such that* $\mathfrak{b}(x(0)) \geq w(0) \geq 0$, *it follows that* $\mathfrak{b}(x(t)) \geq w(t) \geq 0$ *for all* $t \in \mathcal{I}$. *In other words,* $x(0) \in \mathcal{C}$ *implies* $x(t) \in \mathcal{C}$ *for all* $t \in \mathcal{I}$. *If solutions* $x$ *are complete, it holds that* $\mathcal{I} = [0, \infty)$ *which completes the proof.*

We would now like to discuss two distinct cases. If $\mathfrak{b}(x(t)) = 0$ (i.e., $x(t)$ is on the boundary of the safe set), we require that $\dot{\mathfrak{b}}(t) \geq 0$ so that the value of $\mathfrak{b}(x(t))$ cannot decrease. However, if $\mathfrak{b}(x(t)) > 0$ (i.e., $x(t)$ is inside the safe set), we permit $\dot{\mathfrak{b}}(t)$ to be negative. Intuitively, this allows the system to move freely within the safe set and leave any level set of $\mathfrak{b}$ unless its the zero level set.

Theorem 2.1 requires that solutions $x : \mathcal{I} \to \mathbb{R}^n$ to the IVP in equation (2.6) are complete. We recall that a solution $x$ may be maximal but not necessarily complete (i.e., $\mathcal{I}$ may be bounded). How can we now guarantee that a solution is complete? One way is by using lemma 2.2. In fact, if $\mathcal{C}$ is a compact set one can show that $x$ is complete. This can be seen as follows. From the proof of theorem 2.1, we know that $x(t) \in \mathcal{C}$ for all $t \in \mathcal{I}$. If $\mathcal{C}$ is compact, it follows, by contradiction and using lemma 2.2, that $x$ is a complete solution and $\mathcal{I} = [0, \infty)$.

Let us illustrate the concept of a barrier function with two examples.

---

**Example 2.8** Consider a one-dimensional nonlinear system with the dynamics

$$\dot{x} = -(1 - x^2)x.$$

One can already see that trajectories starting from $x_0 < -1$ or $x_0 > 1$ will diverge to infinity, while trajectories starting from $x_0 \in (-1,1)$ will converge to the origin. Consider the safety specification $\mathfrak{b}(x) := a - x^2 \geq 0$ for a constant $a \in [0,1)$ that encodes the safe set $\mathcal{C} = \{x \in \mathbb{R} | a - x^2 \geq 0\} = [-\sqrt{a}, \sqrt{a}]$. Let us check if the system is safe by checking if $\mathfrak{b}$ is a barrier function (i.e., by verifying the condition in equation 2.7). We have that

$$\frac{\partial \mathfrak{b}(x)}{\partial x} H(x) = 2x^2(1 - x^2),$$

which is positive for all $|x| < 1$ (zero for $|x| = 1$ and negative for all $|x| > 1$). Hence it is easy to see that the condition in equation (2.7) holds on $\mathcal{C}$ for any extended class $\mathcal{K}$ function $\alpha$ since $a < 1$ and $-\alpha(\mathfrak{b}(x)) \leq 0$ for all $x \in \mathcal{C}$. By continuity of $\frac{\partial \mathfrak{b}(x)}{\partial x} H(x,t)$, one can then find a small $\epsilon$ such that $\mathfrak{b}$ is a barrier function on the set $\mathcal{D} := \mathcal{C} \oplus \epsilon$.

---

**Example 2.9** Recall the bicycle model from example 2.3, but let us now focus only on the orientation $\theta$ and the $p_y$ position, and neglect the $p_x$ position (e.g., when we assume the vehicle is driving on a road in $p_x$ direction). The dynamics are thus

$$\dot{x} = \begin{bmatrix} \dot{p}_y \\ \dot{\theta} \end{bmatrix} = \begin{bmatrix} v \sin(\theta) \\ \frac{v}{L} \tan(\delta) \end{bmatrix}.$$

The safety specification is that $(p_y(t), \theta(t)) \in [-1,1]^2$ should hold for all times $t \geq 0$ (i.e., the set $[-1,1]^2$ should be forward invariant). This set specifies a safety cone in the $p_x$-axis direction with orientation $\theta \in [-1,1]$, while staying within a distance of 1 in the middle of the road. Assume that we are given a safety controller:[8]

$$\begin{aligned} v &:= -\sin(\theta) y \\ \delta &:= \tan^{-1}(|\theta| y), \end{aligned}$$

and we are asked to check if the safety specification is satisfied. Under the safety controller, the closed-loop dynamics become

$$\dot{x} = \begin{bmatrix} -\sin^2(\theta) y \\ \frac{-\sin(\theta) y^2 |\theta|}{L} \end{bmatrix}.$$

Consider the two barrier function candidates $\mathfrak{b}_1(x) := 1 - p_y^2$ and $\mathfrak{b}_2(x) := 1 - \theta^2$, which encode the given safety specifications. We calculate

[8]The intuition behind the safety controller is as follows. The translational controller $v$ ensures that the vehicle drives backward toward the lane if $\theta y > 0$ and forward toward the lane if $\theta y < 0$. The lateral controller $\delta$ then simply steers the orientation $\theta$ to zero.

$$\frac{\partial \mathfrak{b}_1(x)}{\partial x} H(x) = 2\sin^2(\theta) y^2$$
$$\frac{\partial \mathfrak{b}_2(x)}{\partial x} H(x) = \frac{\theta \sin(\theta) y^2 |\theta|}{L}$$

and note that, as in the previous example, the derivatives are positive on the corresponding safety sets $\mathcal{C}_1 := \{x \in \mathbb{R}^2 | \mathfrak{b}_1(x) \geq 0\}$ and $\mathcal{C}_2 := \{x \in \mathbb{R}^2 | \mathfrak{b}_2(x) \geq 0\}$. As in example 2.8 we can argue that the functions $\mathfrak{b}_1$ and $\mathfrak{b}_2$ are barrier functions.

### 2.1.4 Stability

The notions of forward invariance and barrier functions are useful when we want to show that system trajectories stay within a given set. However, what if we are instead interested in showing convergence of system trajectories to a desired set (e.g., an operating point or region)? What if we are interested in showing that system trajectories are simply bounded? To answer these questions, we next introduce the notion of stability and, to make our exposition easy to follow, we focus on stability with respect to the origin. Therefore, let us assume that the origin is an equilibrium point; that is, that $H(0,t) = 0$ for all times $t \geq 0$. This is without losing generality, as we can use coordinate transformations to shift nonorigin equilibrium points to the origin. Let us now define stability in the sense of Lyapunov.

**Definition 2.6 (Lyapunov Stability)** *The equilibrium point $x = 0$ of the system in equation (2.6) is*

- *Stable if, for each $\epsilon > 0$, there is a constant $\delta > 0$ such that*

$$\|x_0\| \leq \delta \implies \|x(t)\| \leq \epsilon \text{ for all } t \geq 0; \tag{2.8}$$

- *Unstable if it is not stable;*
- *Asymptotically stable if it is stable and there exists a constant $\delta_a > 0$ such that*

$$\|x_0\| \leq \delta_a \implies \lim_{t \to \infty} x(t) = 0. \tag{2.9}$$

Intuitively, stability as per equation (2.8) means that a system trajectory starting $\delta$-close to the equilibrium point will lead to the system trajectory staying $\epsilon$-close to the equilibrium point. The stability condition in equation

(2.8) is given in a challenge-and-answer form in the same way that continuity of functions is defined (i.e., we can pick the "stay" close level $\epsilon$ arbitrarily and have to find a corresponding "start" close level of $\delta$). Naturally, it has to hold that $\delta \leq \epsilon$. It is also important to note that stability implies that system trajectories are bounded as can be seen from equation (2.8). Asymptotic stability as per equation (2.9) is stronger than stability and also requires convergence to the equilibrium point. Let us illustrate these three concepts on the popular example of a pendulum.

---

**Example 2.10** A pendulum as illustrated in figure 2.3 can be described by the rotation angle $\theta$ and the rotational velocity $\dot{\theta}$. A model of the pendulum can be derived as

$$\dot{x} = \begin{bmatrix} \dot{\theta} \\ -\frac{g}{l}\sin(\theta) - \frac{k}{m}\dot{\theta} \end{bmatrix},$$

where $x := \begin{bmatrix} \theta & \dot{\theta} \end{bmatrix}^T$, $g$ is gravity, $l$ is the length of the pendulum, $k$ is the friction coefficient, and $m$ is the mass of the pendulum. Intuition tells us that the system has two equilibrium points: at the down position $(\theta, \dot{\theta}) = (0, 0)$ of the pendulum and at the upright position $(\theta, \dot{\theta}) = (\pi, 0)$, which can formally be checked by verifying that $H(x) = 0$ holds for these points. Intuition also tells us that the equilibrium point at the upright position is unstable. The equilibrium point at the down position is, on the other hand, stable if there is no friction ($k = 0$) and asymptotically stable if there is friction ($k > 0$).

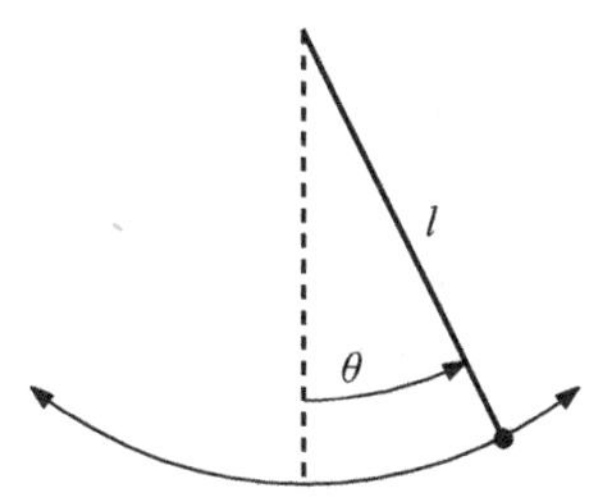


**Figure 2.3**
Geometry of the pendulum.

---

While the discussion in example 2.10 was of a qualitative nature, let us now discuss how we can certify stability of the system in equation (2.6) using a quantitative method. We will use similar concepts as for barrier functions by introducing Lyapunov functions. In fact, Lyapunov functions have been successfully used for decades in nonlinear systems theory.

**Definition 2.7 (Valid Lyapunov Function)** *A continuously differentiable and positive definite*[9] *function $V:\mathbb{R}^n \to \mathbb{R}$ is said to be a* valid Lyapunov function *for equation (2.6) on a set $\mathcal{D}$ if $0 \in \mathcal{D}$ and if, for all $(x,t) \in \mathcal{D} \times \mathbb{R}_{\geq 0}$, it holds that*

$$\frac{\partial V(x)}{\partial x} H(x,t) \leq 0. \tag{2.10}$$

Note first the conceptual difference between Lyapunov and barrier functions. Lyapunov functions are "arbitrary" positive definite functions (i.e, nonnegative functions), whose existence will guarantee stability of the system (as we will formally show in a bit). Lyapunov functions can often be chosen as the energy function of a system. On the other hand, barrier functions describe a desirable set (i.e., a safe set), which we would like to certify as forward invariant. However, note the similarity in the flow condition in equation (2.7) for barrier functions and the flow condition in equation (2.10) for Lyapunov functions. Equation (2.10) considers the inner product of the gradient of the function $V$ with the system dynamics $H$ enforcing that $\dot{V}(x(t)) := \frac{\partial V(x(t))}{\partial x} H(x(t),t) \leq 0$ for almost all $t \in \mathcal{I}$ for a solution $x:\mathcal{I} \to \mathbb{R}^n$ to the IVP in equation (2.6). This ensures that the function value $V(x(t))$ cannot increase. The geometric interpretation is that the angle between the vectors $\frac{\partial V(x(t))}{\partial x}$ and $H(x(t),t)$ has to be obtuse for all times. In other words, since the normal vector $\frac{\partial V(x(t))}{\partial x}$ is pointing outside the current level set $\Omega_\beta := \{x \in \mathbb{R}^n | V(x) \leq \beta\}$ with $\beta := V(x(t))$, the vector field $H(x(t),t)$ is not allowed to also point outside $\Omega_\beta$ so the system stays within $\Omega_\beta$. Let us next show that a valid Lyapunov function certifies stability.

**Theorem 2.2 (Stability via Valid Lyapunov Functions)** *Let the function $H$ that describes the system in equation (2.6) be continuous in its first argument and piecewise continuous in its second argument. Let $x=0$ be an equilibrium point of the system in equation (2.6). Assume that $V$ is a valid Lyapunov function for equation (2.6) on an open and connected set $\mathcal{D}$. Then the equilibrium point $x=0$ is stable.*

*Proof: First, let us pick a constant $r \in (0,\epsilon]$ such that*

$$\mathcal{B}_r := \{x \in \mathbb{R}^n | \|x\| \leq r\} \subset \mathcal{D};$$

*that is, the set $\mathcal{B}_r$ is strictly contained within domain $\mathcal{D}$. Based on the continuity of the function $V$ and since $V(0)=0$, one can find (see, e.g., [151]) a*

[9]By "positive definite," we mean that $V(x)>0$ for all $x \neq 0$ and $V(0)=0$.

*constant* $\beta > 0$ *such that*

$$\Omega_\beta := \{x \in \mathbb{R}^n | V(x) \leq \beta\} \subset \mathbb{B}_r;$$

*that is, the set* $\Omega_\beta$ *is strictly included in the set* $\mathcal{B}_r$. *Note next that* $\dot{w}(t) = 0$ *with initial condition* $w_0 \geq 0$ *admits a unique and complete solution* $w : \mathbb{R}_{\geq 0} \to \mathbb{R}$ *such that* $w(t) = w_0$ *for all* $t \geq 0$. *Each solution* $x : \mathcal{I} \to \mathbb{R}^n$ *to equation (2.5) is now such that*

$$\dot{V}(x(t)) = \frac{\partial V(x(t))}{\partial x} H(x(t), t) \leq 0$$

*for almost all* $t \in \mathcal{I}$. *We can again apply the Comparison Lemma to see that* $V(x(t)) \leq V(x_0)$ *for all times* $t \geq 0$, *so that* $x_0 \in \Omega_\beta$ *implies* $x(t) \in \Omega_\beta$ *for all times* $t \geq 0$. *Again due to continuity of the function* $V$, *there exists a constant* $\delta > 0$ *such that* $\mathcal{B}_\delta \subseteq \Omega_\beta$. *Consequently,* $x_0 \in \mathcal{B}_\delta$ *implies that* $x(t) \in \Omega_\beta$ *for all times* $t \geq 0$ *such that* $x(t) \in \mathcal{B}_\epsilon$ *for all times* $t \geq 0$, *which was to be proven.*

We next illustrate the concept of a valid Lyapunov function in an example.

---

**Example 2.11** Let us continue with the pendulum from example 2.10 and analyze the stability of the equilibrium point at the origin (i.e., for the down equilibrium point of the pendulum). Consider the candidate Lyapunov function

$$V(x) := \frac{g}{l}(1 - \cos(\theta)) + \frac{1}{2}\dot{\theta}^2.$$

Let us next verify the condition in equation (2.10) for this candidate Lyapunov function

$$\begin{aligned} \frac{\partial V(x)}{\partial x} H(x) &= \frac{g}{l}\sin(\theta)\dot{\theta} + \dot{\theta}\Big(-\frac{g}{l}\sin(\theta) - \frac{k}{m}\dot{\theta}\Big) \\ &= -\frac{k}{m}\dot{\theta}^2 \leq 0. \end{aligned}$$

According to theorem 2.2, we now know that the origin is stable, both with friction ($k > 0$) and without ($k = 0$).

---

In example 2.11 and in theorem 2.2, we have shown how we can certify stability, but how can we certify asymptotic stability? For instance, in the case of the pendulum with friction ($k > 0$), we expect the equilibrium point at the origin to be asymptotically stable. One way of showing asymptotic stability is to use La Salle's invariance principle, for which we have to show that solutions with $\dot{V}(x(t)) = 0$ converge to the origin. We refer the interested reader to [151, chapter 4.2]. We will instead replace the condition in equation

(2.10) by a slightly stronger condition. Recall that the condition in equation (2.10) guarantees that the solution at time $t$ cannot leave the level set $\Omega_\beta$ with $\beta := V(x(t))$, i.e., that $x(t') \in \Omega_\beta$ for all $t' \geq t$. A strict decrease in equation (2.10) would instead result in a decrease along these level sets, which we will now use to guarantee asymptotic stability. Lemma 2.5 is given without proof; the interested reader can find the proof in [151, theorem 4.9].

**Lemma 2.5 (Asymptotic Stability)** *Assume that all conditions from theorem 2.2 hold and that additionally, for all $(x,t) \in \mathcal{D} \times \mathbb{R}_{\geq 0}$, it holds that*

$$\frac{\partial V(x)}{\partial x} H(x,t) \leq -W(x), \tag{2.11}$$

*where $W : \mathbb{R}^n \to \mathbb{R}_{\geq 0}$ is a continuous and positive definite function. Then the equilibrium point $x = 0$ is asymptotically stable.*

We highlight that equation (2.11) enforces a strict decrease in the function value of $V(x(t))$ since equation (2.11) enforces $\dot{V}(x(t)) < 0$ for all points except for the origin since the function $W$ is positive semidefinite (i.e., $W(x) > 0$ for all $x \neq 0$ and $W(0) = 0$). To show asymptotic stability of the down position for the pendulum from example 2.11, one can use the Lyapunov function candidate $V(x) = \frac{1}{2}x^T Px + \frac{g}{l}(1 - \cos(\theta))$ with $x := \begin{bmatrix} \theta & \dot{\theta} \end{bmatrix}^T$ for a suitably chosen positive definite matrix $P$. We invite the reader to verify this result.

---

**Remark 2.2** As the reader may already guess, lemma 2.5 reveals a similarity between the derivative conditions $\frac{\partial \mathfrak{b}(x)}{\partial x} H(x,t) \geq -\alpha(\mathfrak{b}(x))$ and $\frac{\partial V(x)}{\partial x} H(x,t) \leq -W(x)$ for barrier and Lyapunov functions, respectively, when the function $W$ is chosen to be the composition of a locally Lipschitz continuous extended class $\mathcal{K}$ function $\alpha$ and $-V$. Indeed, if $\mathfrak{b}$ is a valid barrier function on $\mathcal{D}$, one can show that the set $\mathcal{C}$ is asymptotically stable using the Lyapunov function $V(x) = -\mathfrak{b}(x)$ if $x \in \mathcal{D} \setminus \mathcal{C}$ and $V(x) := 0$ if $x \in \mathcal{C}$ using a similar argument as in lemma 2.5 [330]. This implies that $x(t)$ asymptotically approaches $\mathcal{C}$ if $x(0) \in \mathcal{D} \setminus \mathcal{C}$.

---

Finally, let us present a Lyapunov-like method that guarantees bounded solution $x$ to the IVP in equation (2.6). A solution is bounded if there is a constant $\epsilon > \|x_0\|$ such that $\|x(t)\| \leq \epsilon$ for all $t \geq 0$. We can certify bounded solutions by relaxing the condition in equation (2.10) in the definition of a Lyapunov function to only hold outside a level set $\Omega_\beta$, and by assuming that $\Omega_\beta$ is compact.

**Lemma 2.6 (Certifying Bounded Solutions)** *Let the function $H$ that describes the system in equation (2.6) be continuous in its first argument and* *piecewise continuous in its second argument. Assume that* $V:\mathcal{D}\to\mathbb{R}^n$ *is a continuously differentiable and positive definite function with compact level sets (i.e.,* $\Omega_\beta := \{x\in\mathbb{R}^n | V(x)\le\beta\}$ *is compact). If there is a* $\beta>0$ *such that*

$$\frac{\partial V(x)}{\partial x}H(x,t)\le 0$$

*for all* $(x,t)\in\mathbb{R}^n\setminus\Omega_\beta\times\mathbb{R}_{\ge 0}$*, then solutions* $x$ *to the IVP in equation (2.6) are bounded.*

## 2.2 Selected Topics from Nonlinear Control Theory

In the previous section, we introduced dynamical systems and how a dynamical system can be modeled by an ODE, with its motion described by a corresponding IVP. To analyze existence, uniqueness, and completeness of solutions and forward invariance, safety, and stability, it was assumed that a control law $u:\mathcal{D}\times\mathbb{R}_{\ge 0}\to\mathbb{R}^m$ was already given so that the IVP in equation (2.5) that explicitly depended on the control input $u$ reduced to the IVP in equation (2.6) that had no dependence on the control input $u$. In this section, we will present selected topics from nonlinear control theory to design control laws $u$ such that a system is enforced to be forward invariant or stable. While our exposition is by no means exhaustive, we believe that the selected topics provide the reader with a powerful set of control design tools that are useful for the problem that this book considers (i.e., designing computationally efficient control laws for multi-agent systems under spatio temporal constraints).

In the remainder of this section, we consider again the dynamical system as per equation (2.5), which we give here again for the convenience of the reader:

$$\dot{x}(t)=f(x(t),t)+g(x(t),t)u(x(t),t),\quad x(0)=x_0\in\mathcal{D}.$$

We assume that functions $f$ and $g$ are continuous in their first argument and piecewise continuous in their second argument so that solutions to the IVP in equation (2.5) are guaranteed to exist if the control law $u$ is designed to also be continuous in the first and piecewise continuous in the second argument. We then denote solutions to the IVP in equation (2.5) under such a feedback control law $u:\mathcal{D}\times\mathbb{R}_{\ge 0}\to\mathbb{R}^m$ by $x:\mathcal{I}\to\mathbb{R}^n$, where $\mathcal{I}\subseteq\mathbb{R}_{\ge 0}$ is again the maximum definition interval.

### 2.2.1 Feedback Control, Constraints, and Robustness

Control theory studies the design of control laws for dynamical systems and has great success in academia and industry. Most of this success is explained by the development of feedback control laws (i.e., control laws $u$ that are state dependent and whose value at time $t$ are given by $u(x(t), t)$). Why is feedback so important though, and why can we not just use state-independent feedforward control laws that are computed a priori (i.e., without online information of $x(t)$)? The answer to this question is robustness against uncertainties, or in the words of the authors of a seminal textbook on robust nonlinear control [106]:

> The main purpose of every feedback loop, created by nature or designed by engineers, is to reduce the effect of uncertainty on vital system functions. Indeed, feedback as a design paradigm for dynamic systems has the potential to counteract uncertainty.

For instance, feedback allows for correcting a system in case that there are unknown disturbances (e.g., wind gust), or when there is a mismatch between the nominal model of the system (e.g., $\dot{x} = H(x, t)$), and the real physical system, which may be a perturbed version of the form $\dot{x} = H(x, t) + w(x, t)$. The resulting closed-loop system is illustrated in figure 2.4, and the perturbation $w$ is not restricted to linearly enter the system dynamics. With state information $x(t)$, one can design robust feedback control laws that can account for perturbations such as $w$ by using the information that the state carries. In later chapters, we will see how we can systematically design robust nonlinear feedback control laws.

We are particularly interested in designing feedback control laws that guarantee constraint satisfaction. Indeed, an important aspect of dynamical systems concerns their safety, which can be posed as a constrained control problem in which we would like to satisfy state constraints. In many cases, these state constraints can be transient (i.e., varying with time).

In this book, we focus solely on state feedback so we have to assume to have knowledge of the state $x(t)$ at each time $t$. If only partial measurements of the system state are available, state estimation techniques such as the (extended) Kalman filter can be employed to estimate the state, which we will not discuss here. In this book, we will also not consider the design of optimal feedback control laws (i.e., control laws that are optimal with respect to some cost

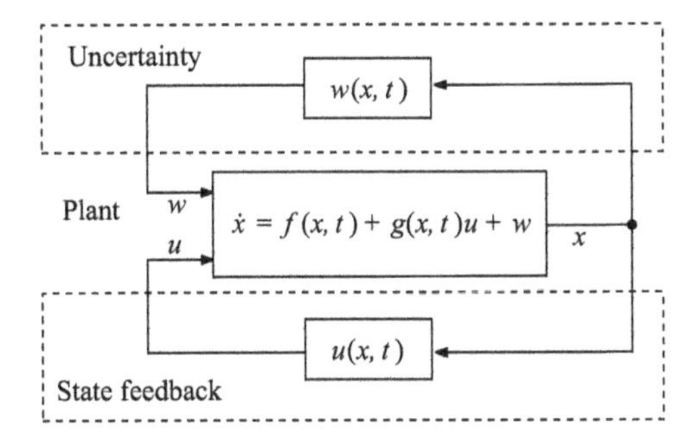


**Figure 2.4**
Closed-loop dynamical system with control input $u$ and unknown disturbance $w$ that enters the system dynamics linearly.

function), and we instead focus on computationally efficient feedback control laws that can satisfy time-varying state constraints.

To provide a first simple example of a feedback control law, let us consider a linear dynamical system of the form $\dot{x} = Ax + Bu$. We would like to design a linear feedback control law $u := Kx$ to asymptotically stabilize the system; that is, with the goal that the closed-loop system $\dot{x} = (A + BK)x$ is asymptotically stable. From linear systems theory [68], it is known that designing the feedback gain matrix $K$ such that the eigenvalues of the matrix $A + BK$ are in the open left halfplane (i.e., all eigenvalues have negative real parts), renders the closed-loop system asymptotically stable. Designing $K$ in this way also provides robustness with respect to perturbations, as mentioned before. The design of control laws for linear systems has been studied in depth; we refer the reader to [68] for an in-depth exposition of tools and algorithms.

As we are generally dealing with nonlinear systems and constrained control problems, linear feedback control laws will not always be sufficient. This motivates the use of nonlinear feedback control laws for which there are various excellent textbooks, such as [151, 134]. We will not present an in-depth treatment of the variety of nonlinear control design tools;[10] instead, we will introduce, in the next three subsections more contemporary nonlinear control techniques and concepts that are suitable for constrained control problems. Indeed, our main goal is to introduce CBFs, CLFs, and funnel control laws that have emerged as powerful feedback control design tools within the last decade.

[10] Commonly used nonlinear control techniques are system linearization, integral control, feedback linearization, sliding mode control, backstepping, passivity-based control, and nonlinear model predictive control.

### 2.2.2 Control Barrier Functions

We now want to introduce and discuss CBFs, which are at least conceptually similar to barrier functions as discussed previously. However, while barrier functions certify forward invariance of a system $\dot{x}=H(x,t)$, CBFs can be used to enforce the forward invariance of a system $\dot{x}=f(x,t)+g(x,t)u$ (i.e., CBFs aid us in finding a controller $u$ that renders the closed-loop system forward invariant). For this purpose, consider again a continuously differentiable function $\mathfrak{b}:\mathbb{R}^n \to \mathbb{R}$, and define the safe set that we would like to render forward invariant as

$$\mathcal{C} := \{x \in \mathbb{R}^n \mid \mathfrak{b}(x) \geq 0\}.$$

We next introduce the concept of a valid CBF that will help us to find a controller $u$ such that set $\mathcal{C}$ is forward invariant.

**Definition 2.8 (Valid Control Barrier Function)** *A continuously differentiable function $\mathfrak{b}:\mathbb{R}^n \to \mathbb{R}$ is said to be a* valid CBF *for equation (2.5) on $\mathcal{D}$ if $\mathcal{C} \subseteq \mathcal{D}$ and if there is a locally Lipschitz-continuous extended class $\mathcal{K}$ function $\alpha:\mathbb{R}\to\mathbb{R}$ such that, for all $(x,t)\in\mathcal{D}\times\mathbb{R}_{\geq 0}$, it holds that*

$$\sup_{u\in\mathbb{R}^m} \frac{\partial \mathfrak{b}(x)}{\partial x}(f(x,t)+g(x,t)u)) \geq -\alpha(\mathfrak{b}(x)). \tag{2.12}$$

Note that compared to the barrier function condition $\frac{\partial \mathfrak{b}(x)}{\partial x}H(x,t) \geq -\alpha(\mathfrak{b}(x))$ in equation (2.7), the CBF condition in equation (2.12) in the CBF definition contains the input $u$ and a supremum operator over all inputs $u\in\mathbb{R}^n$. We are hence asking if, for each pair $(x,t)$, there is a control input $u$ that satisfies the condition $\frac{\partial \mathfrak{b}(x)}{\partial x}(f(x,t)+g(x,t)u)) \geq -\alpha(\mathfrak{b}(x))$. We remark that one can also take the supremum over an input set $\mathcal{U}$ instead of $\mathbb{R}^m$ if one wants to incorporate input constraints. Based on this discussion, we define the set of CBF consistent inputs induced by a valid CBF $\mathfrak{b}$ to be

$$K_{\text{CBF}}(x,t) := \left\{u\in\mathbb{R}^m \mid \frac{\partial \mathfrak{b}(x)}{\partial x}(f(x,t)+g(x,t)u) \geq -\alpha(\mathfrak{b}(x))\right\}.$$

Let us now show under which conditions a control law $u$ that satisfies the constraints imposed by the set-valued function $K_{\text{CBF}}$ renders the set $\mathcal{C}$ forward invariant with respect to the IVP in equation (2.5).[11]

[11]This result was stated in [18, 330], but we provide an alternative proof using the Comparison Lemma instead of Nagumo's theorem so the regularity assumption $\frac{\partial \mathfrak{b}(x)}{\partial x} \neq 0$ if $x\in \text{bd}(\mathcal{C})$ is not required.

**Theorem 2.3 (Forward Invariance via Valid Control Barrier Functions)** *Let the functions $f$, $g$, and $u$ that describe the system in equation (2.5) be continuous in its first argument and piecewise continuous in its second argument. Assume that $\mathfrak{b}$ is a valid CBF for equation (2.5) on $\mathcal{D}$ and that the control law $u$ is such that $u(x,t) \in K_{CBF}(x,t)$ for all $(x,t) \in \mathcal{D} \times \mathbb{R}_{\geq 0}$. If the initial condition $x_0$ is such that $x_0 \in \mathcal{C}$, then each solution $x : \mathcal{I} \to \mathcal{D}$ to the IVP in equation (2.5) under $u$ is such that $x(t) \in \mathcal{C}$ for all $t \in \mathcal{I}$. If solutions $x$ are complete, it holds that $\mathcal{C}$ is forward invariant with respect to the IVP in equation (2.5) under $u$ (i.e., $x_0 \in \mathcal{C}$ implies $x(t) \in \mathcal{C}$ for all $t \geq 0$).*

*Proof: The proof follows similar steps as the earlier proof for valid barrier functions in theorem 2.1. First, note that solutions $x : \mathcal{I} \to \mathbb{R}^n$ to the IVP in equation (2.5) under $u$ exist due to the assumptions made on functions $f$, $g$, and $u$. As $\mathfrak{b}$ is a valid CBF and the control law $u$ is such that $u(x) \in K_{CBF}(x)$, it follows that each solution $x$ is such that*

$$\dot{\mathfrak{b}}(x(t)) = \frac{\partial \mathfrak{b}(x(t))}{\partial x}\big(f(x(t),t) + g(x(t),t)u(x(t),t)\big) \geq -\alpha(\mathfrak{b}(x(t)))$$

*for all $t \in \mathcal{I}$. The rest of the proof follows the same steps as in theorem 2.1 by applying the Comparison Lemma to the solution $w$ of the IVP $\dot{w}(t) = -\alpha(w(t))$, with $0 \leq w(0) \leq \mathfrak{b}(x(0))$, which is such that $w(t) \geq 0$ for all $t \geq 0$.*

We note that the previous result again requires solutions $x : \mathcal{I} \to \mathbb{R}^n$ to the IVP in equation (2.5) under $u$ to be complete. As per our earlier discussion following theorem 2.1, we can again ensure complete solutions if set $\mathcal{C}$ is compact.

From a practical point of view, we have not yet discussed how to obtain a control law $u$ such that $u(x,t) \in K_{\text{CBF}}(x,t)$. We can obtain a minimum norm control law either by solving a convex quadratic optimization problem or even analytically in closed form. For the former, let $u(x,t) := \hat{u}$, where $\hat{u}$ is given by

$$\underset{\hat{u}}{\operatorname{argmin}} \; \|\hat{u}\|^2 \tag{2.13}$$

$$\text{s.t.} \; \frac{\partial \mathfrak{b}(x)}{\partial x}(f(x,t) + g(x,t)\hat{u}) \geq -\alpha(\mathfrak{b}(x)). \tag{2.14}$$

The advantage of such an optimization-based control law is that we could easily replace the objective $\|\hat{u}\|^2$ with the objective $\|\hat{u} - u_d(x,t)\|^2$ for a desired control law $u_d$, which may achieve a performance objective (i.e., tracking a desired trajectory), but may not render the set $\mathcal{C}$ forward invariant (i.e., safe). With the objective $\|\hat{u} - u_d(x,t)\|^2$, the system executes $u_d$ whenever possible, but it may override the control command with a safe control input whenever needed. For the minimum norm control law in equation (2.13), we can also obtain an equivalent closed-form expression. Therefore, define the functions

$$B_0(x,t) := -\frac{\partial \mathfrak{b}(x)}{\partial x} f(x,t) - \alpha(\mathfrak{b}(x)),$$
$$B_1(x,t) := \frac{\partial \mathfrak{b}(x)}{\partial x} g(x,t),$$

so that, inspired by [106, chapters 2.4 and 4.2], the control law $u$ can instead be obtained as follows:

$$u(x,t) := \begin{cases} \frac{B_1(x,t)^T B_0(x,t)}{B_1(x,t) B_1(x,t)^T} & \text{if } B_0(x,t) > 0 \\ 0 & \text{otherwise.} \end{cases}$$

Let us now discuss a simple example to get more insight into the use of CBFs.

---

**Example 2.12** Consider a planar system, such as a mobile robot, with state $x := \begin{bmatrix} p_x & p_y \end{bmatrix}^T$, where $p_x$ and $p_y$ denote the two-dimensional position. For simplicity, let us assume single integrator dynamics $\dot{x} = u$, where the control input $u := \begin{bmatrix} u_x & u_y \end{bmatrix}^T$ directly controls the velocity of the car. This could, for instance, be a simplified model of the motion of an omnidirectional robot as shown in figure 2.1. Let us consider the candidate CBF $\mathfrak{b}(x) := 1 - \|x\|^2$, which defines the safe set $\mathcal{C}$ to be the norm ball. It is easy to see that $\mathfrak{b}$ is a valid CBF for any extended class $\mathcal{K}$ function $\alpha$ since

$$\frac{\partial \mathfrak{b}(x)}{\partial x}(f(x,t) + g(x,t)u)) = -2p_x u_x - 2p_y u_y;$$

that is, we can always find a control input $u$ such that equation (2.12) is satisfied. The same observation applies if we replace the system with $\dot{x} = \begin{bmatrix} 0.75p_x + u_x & 1.25p_y + u_y \end{bmatrix}^T$, for which we show simulation results in figure 2.5 from an initial condition $x_0 := \begin{bmatrix} -0.5 & -0.1 \end{bmatrix}^T$ and for the extended class $\mathcal{K}$ function $\alpha(r) := r$. The MATLAB code to reproduce this example is given as follows:

```
1  clc; clear all; close all;
2
3  t=linspace(0,5,101); % time
4  step=t(2)-t(1); % time increment
5  x(:,1)=[-0.5;-0.1]; % initial condition
6
7  for i=1:length(t)
8      %calculate control input
9      b(i) = 1-norm(x(:,i))^2;
10     db =-2*[x(1,i) x(2,i)];
11     f=[0.75*x(1,i);1.25*x(2,i)];
12     g=eye(2);
```

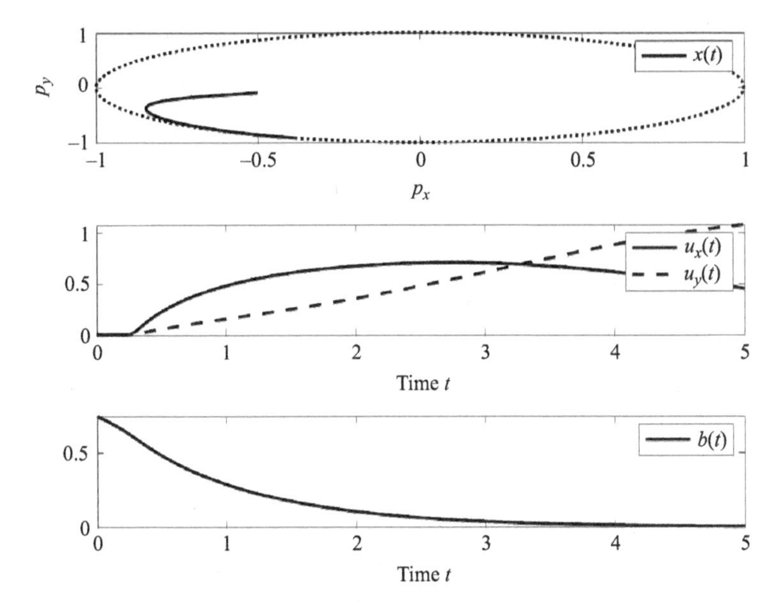


**Figure 2.5**
The evolution of the state $x(t)$, the control input $u(x(t), t)$, and the CBF $\mathfrak{b}(x(t))$ under the CBF controller in example 2.12.

```
13        alpha=b(i);
14        u(:,i) = quadprog(eye(2),zeros(1,2),-db*g,alpha
15            +db*f);
16
17        % Euler method
18        x(:,i+1)=x(:,i)+step*(f+g*u(:,i));
19    end
```

Note that not every CBF candidate $\mathfrak{b}$ is valid as per equation (2.12). Finding valid CBFs can be a challenging task for general dynamical systems, as one is simultaneously looking for a function $\mathfrak{b}$ and a control input $u$ that satisfy the constraint in equation (2.12). For general nonlinear systems, there are techniques to construct valid CBFs using bilinear sum-of-squares programs (see, e.g., [329, 317, 15]), or more recent learning techniques that learn valid CBFs from data (see, e.g., [248, 140, 336, 238, 178]).

In this book, we will focus on dynamical systems for which we can analytically construct valid CBFs as in example 2.12. Therefore, we will place assumptions on the system under consideration. For instance, for higher-order relative degree systems, such as the double integrator from example 2.1 or many mechanical systems, we can analytically construct valid CBFs with the concept of higher-order CBFs [325, 288, 76]. In section 4.5 in chapter 4, we will also show examples of how to construct valid CBFs for the unicycle dynamics previously presented in equation (2.4). We emphasize that we will particularly be interested in constructing CBFs for multi-agent systems.

In this spirit, let us next discuss a second example in which we consider a multirobot system consisting of two robots. The CBF, as presented in example 2.13, will later be referred to as a "centralized CBF," as opposed to "decentralized CBFs."

---

**Example 2.13** Let us next consider a set of two robots with single integrator dynamics. We can stack the states of both robots as $x := \begin{bmatrix} x_1 & x_2 \end{bmatrix}^T$ with their two-dimensional positions $x_i := \begin{bmatrix} p_{x,i} & p_{y,i} \end{bmatrix}^T$ and dynamics $\dot{x}_i = \begin{bmatrix} u_{x,i} & u_{y,i} \end{bmatrix}^T$. For collision avoidance between the two robots, we consider the candidate CBF $\mathfrak{b}(x) := \|x_1 - x_2\|^2 - 1$. It can again be easily seen that $\mathfrak{b}$ is a valid CBF for any extended class $\mathcal{K}$ function $\alpha$ since

$$\begin{aligned}\frac{\partial \mathfrak{b}(x)}{\partial x}(f(x) + g(x)u)) = {} & 2(p_{x,1} - p_{x,2})u_{x,1} - 2(p_{x,1} - p_{x,2})u_{x,2} \\ & + 2(p_{y,1} - p_{y,2})u_{y,1} - 2(p_{y,1} - p_{y,2})u_{y,2}.\end{aligned}$$

---

Let us conclude this section with a remark regarding asymptotic stability of set $\mathcal{C}$, which will be a useful property when the system starts outside $\mathcal{C}$.

---

**Remark 2.3** It can be shown that the set $\mathcal{C}$ is asymptotically stable when solutions to equation (2.5) under the control law $u$ are complete (e.g., when $\mathcal{C}$ is compact [330]). We only discussed stability with respect to an equilibrium point in section 2.1.4. What does asymptotic stability mean with respect to a set? We can define the stability of a set in almost the same way by defining the distance function to the set $\mathcal{C}$ as $d(x, \mathcal{C}) := \inf_{\zeta \in \mathcal{C}} \|x - \zeta\|$. Now, the set $\mathcal{C}$ is

- Stable if, for each $\epsilon > 0$, there exists a constant $\delta > 0$ such that

$$d(x_0, \mathcal{C}) \leq \delta \implies d(x(t), \mathcal{C}) \leq \epsilon \text{ for all } t \geq 0;$$

- Asymptotically stable if the set $\mathcal{C}$ is stable and there is a constant $\delta_a > 0$ such that

$$d(x_0, \mathcal{C}) \leq \delta_a \implies \lim_{t \to \infty} d(x(t), \mathcal{C}) = 0.$$

In other words, when the set $\mathcal{C}$ is asymptotically stable, $x(t)$ approaches $\mathcal{C}$ as $t \to \infty$ when $x_0 \in \mathcal{D} \setminus \mathcal{C}$ (i.e., when the system starts outside the set $\mathcal{C}$).

The proof for this result, presented in [330], uses the Lyapunov function

$$V(x) := \begin{cases} 0 & \text{if } x \in \mathcal{C} \\ -h(x) & \text{if } x \in \mathcal{D} \setminus \mathcal{C}. \end{cases}$$

Note that $V(x) = 0$ for $x \in \mathcal{C}$ and $V(x) > 0$ for $x \in \mathcal{D} \setminus \mathcal{C}$, while $u$ can be selected such that $\frac{\partial V(x)}{\partial x}(f(x,t) + g(x,t)u) \leq \alpha(-V(x)) < 0$ for all $x \in \mathcal{D} \setminus \mathcal{C}$, which resembles the conditions for asymptotic stability of a set given in lemma 2.5.

## 2.2.3 Control Lyapunov Functions

In the same way that we introduced CBFs to enforce forward invariance of the system in equation (2.5) by a suitable choice of $u$, we will now introduce CLFs to enforce the stability of equation (2.5).

**Definition 2.9 (Valid Control Lyapunov Function)** *A continuously differentiable and positive definite function $V: \mathbb{R}^n \to \mathbb{R}$ is said to be a* valid control Lyapunov function *for equation (2.5) on a set $\mathcal{D}$ if $0 \in \mathcal{D}$ and if, for all $(x,t) \in \mathcal{D} \times \mathbb{R}_{\geq 0}$, it holds that*

$$\inf_{u \in \mathbb{R}^m} \frac{\partial V(x)}{\partial x}(f(x,t) + g(x,t)u)) \leq 0. \tag{2.15}$$

Based on this definition, we can define the set of CLF consistent inputs induced by a valid CLF $V$ to be

$$K_{\text{CLF}}(x,t) := \Big\{ u \in \mathbb{R}^m \mid \frac{\partial V(x)}{\partial x}(f(x,t) + g(x,t)u) \leq 0 \Big\}.$$

Based on what we know about CBFs, we can now guarantee stability when the control law $u$ is such that $u(x,t) \in K_{\text{CLF}}(x,t)$, which is formally stated next and presented without proof.

**Theorem 2.4 (Stability via Valid Control Lyapunov Functions)** *Let the functions $f$, $g$, and $u$ that describe the system in equation (2.5) be continuous in its first argument and piecewise continuous in its second argument. Let $x = 0$ be an equilibrium point of the system in equation (2.5). Assume that $V$ is a valid CLF for equation (2.5) on an open and connected set $\mathcal{D}$, and the control law $u$ is such that $u(x,t) \in K_{CLF}(x,t)$ for all $(x,t) \in \mathcal{D} \times \mathbb{R}_{\geq 0}$ with $u(0,t) = 0$. Then the equilibrium point $x = 0$ is stable.*

To obtain a stabilizing minimum norm controller, we remark that we can again formulate a convex quadratic optimization program or obtain a closed-form expression. In other words, $u(x,t) = \hat{u}$ can either be obtained by solving

$$\begin{aligned} &\underset{\hat{u}}{\operatorname{argmin}} \; \|\hat{u}\|^2 \\ &\text{s.t.} \; \frac{\partial V(x)}{\partial x}(f(x,t) + g(x,t)\hat{u}) \leq 0 \end{aligned}$$

or is given in closed form by the expression

$$u(x,t) := \begin{cases} \frac{B_1(x,t)^T B_0(x,t)}{B_1(x,t) B_1(x,t)^T} & \text{if } B_0(x,t) < 0 \\ 0 & \text{otherwise,} \end{cases}$$

where

$$B_0(x,t) := -\frac{V(x)}{\partial x} f(x,t) \quad \text{and} \quad B_1(x,t) := \frac{\partial V(x)}{\partial x} g(x,t).$$

---

**Remark 2.4** We defined CLFs that require a nonstrict decrease in their function values to achieve stability. We remark that, in the literature, often a version with a strict decrease in the spirit of equation (2.11) in lemma 2.5 is considered to accomplish asymptotic stabibility. For time-invariant systems, it was shown in [24] that the existence of a valid CLF with a strict decrease is indeed necessary for a system to be asympotically stabilizable by a control law. In [280], it was further shown under which conditions a continuous and asymptotically stabilizing control low can be explicitly constructed. The elements $u_i(x)$ of this control law $u(x) := \begin{bmatrix} u_1(x) & \dots u_m(x) \end{bmatrix}^T$ are given as

$$u_i(x) := -b_i(x) \frac{a(x) + \sqrt{a^2(x) + \beta^2(x)}}{\beta(x)},$$

with functions $a$, $b_i$, and $\beta$ selected as follows:[12]

$$\begin{aligned} a(x) &:= \frac{\partial V(x)}{\partial x} f(x), \\ b_i(x) &:= \frac{\partial V(x)}{\partial x} g_i(x), \\ \beta(x) &:= \sum_{i=1}^{m} b_i^2(x). \end{aligned}$$

---

[12]We use the notation that $g_i(x)$ denotes the $i$th column of matrix $g(x)$.

### 2.2.4 Funnel Control

The last control technique that we want to present in this chapter is funnel control. For more information about the origin of the name (and meaning) of "funnel control," we refer the reader to figure 2.8, where the dashed lines form a funnel that gets narrower as time progresses. In summary, the goal in funnel control is to constrain a generic error $e:\mathbb{R}^n \times \mathbb{R}_{\geq 0} \to \mathbb{R}^q$ to $q$ such time-varying funnels that may be defined a priori by a control engineer. Then the error $e$ can, for instance, be the tracking error

$$e(x,t) := \left[e_1(x,t) \dots e_q(x,t)\right]^T := x - x_d(t),$$

where $x_d:\mathbb{R}_{\geq 0} \to \mathbb{R}^q$ is a desired trajectory.

Prescribed performance control (PPC) [33, 37] is such a funnel control strategy that we present and follow in this book. The main idea is to transform the original constrained control problem into an unconstrained control problem, which is solved using Lyapunov methods. We next present a short introduction to funnel control and PPC, and in part III, we use specific error functions $e(x,t)$ to enforce transient spatiotemporal logic constraints. Let us first define the performance function $\gamma:\mathbb{R}_{\geq 0} \to \mathbb{R}_{>0}$ that will define the shape of the funnel.

**Definition 2.10 (Performance Function)** *A performance function $\gamma$: $\mathbb{R}_{\geq 0} \to \mathbb{R}_{>0}$ is a continuously differentiable, bounded, positive, and nonincreasing function. In particular, let*

$$\gamma(t) := (\gamma_0 - \gamma_\infty)\exp(-lt) + \gamma_\infty,$$

*where $\gamma_0, \gamma_\infty \in \mathbb{R}_{>0}$ with $\gamma_0 \geq \gamma_\infty$ and $l \in \mathbb{R}_{\geq 0}$.*

Given a performance function $\gamma_i$ and constants $\underline{M}_i, \overline{M}_i \in [0,1]$, which allow additional freedom in designing the funnels, the control task is to synthesize a continuous feedback control law $u$ such that the solution $x:\mathcal{I} \to \mathbb{R}^n$ to the IVP in equation (2.5) under $u$ is such that

$$-\underline{M}_i\gamma_i(t) < e_i(x(t),t) < \overline{M}_i\gamma_i(t) \tag{2.16}$$

for all $t \geq 0$ if $-\underline{M}_i\gamma_i(0) < e_i(x_0,0) < \overline{M}_i\gamma_i(0)$ for each error $i \in \{1,\dots,q\}$; that is, if the error is initially contained within the funnel. Note that the performance function $\gamma_i$ and constants $\underline{M}_i, \overline{M}_i$ are design parameters by which the transient and steady-state behavior of the error $e_i$ can be

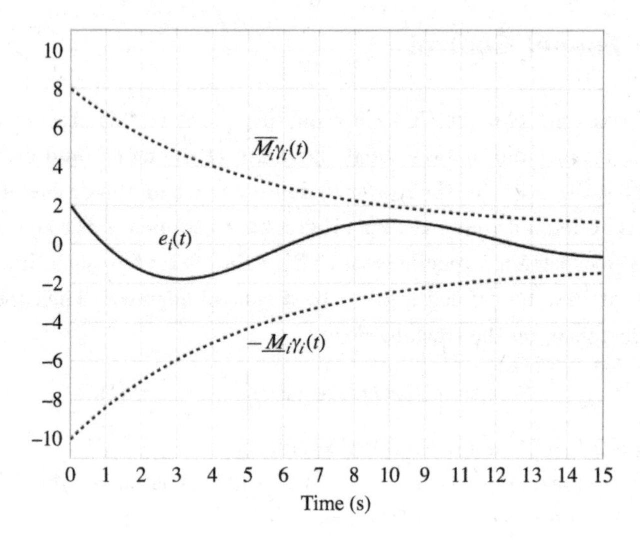


**Figure 2.6**
Evolution of error $e_i(t)$, which is contained in the funnel defined by $\overline{M}_i := 0.8$, $\underline{M}_i := 1$, and $\gamma_i(t) := (\gamma_\infty - \gamma_0)\exp(-lt) + \gamma_0$ with $\gamma_\infty := 10$, $\gamma_0 := 1$, and $l := 0.2$.

prescribed. To illustrate how a performance function $\gamma_i$ can be used to impose a transient behavior on a scalar error $e_i$, we refer the reader to figure 2.6.

---

**Remark 2.5** In definition 2.10, we defined performance functions $\gamma$ that prescribe an exponentially decaying funnel. This is without losing generality, and one could consider performance functions that are not exponentially decaying.

---

Solving the constrained control problem in equation (2.16) may be challenging in general. We therefore transform the constrained control problem in equation (2.16) into an unconstrained control problem via a transformation function $S : (-\underline{M}, \overline{M}) \to \mathbb{R}$, which we define next.

**Definition 2.11 (Transformation Function)** *A transformation function $S : (-\underline{M}, \overline{M}) \to \mathbb{R}$ with $\underline{M}, \overline{M} \in [0, 1]$ is a smooth and strictly increasing function. In particular, let*

$$S(r) := \ln\left(-\frac{r + \underline{M}}{r - \overline{M}}\right).$$

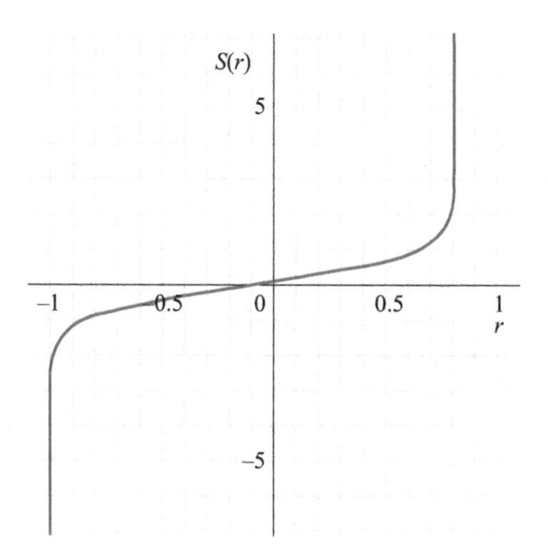


**Figure 2.7**
Transformation function $S$ for the constants $\overline{M}_i := 0.8$ and $\underline{M}_i := 1$.

For the choices of $\underline{M} := 1$ and $\overline{M} := 0.8$ (as used in figure 2.6), we plot the transformation function $S$ in figure 2.7. To transform the constrained control problem in equation (2.16) into an unconstrained control problem, define the normalized error

$$\xi_i(x,t) := \frac{e_i(x,t)}{\gamma_i(t)}.$$

If we now divide equation (2.16) by $\gamma_i(t)$, we obtain the intermediate constrained control problem

$$-\underline{M}_i < \xi_i(x(t),t) < \overline{M}_i. \tag{2.17}$$

By further defining the transformed error

$$\epsilon_i(x,t) := S\big(\xi_i(x,t)\big)$$

and applying the transformation function $S$ to the intermediate constrained control problem in equation (2.17), we obtain the unconstrained control problem

$$-\infty < S\big(\xi_i(x(t),t)\big) < \infty. \tag{2.18}$$

If we now design control law $u$ such that the unconstrained control problem in equation (2.18) is solved (i.e., the transformed error $\epsilon_i(x(t),t)$ is bounded for all times $t \geq 0$), then it follows that the constrained control problem in equation (2.16) is solved. This follows since the transformation function $S$ is strictly increasing, and hence admitting and inverse. To solve the unconstrained control problem, we will view the transformed error as a dynamical system $\dot{\epsilon}_i(x(t),t)$ and use Lyapunov-like methods to ensure that $\epsilon_i(t)$ is bounded by following the result from lemma 2.6. We next show an example to provide more context.

---

**Example 2.14** For illustration purposes, consider the one-dimensional system $\dot{x} = 0.75x + u$ with initial condition $x_0 := 3$. Assume that we want to constrain the system such that equation (2.16) holds for $\underline{M} = \overline{M} := 1$ and for the error $e(x,t) := x$ that we want to constrain to the value of 0.2 in steady state (i.e., we want to regulate $x(t)$ to a region of size 0.2 around the origin). Let $\gamma(t) := (5-0.2)\exp(-t) + 0.2$ (i.e., setting $\gamma_0 := 5$, $\gamma_\infty := 0.2$, and $l := 1$ in definition 2.10). Now, recall that equation (2.16) is a constraint control problem that can be equivalently solved by bounding the transformed error $\epsilon(x(t),t)$. For that purpose, let us derive the dynamics of the transformed error:

$$\dot{\epsilon} = -\frac{2}{(\xi+1)(\xi-1)}\dot{\xi},$$

where we have dropped function arguments for simplicity. By plugging the dynamics of the normalized error $\xi$ into this expression, we obtain

$$\dot{\epsilon} = -\frac{2}{(\xi+1)(\xi-1)}\frac{\dot{x}\gamma - x\dot{\gamma}}{\gamma^2} = -\frac{2}{(\xi+1)(\xi-1)}\frac{\dot{x} - \xi\dot{\gamma}}{\gamma}.$$

For this simplified one-dimensional example, one can now use the control law $u(x,t) = -\epsilon(x,t)$ to achieve that $\epsilon(x(t),t)$ is bounded, consequently achieving that $-\gamma(t) \le x(t) \le \gamma(t)$ if $-\gamma(0) \le x_0 \le \gamma(0)$ initially holds. More formally, this can be shown by using the Lyapunov-like function $V := 0.5\epsilon^2$ and by showing that $\dot{V} = \epsilon\dot{\epsilon} \le 0$ for all $|\epsilon| \ge \mu$ for some value of $\mu$. We are not pursuing a formal proof at this point and instead present simulation results in figure 2.8. The code to reproduce this example is given as follows:

```
1  clc; clear all; close all;
2
3  t=linspace(0,5,101); % time
4  step=t(2)-t(1); % time increment
5  x(1)=3; % initial condition
6
7  for i=1:length(t)
8      %calculate control input
9      gamma(i)=(5-0.2)*exp(-t(i))+0.2;
10     xi(i)= x(i)/gamma(i);
11     epsilon(i)= log(-(xi(i)+1)/(xi(i)-1));
12     u(i) = -epsilon(i);
```

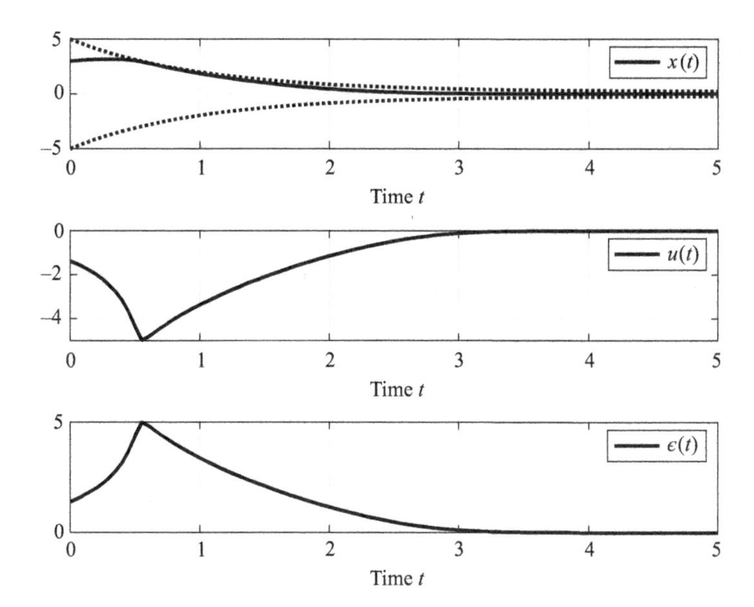


**Figure 2.8**
The state $x(t)$, the control law $u(x(t),t)$, and the transformed error $\epsilon(x(t),t)$ under the funnel control law in example 2.14.

```
13
14      % Newton step
15      x(i+1)=x(i)+step*(0.75*x(i)+u(i));
16  end
```

## 2.3 Multi-Agent Control Systems

In this book, we are interested in dynamical control systems that consist of several subsystems that may be interconnected. For instance, think of a fleet of drones for surveillance, a group of mobile robots in a warehouse, self-driving cars navigating through urban traffic, or a group of robotic manipulators collaboratively grasping and transporting an object. We call such systems "multi-agent control systems" and refer to its subsystems as "agents." In this way, one can argue that multi-agent control systems are essentially high-dimensional dynamical systems. However, when designing control algorithms for multi-agent control systems, we have to address additional challenges in terms of partially available information, limited computation, and increased specification complexity. In fact, agents may have only partial information and may not know the full state of the multi-agent control system due to the agents' limited sensing and communication capabilities. Control under partially available information in networked systems has been studied in detail with the design of distributed control laws using graph theory [209, 57]. We will instead focus on the challenges arising due to limited computation and increased specification complexity, especially when using spatiotemporal logic as a specification formalism. Let us next formally introduce multi-agent control systems and provide some of the notation that we will use throughout this book.

We denote the state of agent $i$ by $x_i \in \mathbb{R}^{n_i}$ (e.g., consisting of an agent's position, velocity, and orientation). The state $x_i$ of agent $i$ may describe different quantities than the state $x_j$ of agent $j$ so the state dimensions need not be the same (i.e., $n_i \neq n_j$). The full state of the multi-agent control system consisting of $N$ agents is defined by the stack of the states of all agents as $x := \begin{bmatrix} x_1^T & \dots & x_N^T \end{bmatrix}^T \in \mathbb{R}^n$, where $n := \sum_{i=1}^N n_i$. Note that the dimensionality $n$ of the multi-agent control system hence directly scales with the number of agents $N$, which can make the control design computationally challenging. The motion of each agent is now described by an ODE; that is, the motion of agent $i$ follows

$$\dot{x}_i = f_i(x_i, t) + g_i(x_i, t)u_i + c_i(x, t),$$

where $u_i \in \mathbb{R}^{m_i}$ is the control input of agent $i$ and the functions $f_i : \mathbb{R}^{n_i} \times \mathbb{R}_{\geq 0} \to \mathbb{R}^{n_i}$ and $g_i : \mathbb{R}^{n_i} \times \mathbb{R}_{\geq 0} \to \mathbb{R}^{n_i \times m_i}$ describe the agents' internal and input dynamics. The function $c_i : \mathbb{R}^n \times \mathbb{R}_{\geq 0} \to \mathbb{R}^{n_i}$ describes dynamical couplings between the agents. Agents may be geometrically coupled via the function $c_i$; that is, $c_i$ can describe the coupled dynamics of two drones that are connected via a cable to transport an object. However, agents may also be coupled by their control inputs $u_i$. For instance, to avoid collisions, each agent may use a collision avoidance controller $u_i^{\text{col}}(x)$ and apply the control action $u_i(x) := u_i^{\text{col}}(x) + v_i$, where $v_i$ is a free control input that we can design to achieve other control objectives. In this case, the dynamics become $\dot{x}_i = f_i(x_i, t) + g_i(x_i, t)v_i + \tilde{c}_i(x, t)$, with the coupling term $\tilde{c}_i(x, t) := g_i(x_i, t)u_i^{\text{col}}(x) + c_i(x, t)$. Finally, note that we have already encountered our first multi-agent control system in example 2.13.

**Example 2.15** Consider a swarm of $N$ drones where each drone may be abstractly modeled as a double integrator (recall example 2.1). However, wind gusts created by the rotors of the drones may now induce couplings between the dynamics of the agents. The dynamics of drone $i$ in the two-dimensional plane, not modeling altitude for convenience, may hence be described as

$$\dot{x}_i(t) = \underbrace{\begin{bmatrix} 0 & 1 & 0 & 0 \\ 0 & 0 & 0 & 0 \\ 0 & 0 & 0 & 1 \\ 0 & 0 & 0 & 0 \end{bmatrix}}_{=:A_i} x_i(t) + \underbrace{\begin{bmatrix} 0 & 0 \\ 1 & 0 \\ 0 & 0 \\ 0 & 1 \end{bmatrix}}_{=:g_i} u_i(t) + c_i(x), \tag{2.19}$$

where we recall that state $x_i$ contains the two-dimensional positions and velocities of the agents. Function $c_i$ may in general be hard to model or could even be unknown. By designing robust control laws $u_i$, we can account for such uncertainty.

Agents in multi-agent control systems are typically coupled by common and shared objectives. In this book, we consider these objectives to be expressed as spatiotemporal logic specifications, as will be introduced more formally in chapter 3. In fact, agents may be required to satisfy *global and local specifications*. By "global," we mean collaborative specifications in which the agents share the same objective (e.g., forming a formation or avoiding collisions). On the other hand, "local" specifications may be objectives individually assigned to agents (e.g., each agent may be required to visit a sequence of goal regions). Local specifications may not be known to other agents. Importantly, the set of local specifications may be conflicting or adversarial, in the sense that it may not be possible to satisfy all of them. For instance, it may not be possible for each agent to visit their individual goal regions while simultaneously satisfying a formation constraint.

The goal in multi-agent control systems research is to design the control inputs $u_i$ so the local and global objectives of the agents are satisfied. Control design for global specifications such as rendezvous, formation control, connectivity maintenance, and collision avoidance have been extensively studied, and these problems have been solved under communication constraints [209, 57]. However, multi-agent control systems may be subject to more complex specifications that go beyond these standard objectives, including reactive (if agent $i$ fails, agent $j$ should take over), repetitive (all agents should repetitively rendezvous), temporal ordering (agent $i$ should arrive at the intersection before agent $j$), and punctuality (agents should form a formation within 2 minutes) specifications [43]. The high-dimensional nature of multi-agent systems and the complexity of these specifications make the control design computationally challenging, which is one of the main bottlenecks.

In this book, we will focus on *designing centralized and decentralized control laws* for multi-agent control systems under complex spatiotemporal logic specifications. By "centralized," we mean that a central control law is designed by having access to functions $f_i$ and $g_i$ (and potentially $c_i$), as well as the states $x_i$ of all agents. Therefore, the goal of a centralized approach is to compute the stacked control input $u := \begin{bmatrix} u_1^T & \dots & u_N^T \end{bmatrix}^T$ using the stacked dynamics

$$\dot{x} = f(x,t) + g(x,t)u + c(x,t), \tag{2.20}$$

where we have used the stacked functions

$$\begin{aligned} f(x,t) &:= \begin{bmatrix} f_1(x_1,t)^T & \dots & f_N(x_N,t)^T \end{bmatrix}^T, \\ g(x,t) &:= \text{diag}(g_1(x_1,t),\dots,g_N(x_N,t)), \\ c(x,t) &:= \begin{bmatrix} c_1(x,t)^T & \dots & c_N(x,t)^T \end{bmatrix}^T. \end{aligned}$$

Note that a centralized control design approach can be viewed as addressing a single-agent control problem. We are particularly interested in designing scalable centralized control laws for such systems under spatiotemporal logic specifications. While we solve a single control problem in a centralized approach, this may still be computationally expensive and require global knowledge of each agent. In a decentralized approach, we instead aim at solving a set of smaller control problems. Typically, each agent solves its own control problem to compute $u_i$.

## 2.4 Notes and References

Most of the material that was covered in this chapter is fairly standard and has appeared in other textbooks. A big portion of the content on systems theory of sections 2.1 and 2.2 was presented in much detail in [129, 23, 68] for linear systems, and in [151, 281, 278, 264, 134, 106] for nonlinear systems. CBFs appeared in recent textbooks [326, 80]. Specifically, the notion of foward invariance is described in detail in [52]. The material on prescribed performance control has not appeared in any textbook before.

The notion of barrier functions for verifying forward invariance of a set was introduced in [235, 234, 233, 236]. The authors also show how barrier functions can be constructed using sum-of-squares programming. Lyapunov functions for stability were presented long ago in the seminal works by Aleksandr Lyapunov; a translated version is available in [194]. Similar to barrier functions, Lyapunov functions can be constructed using sum-of-squares programming [227]. Alternatives to sum-of-square programming are data-driven approaches where a candidate barrier or Lyapunov functions is learned, and where satisfiability modulo theory solvers are used to either prove validity or to produce a counterexample [9, 144]. Reciprocal CBFs were first presented in [319] to guarantee the existence of a control law that renders a given set forward invariant. This notion was later refined in [16, 18, 15], where zeroing CBFs were proposed, which is exactly the notion of CBF that we presented in this chapter. Related to these works is the combination of barrier and Lyapunov functions in [295]. Control Lyapunov functions appeared in [280, 25, 279].

Prescribed performance control is a funnel-based control technique used for deriving analytic control laws that enforce transient and steady-state constraints of a tracking error [37, 33, 34]. The earliest versions of funnel control laws, as well as the precursors of prescribed performance control, were presented in [133]. These works have been extended over the years for nonlinear systems with strict/arbitrary relative degree and even for underactuated systems (see, e.g., [46, 167, 45, 47]). We will cite more related sources, specifically with regard to prescribed performance control, in later chapters.

Section 2.3 provided a brief introduction to multi-agent control systems. We did not cover traditional works that derive distributed control laws under information constraints using graph theory; see [209, 62, 57] for an overview. Multi-agent navigation and collision avoidance were proposed using navigation functions [90, 89, 66, 202], which are based on the seminal works on navigation functions in [246, 155]. Distributed control algorithms that solve the consensus problem have appeared in [220, 244, 271, 221, 263, 88, 168]. The goal in consensus problems is that all agents have to converge to the same value for some common quantity of interest. Solutions to the formation control problem, in which agents are supposed to achieve predefined geometrical shapes, have been addressed in [219, 290, 291, 292, 260, 216, 95]. These works are based on the behavioral model of flocks introduced in [245]. To allow agents to continuously exchange information, solutions to the connectivity maintenance problem have appeared in [139, 334, 333].

# Chapter 3

# Formal Methods and Spatiotemporal Logics

Over the past decades, the formal methods community proposed and studied a broad range of specification languages and system verification/design methods—classically with a focus on discrete software systems due to the community's roots in computer science. These methods can be viewed as multipurpose tools due to the expressivity of formal languages that go beyond notions of stability and forward invariance. We present a selection of the discoveries from the field that we find particularly useful in the study of multi-agent control systems. As we consider continuous-time dynamical systems with a continuous state space, as introduced in chapter 2, we focus on continuous specification languages (i.e., specification languages defined in continuous time and over a continuous state space). Nonetheless, we also provide an introduction to discrete specification languages and verification/design tools and refer the interested reader to [43, 27, 77, 105, 268] for a more detailed treatment of the topic. After reading the chapter, the reader will be familiar with the terminology that we use throughout this book, as well as with the most important concepts and techniques.

In section 3.1, we define important terminology and provide a general introduction to formal specification languages, formal verification, and formal design. To provide some first intuition, we introduce linear temporal logic (LTL), a frequently used discrete specification language. We define the LTL syntax; that is, we show how well-defined LTL formulas are built from atomic symbols by using Boolean and temporal operators, as well as the LTL semantics, which define when an LTL formula is satisfied or violated. In section 3.2, we focus on continuous specification languages and introduce spatiotemporal logics. Particularly, section 3.2.1 presents signal temporal logic (STL), which is a spatiotemporal logic that we use throughout the book because of its expressivity and simplicity. We then introduce the robustness degree and the robust semantics of STL in section 3.2.2, which let us reason about the robustness of an STL formula;[1] that is, about the satisfaction quality or the violation severity. Finally, we introduce timed automata in section 3.3 and show how a timed automaton can encode metric interval temporal logic, which is a logic defined in continuous time but not over a continuous state space. We remark at this point that sections 3.1 and 3.2 are especially relevant for parts II and III of the book, while section 3.3 will only be relevant later, in part IV, when we discuss planning methods for spatiotemporal logic specifications.

## 3.1 Formal Methods

Finding mistakes in the design of a system can be important, even more so to do correct-by-construction system design, and motivates the use of formal methods. To recap from chapter 1, formal methods are mathematical techniques for the specification, design, and verification of software and hardware systems. These techniques are formal, as system specifications are grounded in mathematical logic that allow a formal deduction of system correctness in that logic [79, 323, 1]. In other words, by using formal methods, we can find out if the system is doing what we intend it to do. Formal methods broadly comprise of the following:

- Formal languages to specify a desired system requirement (we refer to requirements formulated in a formal language as "specifications/tasks")
- Automated verification tools that check whether a system implements (i.e., satisfies) the specification
- Automated design tools for correct-by-construction system design

### 3.1.1 Linear Temporal Logic

**Formal languages** are formal in the sense that they have a syntax and semantics that fall within the domain of the system, and they should enable us to

[1] Note that we use the terms "STL specification," "STL task," and "STL formula" interchangeably.

infer useful information about the system. Temporal logics are commonly used as a formal language and were especially developed for the analysis and design of reactive systems (i.e., systems with external inputs). Temporal logics are extensions of Boolean logic (propositions, negations, conjunctions, and disjunctions) by adding temporal operators (next, until, eventually, or always) to reason about temporal properties.

The atomic elements of temporal logics are propositions, which are abstract statements about the world that are either true or false. For instance, for a cleaning robot, we can define the propositions as follows:

$$p_{\text{clean}} := \begin{cases} \top & \text{if the robot is in the cleaning area} \\ \bot & \text{otherwise} \end{cases}$$

$$p_{\text{camera}} := \begin{cases} \top & \text{if the robot's camera is turned on} \\ \bot & \text{otherwise,} \end{cases}$$

where $\top$ and $\bot$ are the symbols for true and false, respectively. We can now use the conjunction $\wedge$ to express that $p_{\text{clean}}$ and $p_{\text{camera}}$ hold via the statement $p_{\text{clean}} \wedge p_{\text{camera}}$ (i.e., that the robot is in the cleaning area and that the camera is turned on). We can further use the negation $\neg$ to express that the camera is turned off, via the statement $\neg p_{\text{camera}}$. Let us introduce another proposition:

$$p_{\text{charge}} := \begin{cases} \top & \text{if robot is in battery charging area} \\ \bot & \text{otherwise.} \end{cases}$$

The statement $\neg p_{\text{camera}} \wedge p_{\text{charge}}$ now expresses that the robot is in the battery-charging area and that the camera is turned off. In addition, we can use the disjunction $\vee$ to express that $p_{\text{clean}}$ or $p_{\text{charge}}$ holds, via the statement $p_{\text{clean}} \vee p_{\text{charge}}$ (i.e., that the robot is in the cleaning area or in the battery-charging area). Note that we can construct other Boolean operators from negations, conjunctions, and disjunctions, such as the implication $\Rightarrow$ and equivalence $\Leftrightarrow$. In particular, for two propositions $p$ and $q$, the statement $p \Rightarrow q$ is logically equivalent to $\neg p \vee q$, while $p \Leftrightarrow q$ is logically equivalent to $(p \Rightarrow q) \wedge (q \Rightarrow p)$.

These Boolean statements, however, do not have the expressivity that we require in robotic systems, or cyber-physical systems in general, where we reason over trajectories and are interested in temporal behavior, which motivates the introduction of temporal operators. The eventually operator, for which we use the symbol $F$, allows to express that a statement will hold eventually; for instance, $F\, p_{\text{clean}}$ means that the robot will eventually be within the cleaning area if the robot is initially not within the cleaning area. The always operator $G$ expresses that a statement always holds; for instance, $G\, p_{\text{clean}}$ means that the robot is always within the cleaning area. The until operator $U$ expresses that a statement holds until another statement starts to hold; for instance, $p_{\text{camera}} \, U \, \neg p_{\text{clean}}$ means that the robot's camera is turned on until the robot is not in the cleaning area.

The informal introduction of temporal logics provided here closely resembles the syntax and semantics of LTL, which is arguably one of the most prominent temporal logics [232, 201]. Even though we will not use LTL in this book, we will next define the concept due to its close resemblance to STL and as a formal introduction into temporal logics.

**LTL Syntax.** Consider a set of atomic propositions $AP$ in which we recall that $p \in AP$ can be either true ($\top$) or false ($\bot$). The syntax of LTL defines a set of rules according to which well-defined LTL formulas can be constructed, and it is expressed in the Backus-Naur form as[2]

$$\phi ::= \top \mid p \mid \neg\phi' \mid \phi' \wedge \phi'' \mid \phi' \, U \, \phi''. \tag{3.1}$$

Admittedly, equation (3.1) looks nonstandard, so how should we interpret the LTL syntax? The symbol ::= means that the left side in (3.1), where $\phi$ is a free variable, is assigned to be one of the expressions from the right side, which are separated by the vertical bars $|$. On the right side, the symbols $\top$ and $p$ and the operators $\neg$, $\wedge$, and $U$ have been discussed before, while $\phi'$ and $\phi''$ are already well-defined LTL formulas. Another interpretation is as follows. The elements $\top$ and $p$ are well-defined LTL formulas that we denote, for instance, by $\phi'$ or $\phi''$. Using these well-defined LTL formulas, we can recursively construct new well-defined LTL formulas using negations ($\neg\phi'$), conjunctions ($\phi' \wedge \phi''$), and the "until" operator ($\phi' \, U \, \phi''$), and repeat the process from there.

The cautious reader now may have observed that we omitted the Boolean disjunction ($\vee$), implication ($\Rightarrow$), and equivalence ($\Leftrightarrow$) operators, as well as the temporal eventually ($F$) and always ($G$) operators from the LTL syntax. This is since we can define them from the existing operators in equation (3.1) as follows:

$$\begin{aligned}
&\phi' \vee \phi'' := \neg(\neg\phi' \wedge \neg\phi'') && \text{(disjunction operator),}\\
&\phi' \Rightarrow \phi'' := \neg\phi' \vee \phi'' && \text{(implication operator),}\\
&\phi' \Leftrightarrow \phi'' := (\phi' \Rightarrow \phi'') \wedge (\phi'' \Rightarrow \phi') && \text{(equivalence operator),}
\end{aligned}$$

[2]The reader familiar with LTL will notice that we omitted the "next" operator from its syntax. By introducing LTL, we intend to gradually prepare the reader for our introduction to STL where the next operator is not defined due to the continuous-time setting. We hence simply omit the "next" operator in our exposition.

$$F\,\phi' := \top\, U\, \phi' \qquad \text{(eventually operator)},$$

$$G\,\phi' := \neg\, F\, \neg\phi' \qquad \text{(always operator)}.$$

**Example 3.1** Let us provide a simple example of a well-defined LTL formula. Recall the previously introduced propositions $p_{\text{clean}}$, $p_{\text{camera}}$, and $p_{\text{charge}}$, and define another proposition as follows:

$$p_{\text{battery}} := \begin{cases} \top & \text{if robot battery level is critical} \\ \bot & \text{otherwise,} \end{cases}$$

which indicates when the robot's battery is almost empty and needs to be recharged. We can now construct the well-defined LTL formula

$$\phi := (F\,p_{\text{clean}}) \wedge (G\,(p_{\text{camera}} \Leftrightarrow p_{\text{clean}})) \wedge (G\,(p_{\text{battery}} \Rightarrow F\,p_{\text{charge}})),$$

which expresses that (1) at some future time, the robot should be in the cleaning region ($F\,p_{\text{clean}}$); (2) at all times, the camera should be turned on if and only if the robot is in the cleaning region ($G\,(p_{\text{clean}} \Leftrightarrow p_{\text{camera}})$); and (3) at all times, if the battery level is critical, the robot should reach the charging area at some future time ($G\,(p_{\text{battery}} \Rightarrow F\,p_{\text{charge}})$). We could alternatively replace the statement $F\,p_{\text{clean}}$ with the statement $G\,F\,p_{\text{clean}}$ to express that the robot should always eventually (i.e., periodically) be within the cleaning region.

**LTL Semantics.** In the previous examples, we have loosely interpreted LTL formulas and informally assigned them a meaning. To formally determine if a well-defined LTL formula is satisfied, we now define the LTL semantics. These semantics are defined over discrete-time Boolean signals. To get some basic intuition first, we refer the reader to figure 3.1, where several time

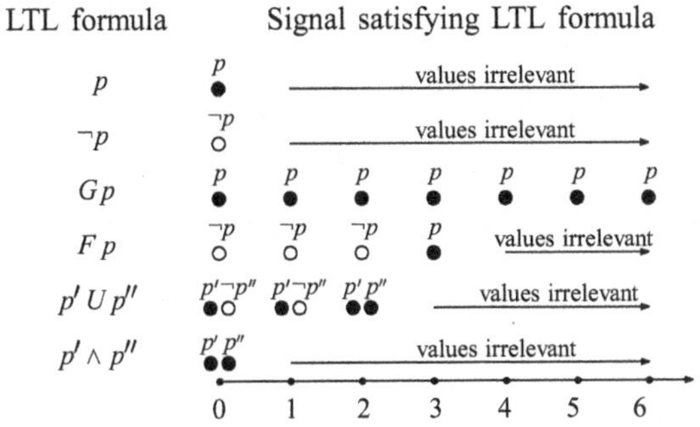


**Figure 3.1**
Intuitive explanation of LTL semantics for basic LTL operators.

sequences of truth values of propositions are shown that satisfy different LTL formulas.

More formally, consider now the discrete-time signal $\sigma : \mathbb{N} \rightarrow \mathbb{B}^{|AP|}$, where we recall that $\mathbb{B} := \{\top, \bot\}$ is the set of Boolean truth values. The signal value $\sigma(t)$ at time $t$ consists of the elements $\sigma_1(t), \ldots, \sigma_{|AP|}(t)$ that indicate the truth value of each proposition $p_1, \ldots p_{|AP|}$. In other words, $\sigma_i(t) = \top$ means that proposition $p_i$ is true at time $t$, while $\sigma_i(t) = \bot$ means that $p_i$ is false at time $t$.

The semantics of LTL are now defined as a relation $\models$ between signals and LTL formulas. In particular, $(\sigma, t) \models \phi$ indicates that signal $\sigma$ satisfies formula $\phi$ at time step $t$, while $(\sigma, t) \not\models \phi$ indicates the opposite. We recursively define this relation and start with defining it for the Boolean operators:

$$
\begin{aligned}
(\sigma, t) \models \top & \quad \text{iff} \quad \text{holds by definition}, \\
(\sigma, t) \models p_i & \quad \text{iff} \quad \sigma_i(t) = \top, \\
(\sigma, t) \models \neg \phi' & \quad \text{iff} \quad (\sigma, t) \not\models \phi', \\
(\sigma, t) \models \phi' \wedge \phi'' & \quad \text{iff} \quad (\sigma, t) \models \phi' \text{ and } (\sigma, t) \models \phi''.
\end{aligned}
$$

The semantics of proposition, negation, and conjunction are straightforward and follow our previously provided intuition. The semantics of disjunction, implication, and equivalence follow immediately from these due to logical equivalences as argued before. For the sake of understanding and completeness, we state them explicitly:

$$
\begin{aligned}
(\sigma, t) \models \phi' \vee \phi'' & \quad \text{iff} \quad (\sigma, t) \models \phi' \text{ or } (\sigma, t) \models \phi'', \\
(\sigma, t) \models \phi' \Rightarrow \phi'' & \quad \text{iff} \quad (\sigma, t) \models \phi' \text{ implies } (\sigma, t) \models \phi'', \\
(\sigma, t) \models \phi' \Leftrightarrow \phi'' & \quad \text{iff} \quad (\sigma, t) \models \phi' \text{ iff } (\sigma, t) \models \phi''.
\end{aligned}
$$

To evaluate the semantics of a Boolean formula $\phi$, we can recursively apply the semantics to the structure of $\phi$. Note that the structure of a formula $\phi$ can be thought of as a tree. Let us illustrate the use of the semantics in an example.

---

**Example 3.2** Consider the Boolean formula $\phi := (p_1 \wedge p_2) \vee (\neg p_3)$ where the disjunction $\vee$ is the parent node from which we have the branches $p_1 \wedge p_2$ and $\neg p_3$ that we can trace down further, see figure 3.2. According to the LTL syntax, we have for this example that

$$(\sigma, t) \models \phi \quad \text{iff} \quad (\sigma, t) \models p_1 \wedge p_2 \quad \text{or} \quad (\sigma, t) \models (\neg p_3)$$

where it holds that

$$
\begin{aligned}
(\sigma, t) \models p_1 \wedge p_2 & \quad \text{iff} \quad (\sigma, t) \models p_1 \quad \text{and} \quad (\sigma, t) \models p_2 \\
(\sigma, t) \models \neg p_3 & \quad \text{iff} \quad (\sigma, t) \not\models p_3.
\end{aligned}
$$

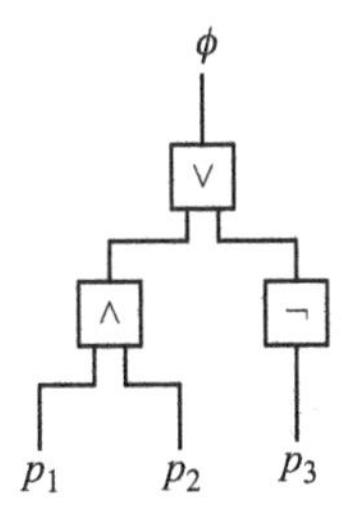


**Figure 3.2**
Tree-like structure of the Boolean formula $\phi := (p_1 \wedge p_2) \vee (\neg p_3)$.

By using the semantics for propositions, we conclude that $(\sigma, t) \models \phi$ iff

- $\sigma_1(t) = \top$ and $\sigma_2(t) = \top$, or
- $\sigma_3(t) = \bot$.

---

It is important to note that the time quantifier $t$ in the semantics has not changed so far and has not been important. However, this will change now, when we present the semantics of the until operator:

$$(\sigma, t) \models \phi' U \phi'' \quad \text{iff} \quad \exists t'' \geq t \text{ s.t. } (\sigma, t'') \models \phi''$$
$$\text{and } \forall t' \text{ s.t. } \; t \leq t' < t'', \; (\sigma, t') \models \phi'.$$

Intuitively, $(\sigma, t) \models \phi' U \phi''$ means that there is a time $(\exists t'' \geq t)$ where $\phi''$ holds, and until then, we have that $\phi'$ holds for all times ($\forall t'$ s.t. $t \leq t' < t''$). For completeness, we present the semantics of the eventually and always operator:

$$(\sigma, t) \models F\phi' \quad \text{iff} \quad \exists t' \geq t \text{ s.t. } (\sigma, t') \models \phi'$$
$$(\sigma, t) \models G\phi' \quad \text{iff} \quad \forall t' \geq t, \; (\sigma, t') \models \phi'.$$

LTL formulas interpreted over these semantics can now specify a diverse range of system specifications; for instance, safety $(G\neg p)$, sequencing $(F(p_1 \wedge F(p_2 \wedge Fp_3)))$, response/reactivity $(p_1 \Rightarrow p_2)$, or surveillance $(GFp)$.

## 3.1.2 Beyond Linear Temporal Logic

To give some historical background, various other temporal logics have been proposed. Computation tree logic (CTL), and its extension CTL* [96], allow for reasoning about more than one system path by introducing path quantifiers to its syntax [78], which we will not consider in this book. Note that the LTL semantics were defined with respect to a single path $\sigma$, while CTL semantics are defined with respect to transition systems (formally defined next) that may admit multiple paths.[3] Probabilistic extensions such as the probabilistic computation tree logic (PCTL) for probabilistic quantification of statements were also proposed [127].

Even though the term "temporal" suggests that temporal logics allow for reasoning about the temporal behavior of a system, it is important to note that this is only true in an abstract way. First, all previously mentioned temporal logics reason over discrete time. In order to determine if a robot trajectory satisfies a temporal logic specification, the trajectory needs to be discretized in time. Second, these temporal logics only allow a qualitative reasoning about temporal properties. For instance, in example 3.1, one can specify that "at some future time, the robot should be within the cleaning area" ($F\, p_{\text{clean}}$), but it is not possible to quantify when exactly, or in which time frame, the robot should be in the cleaning area. Metric interval temporal logic (MITL) (as well as STL), which we will introduce later in this discussion, instead reasons over continuous time and quantitative temporal properties [13].

Further, we want to point out that temporal logics are propositional. For cyber-physical systems, the use of abstract propositions may not always be sufficient, as we may need to capture nuances between the extremes of task satisfaction and violation. We therefore consider spatiotemporal logics that allow a continuous quantification of the atomic elements by the use of predicates. Predicates are real-valued functions of the state space that replace propositions as the atomic elements. For instance, instead of the proposition $p_{\text{clean}}$ from before, we use the predicate

$$\mu_{\text{clean}}(x) := \begin{cases} \top & \text{if } \|x - \begin{bmatrix} 10 & 10 \end{bmatrix}^T \| \leq 1 \\ \bot & \text{otherwise,} \end{cases}$$

which defines the cleaning region of the robot to be

$$O^{\mu_{\text{clean}}} := \{x \in \mathbb{R}^2 | \mu_{\text{clean}}(x) = \top\} = \{x \in \mathbb{R}^2 | \|x - \begin{bmatrix} 10 & 10 \end{bmatrix}^T \| \leq 1\};$$

that is, the cleaning area is centered at $\begin{bmatrix} 10 & 10 \end{bmatrix}^T$ and has a radius of 1. As we shall see later in this chapter, the explicit representation of the state $x$ in

[3] With CTL, one can, for instance, make statements that there is a robot path that satisfies a logical statement or that the statement holds for all robot paths. In contrast, LTL allows one only to reason linearly over a single robot path.

the atomic elements will allow us to define robust semantics that indicate how robustly a temporal logic formula is satisfied. STL is a spatiotemporal logic that uses predicates and closely follows the syntax and semantics of LTL [199]. In addition, STL is defined over continuous time (much like MITL); we will introduce its syntax and semantics later in this chapter.

### 3.1.3 Traditional Formal Verification and Design

**Formal verification** is the process of checking if a system satisfies a given formal specification (e.g., a temporal logic specification) to guarantee the correctness of the system. Various theories and automated tools were developed for the verification of deterministic systems, such as model checking [27, 77] or theorem proving [86, 274, 275, 272]. Nondeterministic system verification was studied using probabilistic model checking [49, 127, 162]. Model checking is arguably the most common technique for system verification that requires obtaining a model of the system, which is often referred to as an abstraction [40, 43, 228]. The idea in model checking is to exhaustively search the state space of the system and to determine, based on this search, if the system specification is satisfied. Such exhaustive search techniques require a discrete system abstraction at their core and render the verification problem computationally intractable for multi-agent systems due to the dimensionality of the state space of a multi-agent system, known as the "curse of dimensionality."

While we will not use discrete verification/design tools in later chapters, we nonetheless provide a brief introduction next to motivate the approach taken in this book and to be able to borrow some of the underlying ideas later. As a system model, labeled finite-state transition systems or variants thereof are typically used.

**Definition 3.1 (Transition System)** *A transition system is as a tuple* $TS := (X, X_0, \delta, AP, L)$*, where* $X$ *is a set of states,* $X_0 \subseteq X$ *is the set of initial states,* $\delta \subseteq X \times X$ *is a transition relation such that* $(x', x'') \in \delta$ *indicates a transition from state* $x'$ *to state* $x''$*,* $AP$ *is the set of atomic propositions, and* $L : X \to \mathbb{B}^{|AP|}$ *is a state-labeling function.*

Naturally, a transition system $TS$ is finite state if the state space $X$ is a finite set. As an example, reconsider the cleaning robot and assume that the two-dimensional workspace of the robot is partitioned into a grid, as shown in figure 3.3(a). Every grid cell corresponds to a state in $X$, and the transition

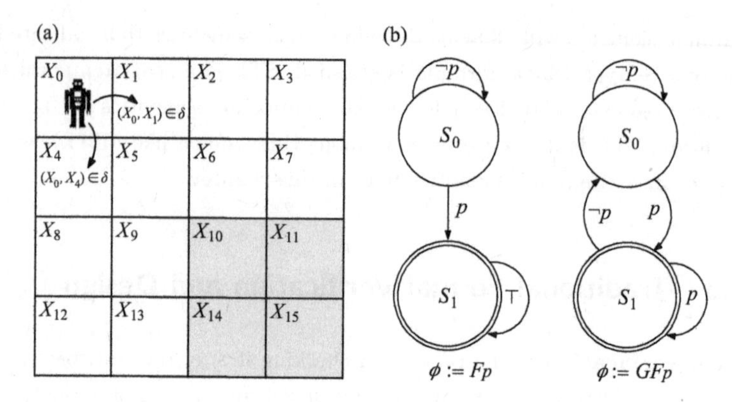


**Figure 3.3**
(a) An illustration of a transition system $TS$ describing the potential motion of a cleaning robot. (b) Büchi automata for the LTL specifications $F\,p$ and $G\,F\,p$.

relation $\delta$ determines if the robot can transition from one cell to another. The labeling function $L$ marks, in this case, that the cells $X_{10}$, $X_{11}$, $X_{14}$, and $X_{15}$ (shown in gray) are within the cleaning area.

A trajectory of the transition system $TS$ is now an infinite state sequence $x_0, x_1, x_2, \ldots \in X$ if $x_0$ is an initial state (i.e., $x_0 \in X_0$) and if consecutive states satisfy the transition relation $\delta$ (i.e., $(x_i, x_{i+1}) \in \delta$ for each $i \in \mathbb{N}$). The path of this infinite state sequence is the labeled state sequence $\sigma := L(x_0), L(x_1), L(x_2) \ldots$. We can now immediately see the connection between a path $\sigma$ of the system and the satisfaction of an LTL specification $\phi$ via the semantics $(\sigma, t) \models \phi$ that we previously defined. However, this will only tell us if a single path $\sigma$ satisfies the specification $\phi$, not if all paths of the system $TS$ satisfy the specification.

The key enabler to check if the system satisfies $\phi$ (i.e., if all paths of $TS$ satisfy $\phi$) is the representation of specification $\phi$ as an automaton (i.e., as another discrete representation). In fact, temporal logic specifications can be translated into finite state automata (e.g., LTL specifications can be translated into Büchi automata [321, 27]).

**Definition 3.2 (Nondeterministic Büchi Automaton)** *A nondeterministic Büchi automaton is a tuple* $BA := (S, S_0, \Delta, AP, \mathcal{A})$, *where* $S$ *is a finite set of states,* $S_0 \subseteq S$ *is the set of initial states,* $\Delta \subseteq S \times \mathbb{B}^{|AP|} \times S$ *is a transition relation so* $(s', d, s'') \in \Delta$ *indicates a transition from state* $s'$ *to state* $s''$ *with input label* $d \in \mathbb{B}^{|AP|}$, $AP$ *is the set of atomic propositions, and* $\mathcal{A} \subseteq S$ *is a set of accepting states.*

For an input signal $\sigma : \mathbb{N} \rightarrow \mathbb{B}^{|AP|}$, a run of the Büchi automaton $BA$ is an infinite state sequence $s := (s_0, s_1, s_2, \ldots)$, with $s_0, s_1, s_2, \ldots \in S$ if $s_0$ is an initial state (i.e., $s_0 \in S_0$) and if consecutive states satisfy the transition relation $\Delta$ (i.e., $(s_i, \sigma(i), s_{i+1}) \in \Delta$ for each $i \in \mathbb{N}$). An "accepting run" is a run for which $\text{inf}(s) \cap \mathcal{A} \neq \emptyset$, where $\text{inf}(s)$ is the set of states that appear infinitely often in $s$. In other words, the condition $\text{inf}(s) \cap \mathcal{A} \neq \emptyset$ requires that the set of accepting states $\mathcal{A}$ is visited infinitely often. Accepting runs are to be distinguished from nonaccepting runs and indicate a specific property of interest of the input signal $\sigma$.

Given an LTL formula $\phi$, a nondeterministic Büchi automaton $BA_\phi$ can always be constructed so that the set of input signals $\sigma$ that result in an accepting run of $BA_\phi$ corresponds exactly to the set of signals that satisfy $\phi$ [27]; see figure 3.3(b) for two simple examples. We will not further discuss the construction of $BA_\phi$ here; we just remark that there are automated tools that perform the translation from LTL specification $\phi$ to Büchi automaton $BA_\phi$ (see e.g., [111]). In section 3.3, we discuss in detail how timed automata, which are automata equipped with clocks, are constructed that can encode MITL specifications.

The last step in this verification process is to construct a product automaton between the transition system $TS$ and the Büchi automaton $BA_{\neg\phi}$. Note particularly the explicit use of the negated formula $\neg\phi$ within this counterexample guided approach. To keep the exposition simple, we will not define this automata product formally (we will do so later for timed automata in section 3.3), but the intuition is that the product automaton exactly characterizes the set of all paths of the system $TS$ that satisfy the specification $\neg\phi$. Consequently, if this set is nonempty, we know that there is a path that violates $\phi$. If such a counterexample exists, we know that the system $TS$ does not satisfy $\phi$. It is straightforward to find such violating paths, if one exists, using graph search techniques such as Dijkstra's algorithm.

Let us briefly also talk about **formal design**, where the goal is to design a control system such that the specification is satisfied by design [43]. Model checking based design techniques follow a similar procedure as the model-based verification process described previously. Here, we instead consider a controllable transition system that is a tuple $TS_c := (X, X_0, \delta_c, AP, L)$, where $X$, $X_0$, $AP$, and $L$ are as in definition 3.1. However, we now need to replace the transition relation by a transition function to eliminate (uncontrollable) nondeterminism. In other words, let $\delta_c : X \times U \rightarrow X$ be a transition function so $x'' = \delta_c(x', u)$ indicates a transition from state $x'$ to state $x''$ under control input $u \in U$. One can then construct the automata product between $TS_c$

and the Büchi automaton $BA_\phi$ and use graph search techniques to extract a sequence of control inputs that result in a path in $TS_c$ that satisfies $\phi$.

The **main bottleneck for multi-agent systems** using discrete verification and design techniques is of computational nature. Particularly, a model checking based approach has exponential complexity. A nondeterministic Büchi automaton $BA_\phi$ consists of $2^{|\phi|+\log(|\phi|)}$ states in the worst case where $|\phi|$ describes the number of Boolean and temporal operators within specification $\phi$ [27]. The automata product between the nondeterministic Büchi automaton $BA_\phi$ and the transition system $TS$ then already consists of up to $|TS| \cdot 2^{|\phi|+\log(|\phi|)}$ states. It is easy to see that the state space $|TS|$ of a multi-agent system may be extremely large. This makes it impossible to handle these large product automata from a computational and memory point of view, especially when it comes to real-time computation. In fact, for $N$ agents that are described by the same transition system $TS_i$, their product transition system $TS$ consists of $|TS_i|^N$ states in the worst case. This motivates the control design approach taken in this book.

## 3.2 Spatiotemporal Logics

Spatiotemporal logics use predicates instead of propositions, which allow for capturing quantitative spatial properties. Spatiotemporal logics thus enable a robustness quantification; that is, we can determine how robustly a specification is satisfied or how severely a specification is violated. Such a robustness quantification will prove useful for feedback control design, discussed later in this book. In this section, we introduce STL as a spatiotemporal logic that is interpreted over continuous time, making it a continuous spatiotemporal logic.

### 3.2.1 Signal Temporal Logic

As opposed to LTL, STL [199, 32] is a predicate-based logic that is defined in continuous time so STL can capture quantitative temporal and spatial system properties. Specifically, the quantification of time allows the expression of more complex specifications that include specific timing requirements or deadlines. For instance, STL specifications can encode combinations of timed reachability ("reach region A within 30 seconds"), timed surveillance ("visit regions B, C, and D every 10–60 seconds while agents form a triangular formation"), timed safety ("always between 5–25 seconds, stay at least 1 meter away from region E"), and many others.

We already briefly introduced predicates in section 3.1.2. More formally, the truth value of a predicate $\mu:\mathbb{R}^n\rightarrow\mathbb{B}$ is obtained after he evaluation of a function $h:\mathbb{R}^n\rightarrow\mathbb{R}$, which we refer to as the "predicate function." In other words, we have

$$\mu(x):=\begin{cases}\top & \text{if } h(x)\geq 0\\ \bot & \text{otherwise.}\end{cases} \tag{3.2}$$

STL is interpreted over real-valued continuous-time signals $x:\mathbb{R}\rightarrow\mathbb{R}^n$, as opposed to LTL, which is interpreted over Boolean-valued discrete-time signals $\sigma:\mathbb{N}\rightarrow\mathbb{B}^{|AP|}$.[4] Note that we let the real numbers $\mathbb{R}$ define the time domain. With minimal change in the following definitions, we could let the nonnegative real numbers $\mathbb{R}_{\geq 0}$ define the time domain (we will comment on this later in this discussion). Let us illustrate the use of predicates briefly in an example.

---

**Example 3.3** Assume the predicate function $h(x):=\epsilon-\|x\|$, which defines the predicate $\mu(x)=\top$ iff $\|x\|\leq\epsilon$. For a given trajectory $x:\mathbb{R}_{\geq 0}\rightarrow\mathbb{R}^n$ that is illustrated in figure 3.4, it trivially follows that $\mu(x(t))=\top$ iff $x(t)$ is within the gray ball of radius $\epsilon$ centered at the origin.

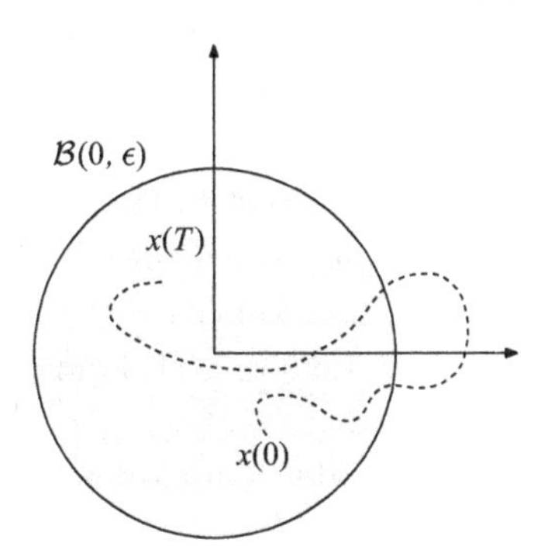


**Figure 3.4**
An example of a predicate function $h(x):=\epsilon-\|x\|$ that defines the predicate $\mu(x)=\top$ iff $\|x\|\leq\epsilon$.

---

[4]We will generally use the symbol $x$ to denote signals, but for convenience and when clear from the context, we will sometimes also use $x$ to denote states, such as in equation (3.2).

We emphasize that predicate function $h$ can be any function encoding, for instance, relative and absolute distances. For control under STL specifications, we will later restrict our attention to continuously differentiable functions $h$.

The syntax of STL defines a set of rules according to which well-defined STL formulas can be constructed, similar to LTL. Therefore, let $M$ denote a set of predicates $\mu \in M$. The STL syntax is expressed in the Backus-Naur form as

$$\phi ::= \top \,|\, \mu \,|\, \neg\phi' \,|\, \phi' \wedge \phi'' \,|\, \phi' U_I \phi'' \,|\, \phi' \underline{U}_I \phi'' , \tag{3.3}$$

where $\phi'$ and $\phi''$ are already well-defined STL formulas and where the true statement ($\top$), negations ($\neg$), and conjunctions ($\wedge$) have the same meaning as in the LTL case. However, $U_I$ now denotes the timed until operator (simply referred to as the until operator in the remainder) with time interval $I \subseteq \mathbb{R}_{\geq 0}$, while $\underline{U}_I$ is the timed "since" operator (which we explain shortly).

While forming well-defined STL formulas follows the same rules and intuition as for LTL, there are three notable differences in the STL syntax: (1) the use of predicates $\mu$ instead of propositions $p$; (2) the appearance of the time interval $I$ in the until operator $U_I$; and (3) the existence of the since operator $\underline{U}_I$. Intuitively, the until operator $\phi' U_I \phi''$ encodes that $\phi'$ has to be true from now on, until $\phi''$ becomes true at some future time within the time interval $I \subseteq \mathbb{R}_{\geq 0}$. The since operator $\phi' \underline{U}_I \phi''$, on the other hand, encodes that $\phi''$ was true at some past time within the time interval $I$, and since then, $\phi'$ is true.

We are now ready to introduce the STL semantics that define when a signal $x : \mathbb{R} \to \mathbb{R}^n$ satisfies an STL specification $\phi$. These semantics are again defined as a relation $\models$ between the signal $x$ and the STL formula $\phi$, and $(x,t) \models \phi$ indicates that the signal $x$ satisfies formula $\phi$ at time step $t$, while $(x,t) \not\models \phi$ indicates the opposite. We say that an STL formula $\phi$ is satisfiable if there exists a signal $x : \mathbb{R} \to \mathbb{R}^n$ such that $(x,0) \models \phi$. For compactness, we state the STL semantics up front, and provide a discussion afterwards.

**Definition 3.3 (STL Semantics)** *For a given signal $x : \mathbb{R} \to \mathbb{R}^n$ and an STL formula $\phi$, the semantics of $\phi$ are recursively defined on its formula structure as*

$$\begin{aligned}
(x,t) \models \top \quad & \textit{iff} \quad \textit{holds by definition,} \\
(x,t) \models \mu \quad & \textit{iff} \quad h(x(t)) \geq 0 \\
(x,t) \models \neg\phi' \quad & \textit{iff} \quad (x,t) \not\models \phi'
\end{aligned}$$

$$(x,t) \models \phi' \wedge \phi'' \quad \textit{iff} \quad (x,t) \models \phi' \textit{ and } (x,t) \models \phi''$$
$$(x,t) \models \phi' U_I \phi'' \quad \textit{iff} \quad \exists t'' \in t \oplus I \textit{ s.t. } (x,t'') \models \phi'' \textit{ and } \forall t' \in (t,t''),\ (x,t') \models \phi'$$
$$(x,t) \models \phi' \underline{U}_I \phi'' \quad \textit{iff} \quad \exists t'' \in t \ominus I \textit{ s.t. } (x,t'') \models \phi'' \textit{ and } \forall t' \in (t'',t),\ (x,t') \models \phi'.$$

For Boolean operators, the semantics compared to LTL differ only in the way that predicates are defined. In the next section, we will particularly show how the spatial quantification via predicates leads to a robustness quantification. The timed until operator $U_I$, however, differs significantly compared to the untimed until operator in LTL. Indeed, for $(x,t) \models \phi' U_I \phi''$ to hold, $\phi''$ must hold in the future within the time interval $I$; that is, there must be a future time $t'' \in t \oplus I$ such that $(x,t'') \models \phi''$, while $\phi'$ needs to hold for all times until then (i.e., within the (open) time interval $(t,t'')$). The since operator is symmetric to the until operator in the sense that the definition is structurally similar, but defined backward in time. For $(x,t) \models \phi' \underline{U}_I \phi''$ to hold, it is required that $\phi''$ held in the past within the time interval $I$; that is, there needs to be a past time $t'' \in t \ominus I$ such that $(x,t'') \models \phi''$, while $\phi'$ has held for all times since then (i.e., within the time interval $(t'',t)$). Importantly, we obtain a temporal quantification by the consideration of the time interval $I$ and by interpreting the satisfaction over continuous-time signals.

For signals $x : \mathbb{R}_{\geq 0} \to \mathbb{R}^n$ that are not defined on the whole real line $\mathbb{R}$, definition 3.3 can be changed by simply intersecting the sets $t \oplus I$ and $t \ominus I$ with the time domain $\mathbb{R}_{\geq 0}$ to account for the shorter time domain.

We can now use logical equivalences to derive the disjunction, implication, and equivalence operators, as in the LTL case. Similarly, we can also define the timed eventually and always operators based on the timed until operator, and we also can define the timed "once" and "historically" operators which we derive from the timed since operator. We summarize these operators here:

$$\begin{aligned}
\phi' \vee \phi'' &:= \neg(\neg\phi' \wedge \neg\phi'') && \text{(disjunction operator)},\\
\phi' \Rightarrow \phi'' &:= \neg\phi' \vee \phi'' && \text{(implication operator)},\\
\phi' \Leftrightarrow \phi'' &:= (\phi' \Rightarrow \phi'') \wedge (\phi'' \Rightarrow \phi') && \text{(equivalence operator)},\\
F_I \phi' &:= \top U_I \phi' && \text{(eventually operator)},\\
\underline{F}_I \phi' &:= \top \underline{U}_I \phi' && \text{(once operator)},\\
G_I \phi' &:= \neg F_I \neg\phi' && \text{(always operator)},\\
\underline{G}_I \phi' &:= \neg \underline{F}_I \neg\phi' && \text{(historically operator)}.
\end{aligned}$$

While their interpretation should be intuitive to the reader, we formally state the semantics of these additional operators here:

$$\begin{aligned}
(x,t) \models \phi' \vee \phi'' &\text{ iff } (x,t) \models \phi' \text{ or } (x,t) \models \phi'',\\
(x,t) \models \phi' \Rightarrow \phi'' &\text{ iff } (x,t) \models \phi' \text{ implies } (x,t) \models \phi'',\\
(x,t) \models \phi' \Leftrightarrow \phi'' &\text{ iff } (x,t) \models \phi' \text{ iff } (x,t) \models \phi'',\\
(x,t) \models F_I \phi' &\text{ iff } \exists t' \in t \oplus I \text{ s.t. } (x,t') \models \phi',\\
(x,t) \models \underline{F}_I \phi' &\text{ iff } \exists t' \in t \ominus I \text{ s.t. } (x,t') \models \phi',\\
(x,t) \models G_I \phi' &\text{ iff } \forall t' \in t \oplus I,\ (x,t') \models \phi',\\
(x,t) \models \underline{G}_I \phi' &\text{ iff } \forall t' \in t \ominus I,\ (x,t') \models \phi'.
\end{aligned}$$

**Example 3.4** Let us now integrate timing requirements in the LTL specification considered in example 3.1 by instead formulating an STL specification. Assume that $\mu_{\text{clean}}$, $\mu_{\text{camera}}$, $\mu_{\text{charge}}$, and $\mu_{\text{battery}}$ are predicates obtained by spatial quantification of the propositions $p_{\text{clean}}$, $p_{\text{camera}}$, $p_{\text{charge}}$, and $p_{\text{battery}}$. Consider now the well-defined STL formula

$$\begin{aligned}
\phi := (F_{[5,10]}\, p_{\text{clean}}) \wedge (G_{[5,10]}\, (p_{\text{camera}} \Leftrightarrow p_{\text{clean}}))\\
\wedge (G_{[0,\infty)}\, (p_{\text{battery}} \Rightarrow F_{[0,10]}\, p_{\text{charge}}))
\end{aligned}$$

that expresses that (1) at some future time within the next 5–10 minutes, the robot should be in the cleaning region ($F_{[5,10]}\, p_{\text{clean}}$); (2) at all times within the next 5–10 minutes, the camera should be turned on if and only if the robot is in the cleaning region ($G_{[5,10]}\, (p_{\text{camera}} \Leftrightarrow p_{\text{clean}})$); and (3) at all times, if the battery level is critical, the robot should reach the charging area at some time within the next 10 minutes ($G_{[0,\infty)}\, (p_{\text{battery}} \Rightarrow F_{[0,10]}\, p_{\text{charge}})$).

## 3.2.2 Robustness Quantification

In the previous subsection, we defined the semantics of STL, which tell us if a signal $x : \mathbb{R} \to \mathbb{R}^n$ satisfies an STL specification $\phi$. In this subsection, we define robustness for spatiotemporal logics to quantify how robustly the signal $x$ satisfies the specification $\phi$. In other words, we would like to think about the following question: if $x$ satisfies $\phi$, how different can a signal $x^* : \mathbb{R} \to \mathbb{R}^n$ be from $x$ while still satisfying $\phi$? More precisely, we will focus on spatial robustness closely following [99], as opposed to temporal robustness [92, 184].

What do we mean by spatial robustness, though? To answer this question, we first define the closeness of two signals $x : \mathbb{R} \to \mathbb{R}^n$ and $x^* : \mathbb{R} \to \mathbb{R}^n$ as

$$d(x, x^*) := \sup_{t \in \mathbb{R}} \|x(t) - x^*(t)\|.$$

Assuming that the signal $x$ satisfies the specification $\phi$ (i.e., $(x,t) \models \phi$), our goal is now to compute a robustness parameter $\epsilon$ such that all signals $x^*$ that are such that $d(x, x^*) < \epsilon$ will also satisfy $\phi$ (i.e., $(x^*, t) \models \phi$). Pictorially, one can think of the set of $x^*$ that satisfy $d(x, x^*) < \epsilon$ as a tube of radius $\epsilon$ around the signal $x$.

There are basically two ways of calculating the robustness value $\epsilon$—namely, the robustness degree and the robust semantics. While the robustness degree is exact and provides the largest possible $\epsilon$, its computation is often intractable. The robust semantics are conservative; that is, $\epsilon$ may not be the largest possible value, but its computation is efficient, as the robust semantics are given in closed form. Due to its simplicity and use in this book, we start with defining the robust semantics.

#### 3.2.2.1 Robust and Quantitative Semantics

Underlying the definition of the robust semantics is a notion of robustness for predicates. To define this notion of predicate robustness, let us first define the set of states that satisfy the predicate $\mu$ as

$$O^\mu := \{x \in \mathbb{R}^n | h(x) \geq 0\},$$

where we recall that the predicate $\mu$ is implicitly defined by predicate function $h$. We can now define the signed distance of the signal value $x(t)$ to the set $O^\mu$ as

$$\text{Dist}(x(t), O^\mu) := \begin{cases} \inf_{x^* \in \text{cl}(\mathbb{R}^n \setminus O^\mu)} \|x^* - x(t)\| & \text{if } x(t) \in O^\mu \\ -\inf_{x^* \in \text{cl}(O^\mu)} \|x^* - x(t)\| & \text{otherwise.} \end{cases}$$

Intuitively, if $x(t)$ satisfies $\mu$, then the predicate robustness is the minimum distance of $x(t)$ to the set of states that violate $\mu$ (i.e., to the set $\mathbb{R}^n \setminus O^\mu$). If $x(t)$ does not satisfy $\mu$, then the predicate robustness is the negated minimum distance of $x(t)$ to the set $O^\mu$. Using this definition, we now recursively define the robust semantics of $\phi$ as a real-valued function $r^\phi(x,t)$.

**Definition 3.4 (STL Robust Semantics)** *For a given signal $x : \mathbb{R} \to \mathbb{R}^n$ and an STL formula $\phi$, the robust semantics of $\phi$ are recursively defined on its formula structure as*

$$\begin{aligned} r^\top(x,t) &:= \infty, \\ r^\mu(x,t) &:= \mathit{Dist}(x(t), O^\mu), \\ r^{\neg\phi'}(x,t) &:= -r^{\phi'}(x,t), \end{aligned}$$

$$r^{\phi' \wedge \phi''}(x,t) := \min(r^{\phi'}(x,t), r^{\phi''}(x,t)),$$
$$r^{\phi' U_I \phi''}(x,t) := \sup_{t'' \in t \oplus I} \min(r^{\phi''}(x,t''), \inf_{t' \in (t,t'')} r^{\phi'}(x,t')),$$
$$r^{\phi' \underline{U}_I \phi''}(x,t) := \sup_{t'' \in t \ominus I} \min(r^{\phi''}(x,t''), \inf_{t' \in (t'',t)} r^{\phi'}(x,t')).$$

The robust semantics are hence recursively built from a robustness quantification of predicates $r^{\mu}(x,t)$ via the signed distance. The negation operator $r^{\neg\phi'}(x,t)$ is simply obtained by negating $r^{\phi'}(x,t)$. The conjunction, until, and since operators are obtained by replacing the "and," "∃," and "∀" operators in the STL semantics in definition 3.3 by "min," "sup," and "inf," respectively, and by applying these operators to the robustness values of the subformulas. To quantify robustness, as originally motivated, we now have the result given in theorem 3.1.

**Theorem 3.1 (Robustness Quantification, Corollary 14 in [99])** *For a given signal $x : \mathbb{R} \to \mathbb{R}^n$ and an STL formula $\phi$, it holds that $(x,t) \models \phi$ iff $(x^*,t) \models \phi$ for all signals $x^* : \mathbb{R} \to \mathbb{R}^n$ that are such that $d(x,x^*) < |r^{\phi}(x,t)|$.*

For completeness, let us also define the robust semantics of the remaining operators that simply follow from definition 3.4:

$$r^{\phi' \vee \phi''}(x,t) := \max(r^{\phi'}(x,t), r^{\phi''}(x,t)),$$
$$r^{\phi' \Rightarrow \phi''}(x,t) := \max(-r^{\phi'}(x,t), r^{\phi''}(x,t)),$$
$$r^{\phi' \Leftrightarrow \phi''}(x,t) := \min(\max(-r^{\phi'}(x,t), r^{\phi''}(x,t)), \max(r^{\phi'}(x,t), -r^{\phi''}(x,t))),$$
$$r^{G_I \phi'}(x,t) := \inf_{t' \in t \oplus I} r^{\phi'}(x,t'),$$
$$r^{\underline{G}_I \phi'}(x,t) := \inf_{t' \in t \ominus I} r^{\phi'}(x,t'),$$
$$r^{F_I \phi'}(x,t) := \sup_{t' \in t \oplus I} r^{\phi'}(x,t'),$$
$$r^{\underline{F}_I \phi'}(x,t) := \sup_{t' \in t \ominus I} r^{\phi'}(x,t').$$

According to theorem 3.1, the robust semantics $r^{\phi}$ provide a valid upper bound on how much a signal $x^*$ can be different from $x$ while both result in a satisfaction of $\phi$. However, one issue that we have not discussed so far is the computation of the signed distance $\text{Dist}(x(t), O^{\mu})$ within the definition of the robust semantics $r^{\mu}$ for predicates $\mu$. While the computation of $\text{Dist}(x(t), O^{\mu})$ for linear predicate functions $h$ is straightforward, this may not always be the case for nonlinear predicate functions $h$. For general nonlinear predicate functions, hence, it is common practice to replace $\text{Dist}(x(t), O^\mu)$ by the value of $h(x(t))$. In other words, instead of $r^\mu := \text{Dist}(x(t), O^\mu)$, we use $h(x(t))$ within definition 3.4. We refer to the semantics obtained in this way as quantitative semantics, which we define next.

**Definition 3.5 (STL Quantitative Semantics)** *For a given signal $x:\mathbb{R}\to\mathbb{R}^n$ and an STL formula $\phi$, the quantitative semantics of $\phi$ are recursively defined on its formula structure as*

$$\begin{aligned}
\rho^{\top}(x,t) &:= \infty,\\
\rho^{\mu}(x,t) &:= h(x(t)),\\
\rho^{\neg\phi'}(x,t) &:= -\rho^{\phi'}(x,t),\\
\rho^{\phi'\wedge\phi''}(x,t) &:= \min(\rho^{\phi'}(x,t), \rho^{\phi''}(x,t)),\\
\rho^{\phi' U_I \phi''}(x,t) &:= \underset{t''\in t\oplus I}{sup}\ \min(\rho^{\phi''}(x,t''), \underset{t'\in(t,t'')}{inf}\ \rho^{\phi'}(x,t')),\\
\rho^{\phi' \underline{U}_I \phi''}(x,t) &:= \underset{t''\in t\ominus I}{sup}\ \min(\rho^{\phi''}(x,t''), \underset{t'\in(t'',t)}{inf}\ \rho^{\phi'}(x,t')).
\end{aligned}$$

As opposed to the robust semantics $r^\phi$, theorem 3.1 does not apply to the quantitative semantics $\rho^\phi$, as we illustrate by means of a counterexample next.

---

**Example 3.5** Consider the two predicate functions $h_1(x) := x$ and $h_2(x) := x^2$ that define equivalent predicates $\mu_1(x) = \mu_2(x)$. In fact, $\mu_1$ and $\mu_2$ describe the region $O^{\mu_1} = O^{\mu_2} := \mathbb{R}_{\geq 0}$. If now $x(t) := 0.5$ for all $t \geq 0$, then it holds that $r^{\mu_1}(x,0) = r^{\mu_2}(x,0) = 0.5$ for the robust semantics, while we have $\rho^{\mu_1}(x,0) = 0.5$ and $\rho^{\mu_2}(x,0) = 0.25$ for the quantitative semantics. Note that the quantitative semantics for $\mu_1$ and $\mu_2$ are different, and the quantitative semantics for $\mu_2$ result in an overly conservative value for the true robustness of 0.5. On the other hand, if $x(t) := 2$ for all $t \geq 0$, then it holds that $r^{\mu_1}(x,0) = r^{\mu_2}(x,0) = 2$ for the robust semantics, while we have $\rho^{\mu_1}(x,0) = 2$ and $\rho^{\mu_2}(x,0) = 4$ for the quantitative semantics. In this case, note that $\rho^{\mu_2}$ results in the value of the true robustness 2 being too large.

---

While it is unfortunate that theorem 3.1 does not apply for the quantitative semantics $\rho^\phi$, it is important to keep in mind that larger values of $\rho^\phi$ indicate a more robust satisfaction of $\phi$. Therefore, we can in general increase the robustness by maximizing the quantitative semantics $\rho^\phi$.

For both the robust and quantitative semantics, however, we can show an important connection to Boolean semantics. This connection is often referred to as the soundness property, which we will state in theorem 3.2.

**Theorem 3.2 (Soundness, Proposition 16 in [99])** *For a given signal $x : \mathbb{R} \rightarrow \mathbb{R}^n$ and an STL formula $\phi$, the following two statements hold for the robust semantics:*

1. $(x,t) \models \phi$ *if* $r^{\phi}(x,t) > 0$,
2. $(x,t) \models \phi$ *implies* $r^{\phi}(x,t) \geq 0$,

*and the following two statements hold for the quantitative semantics:*

1. $(x,t) \models \phi$ *if* $\rho^{\phi}(x,t) > 0$,
2. $(x,t) \models \phi$ *implies* $\rho^{\phi}(x,t) \geq 0$.

The first statement, for both the robust and quantitative semantics in theorem 3.2, is important since it allows for using the robust and quantitative semantics when reasoning over the satisfaction of an STL formula $\phi$. For this reason, we will use the quantitative semantics within the main part of the book as they are computationally easier to deal with. Finally, let us illustrate the concept of the robust and quantitative semantics on two examples. In fact, we have chosen these examples so that the definitions of the robust and the quantitative semantics coincide.

---

**Example 3.6** In figure 3.5, we show a continuous-time signal $x : \mathbb{R}_{\geq 0} \rightarrow \mathbb{R}^n$ and analyze the satisfaction of the three STL formulas:

$$\begin{aligned}
\phi_1 &:= F_{[3,5]}(2 \leq x \leq 3), \\
\phi_2 &:= G_{[2,6]}(0.5 \leq x \leq 3), \\
\phi_3 &:= G_{[2,6]}(0.5 \leq x \leq 2).
\end{aligned}$$

Note that the gray boxes in the graphs indicate the spatial and temporal domains that are of interest for the particular temporal operators, here eventually and always. For these specifications, we can calculate the quantitative semantics as

$$\begin{aligned}
\rho^{\phi_1}(x,0) &= \sup_{t' \in [3,5]} \min(3 - x(t'), x(t') - 2), \\
\rho^{\phi_2}(x,0) &= \inf_{t' \in [2,6]} \min(3 - x(t'), x(t') - 0.5), \\
\rho^{\phi_3}(x,0) &= \inf_{t' \in [2,6]} \min(2 - x(t'), x(t') - 0.5).
\end{aligned}$$

It can be seen that $\phi_1$ and $\phi_2$ are both satisfied by $x$ at time 0 (i.e., $(x,0) \models \phi_1$ and $(x,0) \models \phi_2$). However, formula $\phi_1$ is satisfied more robustly than formula $\phi_2$ since $\rho^{\phi_1}(x,0) > \rho^{\phi_2}(x,0)$. This can intuitively be seen

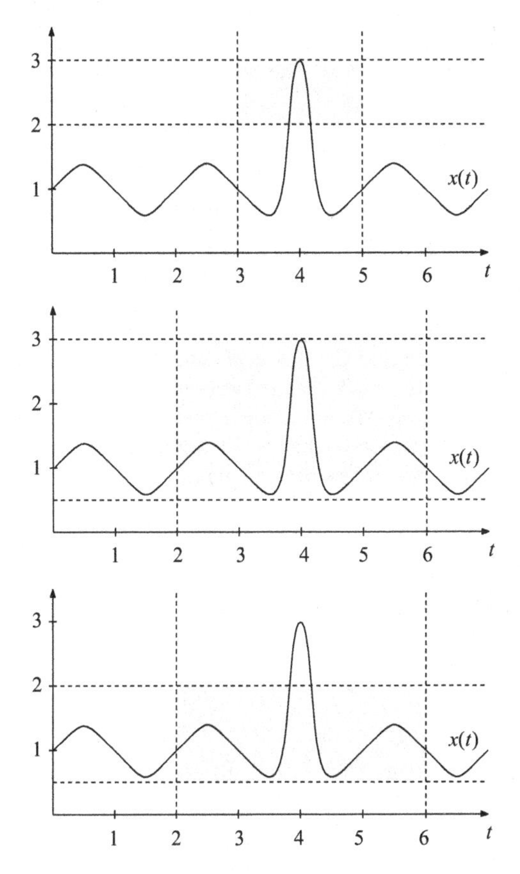


**Figure 3.5**
Top: For the STL formula $\phi_1 := F_{[3,5]}(2 \leq x \leq 3)$, we have that $\rho^{\phi_1}(x,0) = 0.5$ and consequently $(x,0) \models \phi_1$. Middle: For the STL formula $\phi_2 := G_{[2,6]}(0.5 \leq x \leq 3)$, we have that $\rho^{\phi_2}(x,0) = 0.01$, and consequently $(x,0) \models \phi_2$. Bottom: For the STL formula $\phi_3 := G_{[2,6]}(0.5 \leq x \leq 2)$, we have that $\rho^{\phi_3}(x,0) = -1$, and consequently $(x,0) \not\models \phi_3$.

since only small noise added to $x(t)$ at $t = 4$ will lead to $(x,0) \not\models \phi_2$, while small noise will not affect the satisfaction of $\phi_1$. The formula $\phi_3$ is not satisfied since $\rho^{\phi_3}(x,0) = -1$.

**Example 3.7** Consider now the two predicate functions

$$h^{\mu_1}(x) := \|x - x_o\| - r_o$$
$$h^{\mu_2}(x) := r_g - \|x - x_g\|,$$

where $r_o, r_g \in \mathbb{R}_{>0}$ and $x_o, x_g \in \mathbb{R}^n$ encode an obstacle and a goal region, respectively, as illustrated in figure 3.6 (left). The considered STL specification is

$$\phi := G_{[0,\infty)}\mu_1 \wedge F_{[0,9]}G_{[0,\infty)}\mu_2;$$

that is, to always avoid the obstacle defined by $\mu_1$ and to eventually (i.e., within 9 time units) reach the goal region defined by $\mu_2$ and, from then on, to always stay there. We are given a trajectory $x : \mathbb{R}_{\geq 0} \to \mathbb{R}^n$ in figure 3.6 (middle) that does violate the specification (i.e., $(x, 0) \not\models \phi$). In particular, the quantitative semantics are calculated as

$$\rho^{\phi}(x, 0) = \min\Big(\inf_{t' \in [0,\infty)} \text{Dist}(x(t'), O^{\mu_1}),$$
$$\sup_{t'' \in [0,9]} \inf_{t''' \in [t'', t''+\infty)} \text{Dist}(x(t'''), O^{\mu_2})\Big),$$

and from figure 3.6 (right), it can be seen that $\rho^{\phi}(x, 0) = -0.2$.

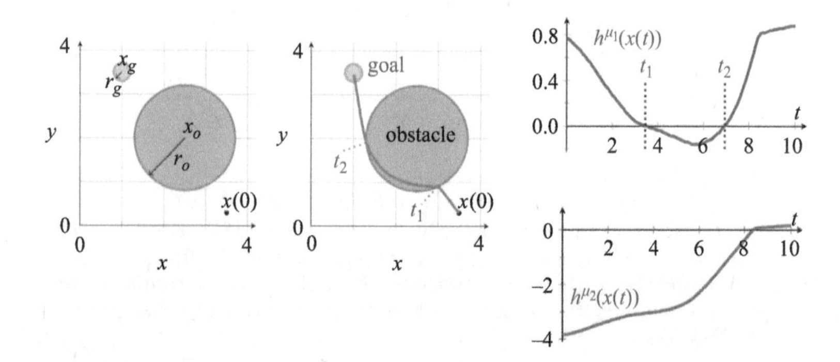


**Figure 3.6**
Left: A reach-avoid specification encoded as $\phi := G_{[0,\infty)}\mu_1 \wedge F_{[0,9]}G_{[0,\infty)}\mu_2$. The regions associated with the predicates $\mu_1$ and $\mu_2$ are illustrated in gray. Middle: A given trajectory $x : \mathbb{R}_{\geq 0} \to \mathbb{R}^n$ that does not satisfy $\phi$ (i.e., $(x, 0) \not\models \phi$). Bottom: The evolution of the predicate functions $h^{\mu_1}(x(t))$ and $h^{\mu_2}(x(t))$ over time. Illustrations provided by Peter Varnai.

#### 3.2.2.2 Robustness Degree

We recall from theorem 3.1 that the robust semantics provide an upper bound on how different a signal $x^*$ can be from a signal $x$ without changing the evaluation of the semantics (i.e., such that $(x,t) \models \phi$ iff $(x^*,t) \models \phi$). The robust semantics are conservative in the sense that there may be a larger upper bound. The robustness degree, on the other hand, finds the largest upper bound [99, definition 7]. Unfortunately, the robustness degree is impossible to compute for most practical specifications, and we thus define the robustness degree here merely for educational purposes. Intuitively, the robustness degree can be understood by partitioning the set of all signals into two sets: one set $\mathcal{L}_t^\phi$ satisfying the specification $\phi$ and the other set $\mathcal{L}_t^{\neg\phi}$ violating $\phi$. Formally, define the set of continuous-time signals that satisfy $\phi$ at time $t$ as

$$\mathcal{L}_t^\phi := \{x : \mathbb{R} \to \mathbb{R}^n | (x,t) \models \phi\}.$$

We can now define the robustness degree as the signed distance of signal $x$ to set $\mathcal{L}_t^\phi$.

**Definition 3.6 (STL Robustness Degree)** *For a given signal $x : \mathbb{R} \to \mathbb{R}^n$ and an STL formula $\phi$, the robustness degree of $\phi$ is defined as*

$$\mathcal{RD}^\phi(x,t) := \begin{cases} \inf_{x^* \in cl(\mathcal{L}_t^{\neg\phi})} d(x,x^*) & \text{if } x \in \mathcal{L}_t^\phi, \\ -\inf_{x^* \in cl(\mathcal{L}_t^\phi)} d(x,x^*) & \text{if } x \notin \mathcal{L}_t^\phi. \end{cases}$$

Intuitively, if the signal $x$ satisfies the specification $\phi$, then the robustness degree is the minimum distance of $x$ to the set of signals that violate $\phi$ (i.e., to the set $\mathcal{L}_t^{\neg\phi}$). If $x$ does not satisfy $\phi$, then the robustness degree is the negated minimum distance of $x$ to set $\mathcal{L}_t^\phi$. Note that the definition of the robustness degree follows a similar construction as the predicate robustness notion $\text{Dist}(x(t), O^\mu)$ that we defined previously, but now by defining the signed distance of the signal $x$ to the set of signals $\mathcal{L}_t^\phi$ using the distance metric $d$. We can now get the same guarantees for the robustness degree as we obtained for the robust semantics in theorem 3.1, which we summarize for completeness in theorem 3.3.

**Theorem 3.3 (Robustness Quantification, Proposition 8 in [99])** *For a given signal $x : \mathbb{R} \to \mathbb{R}^n$ and an STL formula $\phi$, it holds that $(x,t) \models \phi$ iff $(x^*,t) \models \phi$ for all signals $x^* : \mathbb{R} \to \mathbb{R}^n$ that are such that $d(x,x^*) < |\mathcal{RD}^\phi(x,t)|$.*

It can in fact be shown that the robustness degree is strictly larger than the robust semantics in the sense that $|r^\phi(x,t)| \leq |\mathcal{RD}^\phi(x,t)|$ [99, theorem 13].

## 3.3 Timed Automata Theory

While sections 3.1 and 3.2 are relevant for control under spatiotemporal logic specifications in parts II and III of this book, this section will be relevant in part IV, where we discuss planning methods for spatiotemporal logics. For the reader who is primarily interested in parts II and III, we recommend skipping this section for now and return to it before reading part IV. Our goal in this section is to introduce timed automata (specifically timed signal transducers) to represent MITL specifications [13].

### 3.3.1 Metric Interval Temporal Logic

We will now introduce MITL, which is a temporal logic defined over continuous-time signals like STL. However, MITL uses propositions instead of predicates, so robustness cannot be defined as it is for STL.[5] Stated in words, an MITL formula $\varphi$ is an STL formula that is defined over propositions and with additional restrictions on the time interval $I$, which we discuss later in this subsection. This simplifies MITL compared to STL, and STL is strictly more expressive than MITL. Note that we use the symbols $\varphi$ and $\phi$ to distinguish between MITL and STL formulas, respectively. Formally, let $AP$ denote a set of propositions as in LTL. For $p \in AP$, the MITL syntax is

$$\varphi ::= \top \mid p \mid \neg\varphi' \mid \varphi' \wedge \varphi'' \mid \varphi' U_I \varphi'' \mid \varphi' \underline{U}_I \varphi'',$$

where $\varphi'$ and $\varphi''$ are MITL formulas and where $U_I$ and $\underline{U}_I$ are the until and since operators. We restrict the time interval $I$ to belonging to the nonnegative rationals (i.e., $I \subseteq \mathbb{Q}_{\geq 0}$). In addition, we require that $I$ is not a singleton; that is, $I$ cannot be of the form $I := [a, a]$ for $a \in \mathbb{Q}_{\geq 0}$. Note that the former assumption is not restrictive in practice, while the latter assumption excludes punctuality constraints. We can again define disjunction ($\vee$), implication ($\Rightarrow$), equivalence ($\Leftrightarrow$), eventually ($F_I$), once ($\underline{F}_I$), always ($G_I$), and historically ($\underline{G}_I$) operators in the same way as in STL.

Consider now the continuous-time Boolean signal $\bar{\sigma} : \mathbb{R} \to \mathbb{B}^{|AP|}$, which indicates the truth values of the propositions in $AP$ over time (similar to $\sigma$ for LTL). In fact, let $\bar{\sigma}_i(t)$ again denote the truth value of the proposition $p_i \in AP$

[5] Indeed, one can equip propositions in MITL with a notion of robustness as in the original work [99], in which case MITL and STL are almost the same. However, standard MITL does not come with such a robustness notion for propostions.

at time $t$. The relation $(\bar{\sigma}, t) \models \varphi$ indicates that signal $\bar{\sigma}$ satisfies an MITL formula $\varphi$ at time $t$. The semantics of an MTL formula [102, section 4] are then defined as follows:

$$\begin{aligned}
(\bar{\sigma}, t) \models \top \quad &\text{iff} \quad \text{holds by definition,} \\
(\bar{\sigma}, t) \models p_i \quad &\text{iff} \quad \bar{\sigma}_i(t) = \top \\
(\bar{\sigma}, t) \models \neg\varphi' \quad &\text{iff} \quad (\bar{\sigma}, t) \not\models \varphi' \\
(\bar{\sigma}, t) \models \varphi' \wedge \varphi'' \quad &\text{iff} \quad (\bar{\sigma}, t) \models \varphi' \text{ and } (\bar{\sigma}, t) \models \varphi'' \\
(\bar{\sigma}, t) \models \varphi' U_I \varphi'' \quad &\text{iff} \quad \exists t'' \in t \oplus I \text{ s.t. } (\bar{\sigma}, t'') \models \varphi'' \text{ and } \forall t' \in (t, t''), (\bar{\sigma}, t') \models \varphi' \\
(\bar{\sigma}, t) \models \varphi' \underline{U}_I \varphi'' \quad &\text{iff} \quad \exists t'' \in t \ominus I \text{ s.t. } (\bar{\sigma}, t'') \models \varphi'' \text{ and } \forall t' \in (t'', t), (\bar{\sigma}, t') \models \varphi'.
\end{aligned}$$

Note the similarity between the MITL and the STL semantics, which only differ in using Boolean-valued signals instead of real-valued signals, and the way that propositions are evaluated, as opposed to predicates.

## 3.3.2 Timed Automata Representation

An MITL formula $\varphi$ can be translated into a timed automaton that describes all signals $\bar{\sigma}$ that satisfy $\varphi$ [13]. Timed automata were introduced to model the behavior of real-time systems and, in comparison to transition systems and finite automata as introduced in section 3.1.3, timed automata contain clocks and state invariants, as well as guards and resets for transitions [12]. The translation procedure proposed in [13] however, is not constructive. We instead present the method presented in [200, 102], which uses timed signal transducers. Timed signal transducers are timed automata with output labels that enable a compositional construction of a timed signal transducer based on the structure of the formula $\varphi$.

### 3.3.2.1 Timed Signal Transducers

Let $c(t) := \begin{bmatrix} c_1(t) & \dots & c_C(t) \end{bmatrix}^T \in \mathbb{R}^C_{\geq 0}$ be a vector that describes the values of $C$ clocks at time $t$, with an initial value of $c_o(0) := 0$ for each clock $o \in \{1, \dots, C\}$. These clocks have continuous dynamics described by $\dot{c}_o(t) = 1$, as well as discrete dynamics that occur at instantaneous times in form of clock resets. Let $r : \mathbb{R}^C_{\geq 0} \to \mathbb{R}^C_{\geq 0}$ be a reset function such that $r(c) := c'$, where the elements of $c'$ are either $c'_o := c_o$ (no reset) or $c'_o := 0$ (reset). With a slight abuse of notation, we also use $r(c_o) := c_o$ and $r(c_o) := 0$ to refer to resets of an individual clock

$c_o$. Clocks evolve with time when visiting a state of a timed signal transducer, and clocks may be reset during transitions between states. While visiting states or transitioning between states, there may be constraints imposed on the value of the clocks $c$. To describe these constraints, we define clock constraints as Boolean combinations over clock predicates of the form $c_o \leq k$ and $c_o \geq k$ for some constant $k \in \mathbb{Q}_{\geq 0}$.[6] Let $\Phi(c)$ denote the set of all clock constraints over clock variables in $c$. We can now formally define timed signal transducers.

**Definition 3.7 (Timed Signal Transducer [102])** *A timed signal transducer is a tuple* $TST := (S, s_0, \Lambda, \Gamma, c, \iota, \Delta, \lambda, \gamma, \mathcal{A})$*, where* $S$ *is a finite set of states,* $s_0$ *with* $s_0 \cap S = \emptyset$ *is the initial state,* $\Lambda$ *and* $\Gamma$ *are finite sets of input and output variables, respectively;* $\iota : S \to \Phi(c)$ *assigns clock constraints over* $c$ *to each location;* $\Delta$ *is a transition relation so that* $\delta = (s, g, r, s') \in \Delta$ *indicates a transition from* $s \in S \cup s_0$ *to* $s' \in S$*, satisfying the guard constraint* $g \subseteq \Phi(c)$ *and resetting the clocks according to* $r$*;* $\lambda : S \cup \Delta \to BC(\Lambda)$*, and* $\gamma : S \cup \Delta \to BC(\Gamma)$ *are input and output labeling functions where* $BC(\Lambda)$ *and* $BC(\Gamma)$ *denote the sets of all Boolean combinations over* $\Lambda$ *and* $\Gamma$*, respectively, and* $\mathcal{A} \subseteq 2^{S \cup \Delta}$ *is a generalized Büchi acceptance condition.*

Let us now explain how runs of a TST are defined, and specify when a run satisfies the generalized Büchi acceptance condition.

A **run** of a $TST$ over an input signal $d : \mathbb{R}_{\geq 0} \to \mathbb{B}^{|\Lambda|}$ is an alternation of time and discrete steps, resulting in an output signal $y : \mathbb{R}_{\geq 0} \to \mathbb{B}^{|\Gamma|}$. A **time step** of duration $\tau \in \mathbb{R}_{>0}$ is denoted by

$$(s, c(t)) \xrightarrow{\tau} (s, c(t) + \tau),$$

where the following three conditions have to hold for all times $t' \in (0, \tau)$:

1. $d(t + t') \models \lambda(s)$ (state input label satisfied),
2. $y(t + t') \models \gamma(s)$ (state output label satisfied), and
3. $c(t + t') \models \iota(s)$ (state invariant satisfied).

A **discrete step** at time $t$ for a transition $\delta = (s, g, r, s') \in \Delta$ is denoted by

[6] By "Boolean combinations" we mean Boolean formulas formed by using negation and conjunction operators. In this case, we form clock constraints as $g ::= c_o \leq k \mid c_o \geq k \mid \neg g \mid g_1 \wedge g_2$.

$$(s, c(t)) \xrightarrow{\delta} (s', r(c(t))),$$

where the following three conditions have to hold:

1. $d(t) \models \lambda(\delta)$ (transition input label satisfied),
2. $y(t) \models \gamma(\delta)$ (transition output label satisfied), and
3. $c(t) \models g$ (guard satisfied).

Each run starts with a discrete step from the initial state $(s_0, c(0))$. Formally, a run of a $TST$ over $d$ is now the sequence

$$(s_0, c(0)) \xrightarrow{\delta_0} (s_1, r_0(c(0))) \xrightarrow{\tau_1} (s_1, r_0(c(0)) + \tau_1) \xrightarrow{\delta_1} \ldots.$$

Due to the alternation of time and disrete steps, signals $d$ and $y$ can be a concatenation of sequences of points and open time intervals, not allowing for Zeno signals [102].

Let us next define when a run of a $TST$ over an input signal $d$ is **accepting**. Therefore, we associate a function $q: \mathbb{R}_{\geq 0} \rightarrow S \cup \Delta$ with a run as $q(0) := \delta_0$, $q(t) = s_1$ for all $t \in (0, \tau_1), \ldots$. Recall that $\mathcal{A}$ is a generalized Büchi acceptance condition. This means that a run over $d$ is accepting if, for each element $A \in \mathcal{A}$, $\inf(q) \cap A \neq \emptyset$, where $\inf(q)$ contains the states in $S$ that are visited for an unbounded time duration and the transitions in $\Delta$ that are taken infinitely many times. We define the language of $TST$ to be $L(TST) := \{d: \mathbb{R}_{\geq 0} \rightarrow \mathbb{R}^{|\Lambda|} | TST$ has an accepting run over $d\}$, and our next goal will be to construct an automaton $TST_\varphi$ such that $L(TST_\varphi)$ contains exactly those signals that satisfy $\varphi$.

#### 3.3.2.2 Timed Signal Transducers for Metric Interval Temporal Logic

In this section, we construct a timed signal transducer $TST_\varphi$ that encodes the MITL formula $\varphi$. In the first step, the formula $\varphi$ is translated into a logically equivalent formula that only uses the temporal operators $U_{(0,\infty)}$, $\underline{U}_{(0,\infty)}$, $F_{(0,b)}$, and $\underline{F}_{(0,b)}$ for rational constants $b$. This translation is always possible using the rewriting rules in [102, lemmas 4.1, 4.2, 4.3, and 4.4], which we will recall at the end of this section. Assume in the remainder that $\varphi$ corresponds to this logically equivalent formula.

In the second step, we design timed signal transducers for exactly those temporal operators $U_{(0,\infty)}$, $\underline{U}_{(0,\infty)}$, $F_{(0,b)}$, and $\underline{F}_{(0,b)}$ that are contained in the logically equivalent formula $\varphi$; see figure 3.7 for the definition of the timed signal transducers. We remark that all states and transitions except for the state indicated by the dashed circle in $U_{(0,\infty)}$ are accepting and included in $\mathcal{A}$.

Let us explain the functionality of these timed signal transducers by discussing the until operator $\varphi' U_{(0,\infty)} \varphi''$ (figure 3.7(a)). The input variables $\Lambda := \{d_1, d_2\}$ are placeholders and encode the satisfaction of the MITL formulas $\varphi'$ and $\varphi''$.[7] In the simplest case, $\varphi'$ and $\varphi''$ are two propositions, $p_1$ and $p_2$. The timed signal transducer prescribes now at what times $\varphi'$ and $\varphi''$ have to be satisfied so that the MITL formula $\varphi' U_{(0,\infty)} \varphi''$ is satisfied at time zero (i.e., that $(\bar{\sigma}, 0) \models \varphi' U_{(0,\infty)} \varphi''$). Note that there are four initial transitions, and that the output variable is $\Gamma := \{y\}$. We have to select an initial transition that has an output label $y$ because the output signal $y(t)$ encodes $(\bar{\sigma}, t) \models \varphi' U_{(0,\infty)} \varphi''$ and we are interested in satisfaction at time zero. Let us now pick the initial transition leading to the top-left state, in which it has to hold that $d_1 \wedge d_2$. By definition, $\varphi' U_{(0,\infty)} \varphi''$ is satisfied since $d_2$ (encoding $\varphi''$) holds right when entering this state. Let us next pick instead the initial transition leading to the bottom-left state, where it has to hold that $d_1 \wedge \neg d_2$. Since $\neg d_2$ holds in this state, the formula $\varphi' U_{(0,\infty)} \varphi''$ is not satisfied yet, and a transition out of this state has to be taken. Recall that this state is not accepting so that, in order to produce an accepting run, this state has to be left. Now any transition out of this state leads to $d_2$ so that $\varphi' U_{(0,\infty)} \varphi''$ is satisfied. If, in contrast, we want $\varphi' U_{(0,\infty)} \varphi''$ to be violated (i.e., $(\bar{\sigma}, 0) \not\models \varphi' U_{(0,\infty)} \varphi''$), we have to pick an initial transition with output label $\neg y$. We leave it to the reader to verify this case, as well as to check the remaining temporal operators. In addition, we need timed signal transducers for the Boolean operators, as shown in figure 3.8.

Now that we have constructed the basic timed signal transducers, we can compose them to construct a timed signal transducer $TST_\varphi$ for $\varphi$. Therefore, we first construct the formula tree of the MITL formula $\varphi$ at hand; we provide an example of such a tree in figure 3.9. Note that each box in the formula tree corresponds to one timed signal transducers. We now apply two types of composition operators to these timed signal transducers, which are the input-output composition $\triangleright$ and the synchronous product $||$. We formally define these operators at the end of this section. By doing so, we obtain a timed signal transducer $TST_\varphi := (S, s_0, \Lambda, \Gamma, c, \iota, \Delta, \lambda, \gamma, \mathcal{A})$, with input labels $\Lambda := AP$ and output label $\Gamma := \{y\}$. We now have that $TST_\varphi$ has accepting runs over $\bar{\sigma}$ (i.e., $\bar{\sigma} \in L(TST_\varphi)$), with $y(0) = \top$ iff $(\bar{\sigma}, 0) \models \varphi$ [102, theorem 6.7]. Note

[7] In fact, the input signals $d_1(t)$ and $d_2(t)$ encode $(\bar{\sigma}, t) \models \varphi'$ and $(\bar{\sigma}, t) \models \varphi''$, respectively.

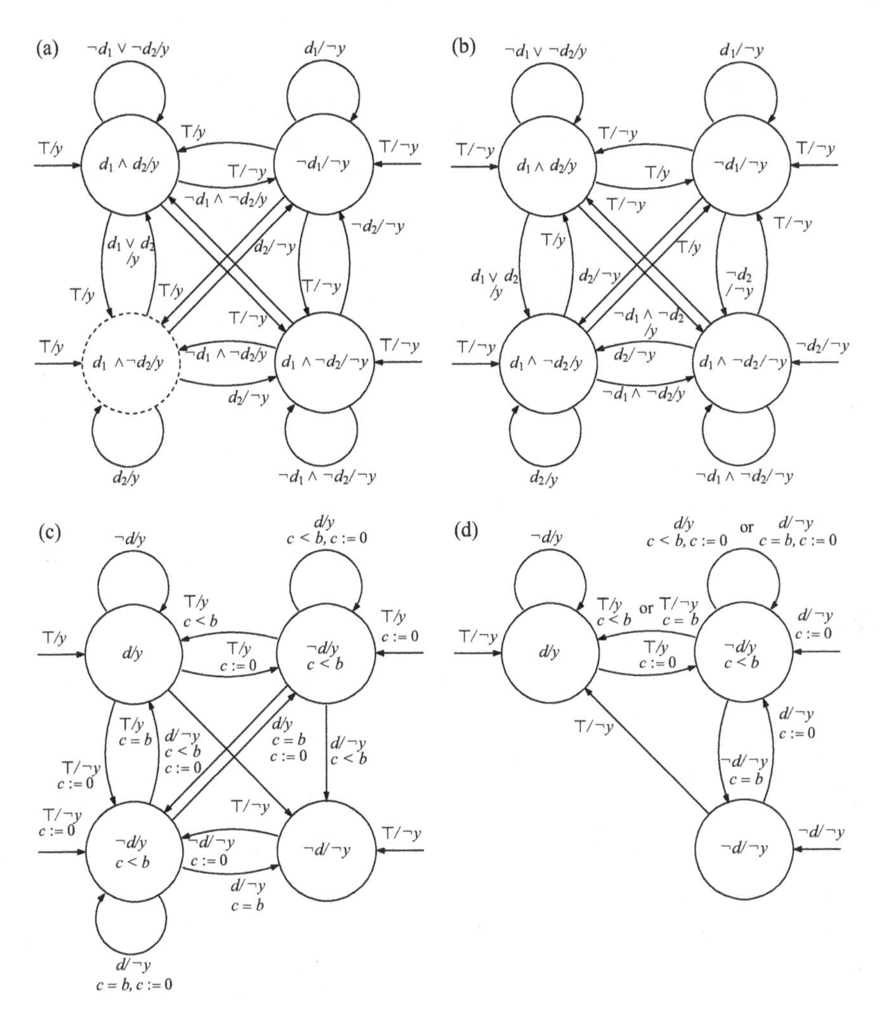


**Figure 3.7**
Timed signal transducers for temporal operators $U_{(0,\infty)}$ (a), $\underline{U}_{(0,\infty)}$ (b), $F_{(0,b)}$ (c), and $\underline{F}_{(0,b)}$ (d). Reprinted with permission from [176, 182].

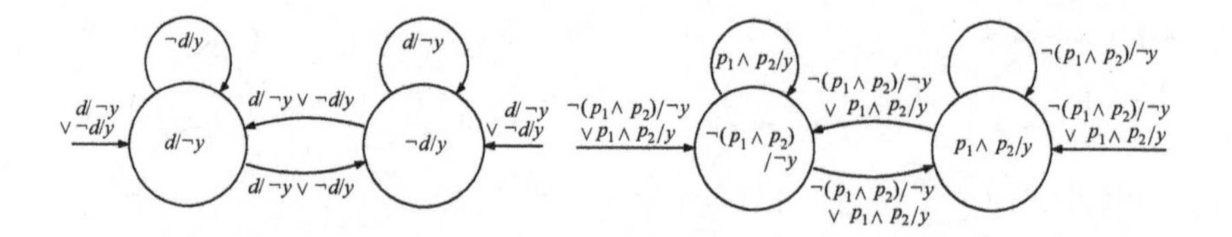


**Figure 3.8**
Timed signal transducers for Boolean operators $\neg$ (left) and $\wedge$ (right). Reprinted with permission from [176, 182].

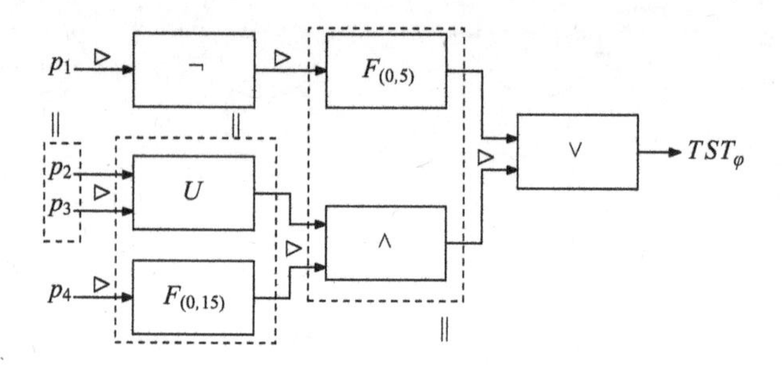


**Figure 3.9**
Formula tree for $\varphi := F_{(0,5)}\neg p_1 \vee (p_2 U_{(0,\infty)} p_3 \wedge F_{(0,15)} p_4)$. Reprinted with permission from [176, 182].

that $y(0) = \top$ (meaning that $\gamma(\delta_0) = y$, where $\delta_0$ is the initial transition) indicates satisfaction of $\varphi$ at time $t = 0$, while $y(0) = \bot$ (i.e., $\gamma(\delta_0) = \neg y$), indicates $(\bar{\sigma}, 0) \not\models \varphi$. We summarize this result in theorem 3.4.

**Theorem 3.4 (Soundness of $TST_\varphi$)** *Given an MITL formula $\varphi$, the timed signal transducer $TST_\varphi$ has an accepting run over the signal $\bar{\sigma} : \mathbb{R}_{\geq 0} \to \mathbb{B}^{|AP|}$; that is, $\bar{\sigma} \in L(TST_\varphi)$, with $y(0) = \top$ iff $(\bar{\sigma}, 0) \models \varphi$.*

In the remainder of this section, we summarize the previously omitted rewriting rules from [102, lemmas 4.1, 4.2, 4.3, and 4.4] and formally define the synchronous product and the input-output composition.

**Rewriting Rules.** We can express the timed until operator $U_I$ as a combination of the untimed until operator $U_{[0,\infty)}$ and the timed eventually operator $F_J$.

**Lemma 3.1 ([102], Lemma 4.1)** *For any rational constants $a$, $b$, and $c$ such that $0 \leq a < b < \infty$ and $0 < c < \infty$, we have*

$$\begin{aligned}
\varphi' U_{(a,b)} \varphi'' &\Leftrightarrow \varphi' U_{(a,\infty)} \varphi'' \wedge F_{(a,b)} \varphi'', \\
\varphi' U_{[a,b)} \varphi'' &\Leftrightarrow \varphi' U_{[a,\infty)} \varphi'' \wedge F_{[a,b)} \varphi'', \\
\varphi' U_{(a,b]} \varphi'' &\Leftrightarrow \varphi' U_{(a,\infty)} \varphi'' \wedge F_{(a,b]} \varphi'', \\
\varphi' U_{[a,b]} \varphi'' &\Leftrightarrow \varphi' U_{[a,\infty)} \varphi'' \wedge F_{[a,b]} \varphi'', \\
\varphi' U_{(c,\infty)} \varphi'' &\Leftrightarrow G_{(0,c]} (\varphi' \wedge \varphi' U_{[0,\infty)} \varphi''), \\
\varphi' U_{[c,\infty)} \varphi'' &\Leftrightarrow G_{(0,c)} \varphi' \wedge G_{(0,c]} (\varphi'' \vee (\varphi' \wedge \varphi' U_{[0,\infty)} \varphi'')).
\end{aligned}$$

We can next express the timed eventually operator $F_I$ by using the timed eventually operator $F_{(0,b)}$.

**Lemma 3.2 ([102], Lemma 4.3)** *For any rational constants $a$, $b$, and $c$ such that $0 < c \leq b - a$, we have*

$$\begin{aligned}
F_{(a+c,b+c)} \varphi' &\Leftrightarrow F_{(0,c)} G_{(0,c)} F_{(a,b)} \varphi', \\
F_{[a+c,b+c)} \varphi' &\Leftrightarrow F_{[0,c)} G_{(0,c]} F_{[a,b)} \varphi', \\
F_{(a+c,b+c]} \varphi' &\Leftrightarrow F_{(0,c]} G_{[0,c)} F_{(a,b]} \varphi', \\
F_{[a+c,b+c]} \varphi' &\Leftrightarrow F_{[0,c]} G_{[0,c]} F_{[a,b]} \varphi'.
\end{aligned}$$

We can next express the timed since operator $\underline{U}_I$ as a combination of the untimed since operator $\underline{U}_{[0,\infty)}$ and the timed once operator $\underline{F}_J$.

**Lemma 3.3 ([102], Lemma 4.2)** *For any rational constants $a$, $b$, and $c$ such that $0 \leq a < b < \infty$ and $0 < c < \infty$, we have*

$$\begin{aligned}
\varphi' \underline{U}_{(a,b)} \varphi'' &\Leftrightarrow \varphi' \underline{U}_{(a,\infty)} \varphi'' \wedge \underline{F}_{(a,b)} \varphi'', \\
\varphi' \underline{U}_{[a,b)} \varphi'' &\Leftrightarrow \varphi' \underline{U}_{[a,\infty)} \varphi'' \wedge \underline{F}_{[a,b)} \varphi'', \\
\varphi' \underline{U}_{(a,b]} \varphi'' &\Leftrightarrow \varphi' \underline{U}_{(a,\infty)} \varphi'' \wedge \underline{F}_{(a,b]} \varphi'', \\
\varphi' \underline{U}_{[a,b]} \varphi'' &\Leftrightarrow \varphi' \underline{U}_{[a,\infty)} \varphi'' \wedge \underline{F}_{[a,b]} \varphi'', \\
\varphi' \underline{U}_{(c,\infty)} \varphi'' &\Leftrightarrow \underline{G}_{(0,c]} (\varphi' \wedge \varphi' \underline{U}_{[0,\infty)} \varphi''), \\
\varphi' \underline{U}_{[c,\infty)} \varphi'' &\Leftrightarrow \underline{G}_{(0,c)} \varphi' \wedge \underline{G}_{(0,c]} (\varphi'' \vee (\varphi' \wedge \varphi' \underline{U}_{[0,\infty)} \varphi'')).
\end{aligned}$$

Finally, we can express the timed once operator $\underline{F}_I$ by using the timed once operator $\underline{F}_{(0,b)}$.

**Lemma 3.4 ([102], Lemma 4.4)** *For any rational constants $a$, $b$, and $c$ such that $0 < c \leq b - a$, we have*

$$\underline{F}_{(a+c,b+c)}\varphi' \Leftrightarrow \underline{F}_{(0,c)}\underline{G}_{(0,c)}\underline{F}_{(a,b)}\varphi',$$
$$\underline{F}_{[a+c,b+c)}\varphi' \Leftrightarrow \underline{F}_{[0,c)}\underline{G}_{(0,c]}\underline{F}_{[a,b)}\varphi',$$
$$\underline{F}_{(a+c,b+c]}\varphi' \Leftrightarrow \underline{F}_{(0,c]}\underline{G}_{[0,c)}\underline{F}_{(a,b]}\varphi',$$
$$\underline{F}_{[a+c,b+c]}\varphi' \Leftrightarrow \underline{F}_{[0,c]}\underline{G}_{[0,c]}\underline{F}_{[a,b]}\varphi'.$$

**Synchronous Product and Input-Output Composition.** In definitions 3.8 and 3.9, we define the synchronous product and the input-output composition.

**Definition 3.8 (Synchronous Product $TST := TST_1 || TST_2$ [102])** *Given*

$$TST_i := (S_i, s_{0,i}, \Lambda_i, \Gamma_i, c_i, \iota_i, \Delta_i, \lambda_i, \gamma_i, \mathcal{A}_i),$$

*with* $i \in \{1, 2\}$, *their synchronous product is*

$$TST := TST_1 || TST_2 = (S, s_0, \Lambda, \Gamma, c, \iota, \Delta, \lambda, \gamma, \mathcal{A}),$$

*with* $S := S_1 \times S_2$, $s_0 := s_{0,1} \times s_{0,2}$, $\Lambda := \Lambda_1 \cup \Lambda_2$, $\Gamma := \Gamma_1 \cup \Gamma_2$, $c := \begin{bmatrix} c_1^T & c_2^T \end{bmatrix}^T$, *and* $\iota(s_1, s_2) := \iota_1(s_1) \wedge \iota(s_2)$. *The transition relation* $\Delta$ *is defined as follows:*

- $((s_1, s_2), g, r, (s_1', s_2')) \in \Delta$ *where* $(s_1, g_1, r_1, s_1') \in \Delta_1$, $(s_2, g_2, r_2, s_2') \in \Delta_2$, $g := g_1 \wedge g_2$, *and* $r := \begin{bmatrix} r_1^T & r_2^T \end{bmatrix}^T$ *(simultaneous transitions),*
- $((s_1, s_2), g_1 \wedge \iota_2(s_2), r_1, (s_1', s_2)) \in \Delta$ *where* $(s_1, g_1, r_1, s_1') \in \Delta_1$ *(left-sided transitions), and*
- $((s_1, s_2), \iota_1(s_1) \wedge g_2, r_2, (s_1, s_2')) \in \Delta$ *where* $(s_2, g_2, r_2, s_2') \in \Delta_2$ *(right-sided transitions);*

*and the input labeling function, defined as follows:*

- $\lambda(s_1, s_2) := \lambda_1(s_1) \wedge \lambda_2(s_2)$ *(state labels),*
- $\lambda((s_1, s_2), g, r, (s_1', s_2')) := \lambda_1(s_1, g_1, r_1, s_1') \wedge \lambda_2(s_2, g_2, r_2, s_2')$ *(simultaneous transitions),*
- $\lambda((s_1, s_2), g_1 \wedge \iota_2(s_2), r_1, (s_1', s_2)) := \lambda_1(s_1, g_1, r_1, s_1') \wedge \lambda_2(s_2)$ *(left-sided transitions), and*
- $\lambda((s_1, s_2), \iota_1(s_1) \wedge g_2, r_2, (s_1, s_2')) := \lambda_1(s_1) \wedge \lambda_2(s_2, g_2, r_2, s_2')$ *(right-sided transitions);*

*while the output labeling function is constructed the same way as the input labeling function. The Büchi acceptance condition is* $\mathcal{A} := \{\mathcal{A}_1 \times (S_2 \cup \Delta_2), (S_1 \cup \Delta_1) \times \mathcal{A}_2\}$.

**Definition 3.9 (Input-Output Composition $TST := TST_1 \triangleright TST_2$ [102])** *Given*

$$TST_i := (S_i, s_{0,i}, \Lambda_i, \Gamma_i, c_i, \iota_i, \Delta_i, \lambda_i, \gamma_i, \mathcal{A}_i),$$

*with $i \in \{1,2\}$, the input-output composition where the output of $TST_1$ is the input of $TST_2$ is*

$$TST := TST_1 \triangleright TST_2 = (S, s_0, \Lambda, \Gamma, c, \iota, \Delta, \lambda, \gamma, \mathcal{A}),$$

*with $S := \{(s_1, s_2) \in S_1 \times S_2 | \text{if } d \models \gamma_1(s_1) \text{ implies } d \models \lambda_2(s_2)\}$, $\Lambda := \Lambda_1$, $\Gamma := \Gamma_2$, and $s_0$, $c$, $\iota$; and $\mathcal{A}$ as defined in the synchronous product. The transition relation $\Delta$ is defined as follows:*

- *$((s_1, s_2), g, r, (s_1', s_2')) \in \Delta$ where $\delta_1 := (s_1, g_1, r_1, s_1') \in \Delta_1$, $\delta_2 := (s_2, g_2, r_2, s_2') \in \Delta_2$, $g = g_1 \wedge g_2$, and $r := \begin{bmatrix} r_1^T & r_2^T \end{bmatrix}^T$ if $d \models \gamma_1(\delta_1)$ implies $d \models \lambda_2(\delta_2)$ (simultaneous transitions),*
- *$((s_1, s_2), g_1 \wedge \iota_2(s_2), r_1, (s_1', s_2)) \in \Delta$ where $\delta_1 := (s_1, g_1, r_1, s_1') \in \Delta_1$ if $d \models \gamma_1(\delta_1)$ implies $d \models \lambda_2(s_2)$ (left-sided transitions), and*
- *$((s_1, s_2), \iota_1(s_1) \wedge g_2, r_2, (s_1, s_2')) \in \Delta$ where $\delta_2 := (s_2, g_2, r_2, s_2') \in \Delta_2$ if $d \models \gamma_1(s_1)$ implies $d \models \lambda_2(\delta_2)$ (right-sided transitions);*

*and the input and output labeling functions are defined as*

- *$\lambda(s_1, s_2) := \lambda_1(s_1)$ and $\gamma(s_1, s_2) := \gamma_2(s_2)$ (state labels),*
- *$\lambda((s_1, s_2), g, r, (s_1', s_2')) := \lambda_1(s_1, g_1, r_1, s_1')$ and $\gamma((s_1, s_2), g, r, (s_1', s_2')) := \gamma_2(s_2, g_2, r_2, s_2')$ (simultaneous transitions),*
- *$\lambda((s_1, s_2), g_1 \wedge \iota_2(s_2), r_1, (s_1', s_2)) := \lambda_1(s_1, g_1, r_1, s_1')$ and $\gamma((s_1, s_2), g_1 \wedge \iota_2(s_2), r_1, (s_1', s_2)) := \gamma_2(s_2)$ (left-sided transitions), and*
- *$\lambda((s_1, s_2), \iota_1(s_1) \wedge g_2, r_2, (s_1, s_2')) := \lambda_1(s_1)$ and $\gamma((s_1, s_2), \iota_1(s_1) \wedge g_2, r_2, (s_1, s_2')) := \gamma_2(s_2, g_2, r_2, s_2')$ (right-sided transitions).*

## 3.4 Notes and References

In section 3.1, we provided a brief introduction into traditional formal verification under temporal logic specifications via modelchecking techniques. Abstractions of physical systems into finite-state transition systems suitable for model checking are further discussed in [14, 228, 287, 41, 40, 115]. Computational techniques to obtain such abstractions in practice are obtained via reachability analysis [113, 114, 161]. For correctness, it is important to note that abstractions have to be designed such that the abstracted system preserves certain properties of the original system. Intuitively, the abstraction should include the behavior of the physical system so the behavior of the abstracted system implies the behavior of the original system. This can be guaranteed via the notion of a bisimulation relation. While we discussed the translation of metric interval temporal logic into a timed signal transducer in section 3.3, many techniques for doing this exist [111, 27, 13, 102]. Beyond model checking for system verification, alternative techniques are based on theorem proving [86, 274, 275, 272] and barrier certificates [235, 233, 236, 50, 51].

Formal design techniques under temporal logic specifications invert the ideas from model checking. There is a broad body of literature on formal design for single-agent systems [e.g., 39, 43, 98, 97, 192, 231, 157, 152, 322, 197, 257, 258, 332]. While these works focus on untimed temporal logic specifications such as LTL, formal design under timed temporal logic specifications such as MITL was considered in [339, 338, 215, 107, 186, 21, 312]. The major bottleneck of these techniques is the state space explosion, which we mentioned in the foreword and discuss further in chapter 8, motivating more efficient probabilistically optimal and complete sampling-based planning techniques [306, 307, 145, 146, 142, 143]. The state space explosion especially becomes an issue in multi-agent systems for which decentralized solutions were proposed [122, 123, 121, 153, 300, 103, 124, 192, 267, 302]. Decentralized solutions mitigate the state space explosion problem, but they still do not scale beyond the number of a few agents, which motivates the approach taken in this book.

Besides LTL, STL, and MITL, as introduced in this chapter, there are various extensions of temporal logics tailored for multi-agent systems, such as counting LTL [217, 257], consensus STL [331], and capability temporal logic [164]. To reason about spatial properties of graphs that model agent interactions directly, multi-agent spatiotemporal logics have been proposed. In Spatial-Temporal Logic (SpaTeL) [125], STL predicates are replaced with spatial formulas over transition systems, while Spatio-Temporal Reach and Escape Logic (STREL) [29, 193] uses spatial reach and escape operators defined over the graph. Spatial Aggregation Signal Temporal Logic (SaSTL) [195] reasons over spatial aggregation and spatial counting characteristics. We remark here that in these works, "spatiotemporal" refers to reasoning over multi-agent graphs, while for this book, we think of "spatiotemporal" more directly in terms of the continuous state space of all agents.

While we have presented robustness notions via the robust semantics and the robustness degree from [99] in section 3.2, various other notions of spatial robustness have been proposed. For instance, the spatially robust semantics from [92], which closely resemble the robust semantics from [99]. Importantly, both works consider spatial robustness by quantifying how much a signal can be perturbed spatially in terms of the metric $d$. Other notions of spatial robustness were introduced in [56, 138, 247, 249, 285, 207, 126, 208, 305, 112]. Robust formal design techniques were proposed based on these notions to find robust system trajectories (e.g., [241, 243, 93, 242, 101, 255, 256, 42, 100, 56] present mixed-integer linear program encodings). Extensions to multi-agent systems were proposed in [189, 190]. However, mixed integer linear programs are NP-hard and computationally challenging despite the existence of efficient solver heuristics. Optimization-based approaches, essentially maximizing robust semantics, have appeared in [207, 259, 126, 208, 225, 226, 158], but are similarly challenging, as the resulting optimization problems are generally nonconvex. More recently, temporal robustness notions were considered, as opposed to spatial robustness [92, 184, 10, 71]. Formal design techniques for temporally robust trajectory design based on mixed-integer linear program encodings were proposed in [250, 251, 252].

# II

# Multi-Agent Control Barrier Functions for Spatiotemporal Constraints

# Chapter 4

# Centralized Time-Varying Control Barrier Functions

In parts II and III of this book, we design efficient feedback control laws for multi-agent control systems under signal temporal logic (STL) specifications. Part II presents control barrier functions (CBFs) for this purpose, while part III presents funnel controllers. We note that the general control problem under STL specifications is NP-hard.[1] We will thus consider fragments of STL (i.e., subsets of STL), to constrain the transient behavior of the agents. These fragments are tractable in the sense that we can design provably correct feedback control laws for these specifications. We then show in part IV how to design planning algorithms for the full fragment of STL using the feedback control laws from parts II and III.

In this chapter, we consider *collaborative* STL specifications, in the sense that the specifications can be satisfied by a coordinated action of the agents (e.g., formations of agents). We present the notion of *time-varying control barrier functions (CBFs)*, and show how an STL specification can be encoded into a time-varying CBF. Particularly, we show how the spatial and temporal

[1] It is known from the Cook-Levin theorem that the Boolean satisfiability problem is NP-complete. The Boolean satisfiability problem is to decide if a propositional Boolean formula is satisfiable, i.e., to find an evaluation of all propositions that satisfy the Boolean formula. Satisfiability of LTL, on the other hand, is already known to be PSPACE-complete [277]. Satisfiability of MITL is in general EXSPACE-complete, and only certain fragments of MITL are known to be PSPACE-complete [13, 12]. As we previously noted, STL is at least as expressive (and hence complex) as MITL due to the use of predicates. In this book, we are not only interested in satisfiability problems, but in control problems of multi-agent systems which additionally increases complexity.

properties of the STL specification at hand are encoded into a CBF. While such time-varying CBFs can often be designed analytically, we present an optimization-based approach for more complex specifications. We then design *centralized* CBF–based feedback control laws that lead to the satisfaction of the specification. As eluded to earlier in this book, "centralized" here means that the control input of each agent is computed centrally as a stack of the agents' control inputs. CBFs, once designed, allow computationally efficient control and can be used even for higher-dimensional, multi-agent systems.

## 4.1 Time-Varying Control Barrier Functions

Let $t_0 \geq 0$ be the initial time of our system. We note that we considered $t_0 := 0$ previously, but choose a more general setting in this section. At time $t \geq t_0$, let $x(t) \in \mathbb{R}^n$ denote the state of a multi-agent control system:

$$\dot{x}(t) = f(x(t)) + g(x(t))u(x(t), t), \quad x(t_0) \in \mathcal{D} \tag{4.1}$$

given in stacked form, e.g., as in equation (2.20), where we recall that the set $\mathcal{D}$ denotes the domain in which the system operates. The functions $f : \mathcal{D} \to \mathbb{R}^n$ and $g : \mathcal{D} \to \mathbb{R}^{n \times m}$ are continuous, and the control law $u : \mathcal{D} \times \mathbb{R}_{\geq 0} \to \mathbb{R}^m$ is to be designed by us. For simplicity and without loss of generality, the system is time invariant. Solutions to the IVP in equation (4.1) are hence functions $x : \mathcal{I} \to \mathcal{D}$, where $t_0$ is the smallest value contained within the left-closed time interval $\mathcal{I}$. For the synthesis of valid CBFs for STL specifications, we will later pose a controllability assumption on $g$, while the theory on time-varying CBFs presented in this section applies to general systems as in equation (4.1).

We will now extend CBFs, as introduced in section 2.2.2, to time-varying CBFs. Consider, therefore, the continuously differentiable function $\mathfrak{b} : \mathbb{R}^n \times [t_0, t_1] \to \mathbb{R}$, and define the time-varying set

$$\mathcal{C}(t) := \{x \in \mathbb{R}^n \mid \mathfrak{b}(x, t) \geq 0\},$$

which explicitly depends on time. For reasons that will become apparent later in this chapter, we have chosen the compact time interval $[t_0, t_1] \subset \mathbb{R}_{\geq 0}$, where $t_0$ and $t_1$ are start and end times, respectively.[2] In this time-varying case, we not only define valid CBFs as in section 2.2.2, but we also define candidate

[2]More generally, one can again consider the time interval $[0, \infty)$.

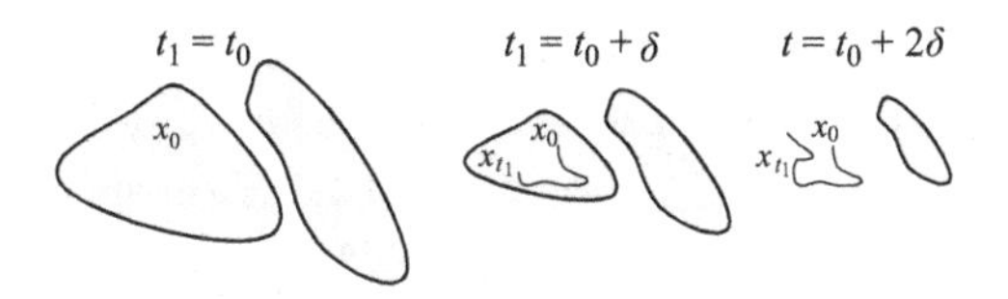


**Figure 4.1**
A set $\mathcal{C}(t)$ that is the union of two not-connected sets and defined via a function $\mathfrak{b}(x,t)$ that is not a candidate CBF.

CBFs. It will hold that $\mathfrak{b}$ being a candidate CBF is a necessary condition for $\mathfrak{b}$ to be a valid CBF. We next state the conditions under which $\mathfrak{b}$ is a candidate CBF.

**Definition 4.1 (Candidate Time-Varying Control Barrier Function)** *A continuously differentiable function $\mathfrak{b}:\mathbb{R}^n \times [t_0,t_1] \to \mathbb{R}$ is said to be a candidate time-varying CBF on $\mathcal{D} \times [t_0,t_1]$ if $\mathcal{D} \supseteq \mathcal{C}(t)$ for all $t \in [t_0,t_1]$, and if, for each $x_0 \in \mathcal{C}(t_0)$, there is an absolutely continuous function $x:[t_0,t_1] \to \mathbb{R}^n$, with $x(t_0):=x_0$, such that $x(t) \in \mathcal{C}(t)$ for all $t \in [t_0,t_1]$.*

We remark that the domain $\mathcal{D}$ of the system has to contain the set $\mathcal{C}(t)$ for all times (i.e., $\mathcal{D} \supseteq \mathcal{C}(t)$ for all $t \in [t_0,t_1]$). This is a natural assumption that resembles the assumption that $\mathcal{C} \subseteq \mathcal{D}$ in the case of static CBFs in section 2.2.2. If the function $\mathfrak{b}$ is now a candidate time-varying CBF, it also holds that the associated set $\mathcal{C}(t)$ is nonempty for each $t \in [t_0,t_1]$, which was similar in the case of static CBFs. However, here the set $\mathcal{C}(t)$ is also such that there is a function $x:[t_0,t_1] \to \mathbb{R}^n$ with $x(t) \in \mathcal{C}(t)$ for all $t \in [t_0,t_1]$. The function $x$ needs to be absolutely continuous so that it may be a solution to equation (4.1) under a suitable control signal $u(x(t),t)$, meaning that there potentially exists a trajectory $x$ of equation (4.1) that evolves within $\mathcal{C}(t)$.

---

**Remark 4.1** The assumption of the existence of such an absolutely continuous function $x:[t_0,t_1] \to \mathbb{R}^n$ can be illustrated as follows: Assume that $\mathcal{C}(t)$ is the union of two sets, $\mathcal{C}_1(t)$ and $\mathcal{C}_2(t)$, that are not connected for all $t \in [t_0,t_1]$; that is, $\mathcal{C}(t):=\mathcal{C}_1(t) \cup \mathcal{C}_2(t)$ with $\mathcal{C}_1(t) \cap \mathcal{C}_2(t)=\emptyset$ for all $t \in [t_0,t_1]$ (see figure 4.1). If now $\mathcal{C}_1(t)$ shrinks so that eventually $\mathcal{C}_1(t)=\emptyset$ for all $t \geq t'$ where $t' \in (t_0,t_1]$ is some constant (think of $t':=2\delta$ in figure 4.1), while $\mathcal{C}_2(t) \neq \emptyset$ for all $t \in [t_0,t_1]$, then $\mathcal{C}(t)$ is nonempty for all $t \in [t_0,t_1]$. However, if $x_0 \in \mathcal{C}_1(t_0)$, there cannot be a solution $x:[t_0,t_1] \to \mathbb{R}^n$ to equation (4.1) with initial condition $x(t_0)=x_0$ so that $x(t) \in \mathcal{C}(t)$ for all $t \in [t_0,t_1]$.

---

A candidate time-varying CBF, however, does not guarantee that we can construct a control law $u$ so that the set $\mathcal{C}(t)$ is forward invariant with respect to the initial value problem (IVP) in equation (4.1).[3] To guarantee this property, we define the notion of a valid time-varying CBF next.

**Definition 4.2 (Valid Time-Varying Control Barrier Function)** *A candidate time-varying CBF $\mathfrak{b}:\mathbb{R}^n \times [t_0,t_1] \to \mathbb{R}$ is said to be a valid time-varying CBF for equation (4.1) on $\mathcal{D}\times[t_0,t_1]$ if there is a locally Lipschitz-continuous extended class $\mathcal{K}$ function $\alpha$ such that, for all $(x,t)\in\mathcal{D}\times[t_0,t_1]$, it holds that*

$$\sup_{u\in\mathbb{R}^m} \frac{\partial\mathfrak{b}(x,t)}{\partial x}(f(x)+g(x)u)+\frac{\partial\mathfrak{b}(x,t)}{\partial t}\geq -\alpha(\mathfrak{b}(x,t)). \tag{4.2}$$

The difference between the constraint in equation (4.2) in definition 4.2 and the constraint in equation (2.12) in definition 2.8, where we defined time-invariant CBFs, is the appearance of the term $\frac{\mathfrak{b}(x,t)}{\partial t}$ (i.e., the partial derivative of $\mathfrak{b}$ with respect to time). We remark that $\frac{\mathfrak{b}(x,t)}{\partial t}$ tightens the constraint in equation (4.2) if $\frac{\mathfrak{b}(x,t)}{\partial t}<0$. In fact, this is the case whenever $\mathfrak{b}(x,t)$ is decreasing in $t$. This constraint tightening makes it more difficult to satisfy the time-varying constraint in equation (4.2) than the time-invariant constraint in equation (2.12). This may result in larger control inputs, but it can also lead to infeasibility. For instance, the constraint in equation (2.12) is feasible if $\frac{\mathfrak{b}(x)}{\partial x}=0$ and $\mathfrak{b}(x)\geq 0$, while the constraint in equation (4.2) can be infeasible if $\frac{\mathfrak{b}(x,t)}{\partial x}=0$, $\mathfrak{b}(x,t)\geq 0$, and $\frac{\mathfrak{b}(x,t)}{\partial t}<0$.

Similar to theorem 2.3 in section 2.2.2, we can now use the time-varying CBF $\mathfrak{b}$ to render the set $\mathcal{C}(t)$ forward invariant. We hence again define the set of safe control inputs induced by a valid time-varying CBF $\mathfrak{b}$ to be

$$K_{\text{CBF}}(x,t):=\left\{u\in\mathbb{R}^m \mid \frac{\partial\mathfrak{b}(x,t)}{\partial x}(f(x)+g(x)u)+\frac{\partial\mathfrak{b}(x,t)}{\partial t}\geq -\alpha(\mathfrak{b}(x,t))\right\}.$$

From this notation, it can easily be seen that theorem 2.3, stated for static CBFs, holds in a slightly modified manner for time-varying CBFs as well, as we have also assumed that $\mathcal{C}(t)\subseteq\mathcal{D}$ for all $t\in[t_0,t_1]$.

[3] Note that forward invariance as introduced in definition 2.4 requires that solutions $x:\mathcal{I}\to\mathcal{D}$ to the IVP in equation (4.1) are complete. We remark that we do not need to assume completeness in this case, as we only consider the restricted time domain $[t_0,t_1]$, and hence only need that $\mathcal{I}\supseteq[t_0,t_1]$.

**Theorem 4.1 (Forward Invariance via Valid Time-Varying Control Barrier Functions)** *Let the functions $f$, $g$, and $u$ that describe the system in equation (4.1) be continuous in its first argument and piecewise continuous in its second argument. Assume that $\mathfrak{b}$ is a valid time-varying CBF for equation (4.1) on $\mathcal{D} \times [t_0, t_1]$ and that the control law $u$ is such that $u(x,t) \in K_{CBF}(x,t)$ for all $(x,t) \in \mathcal{D} \times [t_0, t_1]$. If the initial condition $x(t_0)$ is such that $x(t_0) \in \mathcal{C}(t_0)$, then each solution $x : \mathcal{I} \to \mathbb{R}^n$ to the IVP in equation (4.1) under $u$ is such that $x(t) \in \mathcal{C}(t)$ for all $t \in \mathcal{I} \cap [t_0, t_1]$. If solutions $x$ are defined on $[t_0, t_1]$ (i.e., $\mathcal{I} \supseteq [t_0, t_1]$), it holds that $\mathcal{C}$ is forward invariant with respect to the IVP in equation (4.1) under $u$ over the time interval $[t_0, t_1]$; that is, $x(t_0) \in \mathcal{C}(t_0)$ implies $x(t) \in \mathcal{C}(t)$ for all $t \in [t_0, t_1]$.*

*Proof:* *The proof follows similar steps as the earlier proofs for valid barrier functions in theorem 2.1 and valid CBFs in theorem 2.3. First, note that solutions $x : \mathcal{I} \to \mathbb{R}^n$ to the IVP in (4.1) under $u$ exist due to the assumptions made on the functions $f$, $g$, and $u$. As $\mathfrak{b}$ is a valid time-varying CBF and the control law $u$ is such that $u(x,t) \in K_{CBF}(x)$, it follows that each solution $x$ is such that*

$$\begin{aligned}\dot{\mathfrak{b}}(x(t),t) = &\frac{\partial \mathfrak{b}(x(t),t)}{\partial x}\left(f(x(t),t) + g(x(t),t)u(x(t),t)\right)\\ &+ \frac{\partial \mathfrak{b}(x(t),t)}{\partial t} \geq -\alpha(\mathfrak{b}(x(t),t))\end{aligned}$$

*for all $t \in \mathcal{I} \cap [t_0, t_1]$. Again using the Comparison Lemma, we see that $\mathfrak{b}(x(t), t) \geq 0$ for all $t \in \mathcal{I} \cap [t_0, t_1]$ if $\mathfrak{b}(x(0), 0) \geq 0$. Hence, $x(t_0) \in \mathcal{C}$ implies $x(t) \in \mathcal{C}$ for all $t \in \mathcal{I} \cap [t_0, t_1]$, and $x(t) \in \mathcal{C}(t)$ for all $t \in [t_0, t_1]$ if $x$ is defined on $[t_0, t_1]$.*

Recall that the time interval of interest is $[t_0, t_1]$, and note that we considered the intersection of $\mathcal{I}$ and $[t_0, t_1]$ in theorem 4.1. Using lemma 2.2 and following a similar argument as in our earlier discussions of valid barrier functions and valid CBFs, we can achieve the requirement that solutions $x : \mathcal{I} \to \mathcal{D}$ to the IVP in equation (4.1) under $u$ is defined on $[t_0, t_1]$ if $\mathcal{C}(t)$ is a compact set for all $t \in [t_0, t_1]$. Moving forward, we will ensure this property by construction of $\mathcal{C}(t)$.

## 4.2 Encoding Signal Temporal Logic

In this section, we show how to encode an STL formula $\phi$ into a candidate time-varying CBF $\mathfrak{b} : \mathbb{R}^n \times \mathbb{R}_{\geq 0} \to \mathbb{R}$. Effectively, the function $\mathfrak{b}$ will be a spatiotemporal constraint, meaning that a signal $x : \mathbb{R}_{\geq 0} \to \mathbb{R}^n$ that is such that

$\mathfrak{b}(x(t),t) \geq 0$ for all times $t \geq 0$ satisfies $\phi$ (i.e., it is such that $(x,0) \models \phi$). Note that $\mathfrak{b}$ is now defined over $\mathbb{R}_{\geq 0}$ instead of $[t_0, t_1]$. The reason for this is that $\mathfrak{b}$ will only be piecewise continuously differentiable to encode the STL specification $\phi$, and consequently we will have to apply theorem 4.1 multiple times.

The encoding in $\mathfrak{b}$ will be a sufficient condition for the satisfaction of $\phi$, but not a necessary one. In subsequent sections, we then show under which conditions on the system in equation (4.1) the time-varying CBF $\mathfrak{b}$ is valid.

To be able to construct a function $\mathfrak{b}$ such that it (1) encodes $\phi$ and (2) is a valid time-varying CBF, we consider the following fragment of STL:

$$\psi ::= \top \mid \mu \mid \psi_1 \wedge \psi_2 \tag{4.3a}$$

$$\phi ::= G_{[a,b]}\psi \mid F_{[a,b]}\psi \mid \psi_1 \, U_{[a,b]} \, \psi_2 \mid \phi_1 \wedge \phi_2, \tag{4.3b}$$

where $\psi_1$ and $\psi_2$ are Boolean formulas constructed according to equation (4.3a), while $\phi_1$ and $\phi_2$ are temporal formulas constructed according to equation (4.3b). Note that we assume a special structure for the formulas in the fragment in equation (4.3), and we do not permit negations and disjunctions. In principle, negations and disjunctions could be considered in this fragment and we would still be able to encode the resulting formula $\phi$ into a candidate time-varying CBF $\mathfrak{b}$. However, time-varying CBFs $\mathfrak{b}$ constructed for such formulas may generally not be valid, as negations and disjunctions can lead to a set of local minima and maxima. The intuition for the choice of the fragment in equation (4.3) is that it will enable us to construct functions $\mathfrak{b}(x,t)$ that are concave in $x$ if the predicate functions $h$ of predicates $\mu$ also are concave.

Before we formally present the CBF encoding of the fragment of STL formulas in equation (4.3), we illustrate the main idea in two examples.

---

**Example 4.1** Consider the STL formula

$$\phi := F_{[5,15]}\left(\|x - \begin{bmatrix}10 & 0\end{bmatrix}^T\|^2 \leq 5^2\right),$$

consisting of an eventually operator and a single predicate. Note that the corresponding predicate function is $h(x) := 5^2 - \|x - \begin{bmatrix}10 & 0\end{bmatrix}^T\|^2$. Let $t_0 := 0$ and $t_1 := 15$ and consider, without loss of generality, an initial condition of $x(0) := \begin{bmatrix}0 & 0\end{bmatrix}^T$. Define now the candidate time-varying CBF

$$\mathfrak{b}(x,t) := \gamma(t) - \|x - \begin{bmatrix}10 & 0\end{bmatrix}^T\|^2,$$

where the temporal properties of $\mathfrak{b}$ are defined by the function

$$\gamma(t) := -\frac{75}{15}t + 100.$$

Note that $\mathfrak{b}(x(0),0)=0$ so that the initial condition $x(0)$ is contained within set $\mathcal{C}(t)$ (i.e., $x(0)\in\mathcal{C}(0)$), as required in theorem 4.1. If the solution $x:[t_0,t_1]\to\mathbb{R}^n$ to equation (4.1) under a suitable control law $u$ is now such that $\mathfrak{b}(x(t),t)\geq 0$ for all $t\in[t_0,t_1]$, then the formula $\phi$ is satisfied and $(x,0)\models\phi$ holds. This follows since $\gamma(t_1)=5^2$, so that $\mathfrak{b}(x,t_1)=5^2-\|x-\begin{bmatrix}10 & 0\end{bmatrix}^T\|^2=h(x)$. This means that $\mathfrak{b}(x(t_1),t_1)=h(x(t_1))\geq 0$ implies that $\|x(t_1)-\begin{bmatrix}10 & 0\end{bmatrix}^T\|^2\leq 5^2$, which yields $(x,0)\models\phi$ according to the description of the STL semantics in definition 3.3. Note that $\mathfrak{b}(x(t),t)\geq 0$ for all $t\in[t_0,t_1]$ can be achieved by a suitable control law $u$ (which we explain in the next section) if $\mathfrak{b}$ is a valid, time-varying CBF.

---

In order to use conjunctions as in the semantics in equation (4.3), we use a smooth underapproximation of the "min" operator.[4] Assume, therefore, that we are given $p$ candidate time-varying CBFs $\mathfrak{b}_l:\mathbb{R}^n\times\mathbb{R}_{\geq 0}\to\mathbb{R}$, where $l\in\{1,\dots,p\}$, that encode formulas from equation (4.3) that we would like to combine via a conjunction. The smooth underapproximation that we use is

$$\mathfrak{b}(x,t):=-\frac{1}{\eta}\ln\Big(\sum_{l=1}^{p}\exp(-\eta\mathfrak{b}_l(x,t))\Big),$$

where $\eta>0$ is a design parameter. It holds that $\min_{l\in\{1,\dots,p\}}\mathfrak{b}_l(x,t)\approx\mathfrak{b}(x,t)$, where the accuracy of this approximation increases with $\eta$, as stated in lemma 4.1.

**Lemma 4.1 (Log-Sum Approximation)** *For $\eta>0$, a set of $p$ candidate time-varying CBFs $\mathfrak{b}_l:\mathbb{R}^n\times\mathbb{R}_{\geq 0}\to\mathbb{R}$, where $l\in\{1,\dots,p\}$, and $\mathfrak{b}:\mathbb{R}^n\times\mathbb{R}_{\geq 0}\to\mathbb{R}$, defined as $\mathfrak{b}(x,t):=-\frac{1}{\eta}\ln(\sum_{l=1}^{p}\exp(-\eta\mathfrak{b}_l(x,t)))$, it holds that*

$$\mathfrak{b}(x,t)\leq\min_{l\in\{1,\dots,p\}}\mathfrak{b}_l(x,t)\leq\mathfrak{b}(x,t)+\frac{\ln(p)}{\eta}\tag{4.4}$$

*for each $(x,t)\in\mathbb{R}^n\times\mathbb{R}_{\geq 0}$. Consequently, it holds that*

$$\lim_{\eta\to\infty}\mathfrak{b}(x,t)=\min_{l\in\{1,\dots,p\}}\mathfrak{b}_l(x,t).$$

*Proof:* For $\zeta_1,\dots,\zeta_p\in\mathbb{R}$, we know from [54, p. 72] that

$$\max_{l\in\{1,\dots,p\}}\zeta_l\leq\ln\Big(\sum_{l=1}^{p}\exp(\zeta_l)\Big)\leq\max_{l\in\{1,\dots,p\}}\zeta_l+\ln(p).$$

---

[4]One could instead use the min operator directly. However, the min operator is not differentiable and requires nonsmooth analysis for control design [117, 315], which we want to avoid.

*We also know that* $\max_{l\in\{1,\dots,p\}} \zeta_l = -\min_{l\in\{1,\dots,p\}}(-\zeta_l)$, *so that, by substituting* $\zeta_l := -\eta \mathfrak{b}_l(x,t)$, *we get that*

$$-\eta \min_{l\in\{1,\dots,p\}} \mathfrak{b}_l(x,t) \le \ln\Big(\sum_{l=1}^{p} \exp(-\eta \mathfrak{b}_l(x,t))\Big) \le -\eta \min_{l\in\{1,\dots,p\}} \mathfrak{b}_l(x,t) + \ln(p)$$

*by which both results to be shown follow trivially.*

Lemma 4.1 is useful since $\mathfrak{b}(x,t) \ge 0$ implies $\mathfrak{b}_l(x,t) \ge 0$ for each $l \in \{1,\dots,p\}$; that is, the conjunction operator can be encoded by using the smooth approximation. This idea is illustrated in example 4.2.

---

**Example 4.2** Consider the STL formula $\phi := \phi_1 \wedge \phi_2$, with

$$\phi_1 := F_{[5,15]}\left(\|x - \begin{bmatrix}10 & 0\end{bmatrix}^T\|^2 \le 5^2\right)$$
$$\phi_2 := G_{[5,15]}\left(\|x - \begin{bmatrix}10 & 5\end{bmatrix}^T\|^2 \le 10^2\right)$$

consisting of an eventually operator, an always operator, and two predicates. Note that the corresponding predicate functions are

$$h_1(x) := 5^2 - \|x - \begin{bmatrix}10 & 0\end{bmatrix}^T\|^2,$$
$$h_2(x) := 10^2 - \|x - \begin{bmatrix}10 & 5\end{bmatrix}^T\|^2.$$

Again, let $t_0 := 0$ and $t_1 := 15$, and consider, without loss of generality, an initial condition of $x(0) := \begin{bmatrix}0 & 0\end{bmatrix}^T$. Set $\eta := 1$ and define the candidate time-varying CBF $\mathfrak{b}(x,t) := -\ln\left(\exp(-\mathfrak{b}_1(x,t)) + \exp(-\mathfrak{b}_2(x,t))\right)$, where

$$\mathfrak{b}_1(x,t) := \gamma_1(t) - \|x - \begin{bmatrix}10 & 0\end{bmatrix}^T\|^2,$$
$$\mathfrak{b}_2(x,t) := \gamma_2(t) - \|x - \begin{bmatrix}10 & 5\end{bmatrix}^T\|^2,$$

and the temporal properties of $\mathfrak{b}_1$ and $\mathfrak{b}_2$ are defined by the functions

$$\gamma_1(t) := -6.3333t + 120,$$
$$\gamma_2(t) := 110\exp(-0.4796t) + 90.$$

Note that $\mathfrak{b}(x(0),0) \ge 0$, so that the initial condition $x(0)$ is again contained within the set $\mathcal{C}(0)$. By the choices of $\gamma_1$ and $\gamma_2$, note that $\mathfrak{b}_1(x,t_1) = 5^2 - \|x - \begin{bmatrix}10 & 0\end{bmatrix}^T\|^2 = h_1(x)$ and $\mathfrak{b}_2(x,t') \le h_2(x)$ for all $t' \in [5,15]$. If the solution $x : [t_0,t_1] \to \mathbb{R}^n$ to equation (4.1) under a suitable control law $u$ is now such that $\mathfrak{b}(x(t),t) \ge 0$ for all $t \in [t_0,t_1]$, e.g., when $\mathfrak{b}$ is a valid CBF, then the formula $\phi$ is satisfied and $(x,0) \models \phi$. This can be seen as follows: according to lemma 4.1, it follows that $\mathfrak{b}_1(x(t_1),t_1) \ge 0$ and $\mathfrak{b}_2(x(t'),t') \ge 0$ for all $t' \in [5,15]$, which implies that $\|x(t_1) - \begin{bmatrix}10 & 0\end{bmatrix}^T\|^2 \le 5^2$ and $\|x(t') - \begin{bmatrix}10 & 5\end{bmatrix}^T\|^2 \le 10^2$ for all $t' \in [5,15]$. This leads to $(x,0) \models \phi$ according to the definition of the STL semantics in definition 3.3.

---

**Table 4.1** Step A conditions

| Step A Single Temporal Operators in Equation (4.3b) *without* Conjunctions | |
|---|---|
| $G_{[a,b]}\mu_1$ | $\forall t' \in [a,b]$, $\mathfrak{b}_1(x,t') \leq h_1(x)$, $p := 1$ |
| $F_{[a,b]}\mu_1$ | $\exists t' \in [a,b]$ such that $\mathfrak{b}_1(x,t') \leq h_1(x)$, $p := 1$ |
| $\mu_1 \, U_{[a,b]} \, \mu_2$ | $\exists t' \in [a,b]$ such that $\mathfrak{b}_2(x,t') \leq h_2(x)$, and $\forall t'' \in [0,t']$, $\mathfrak{b}_1(x,t'') \leq h_1(x)$, $p := 2$ |

We generalize these two examples next, and establish a formal connection between the function $\mathfrak{b}: \mathbb{R}^n \times \mathbb{R}_{\geq 0} \to \mathbb{R}$ (later shown to be a candidate and a valid time-varying CBF) and the STL semantics of $\phi$ in definition 3.3. In particular, we formulate conditions that the function $\mathfrak{b}$ has to satisfy to account for the semantics of an STL formula $\phi$ in the fragment in equation (4.3). These conditions on $\mathfrak{b}$ will be discussed in three steps (steps A, B, and C). We note that, besides these conditions, it further needs to hold that the function $\mathfrak{b}$ is such that $\mathfrak{b}(x(0),0) \geq 0$.

Before we start, let us denote the predicate functions that correspond to the predicates $\mu_1, \mu_2, \ldots$ by $h_1, h_2, \ldots$ In steps A and B, we present conditions for the CBF $\mathfrak{b}$ for single temporal operators (i.e., the formula $\phi$ contains only one always, eventually, or until operator). The conditions from step A, shown in table 4.1, particularly present conditions for single temporal operators in equation (4.3b) that do not contain conjunctions of predicates.

For the always and eventually operators $G_{[a,b]}\mu_1$ and $F_{[a,b]}\mu_1$, respectively, we need only one candidate time-varying CBF such that we select $p := 1$ in $\mathfrak{b}(x,t) := -\frac{1}{\eta}\ln(\sum_{l=1}^{p} \exp(-\eta \mathfrak{b}_l(x,t)))$. Intuitively, we require that $\mathfrak{b}_1(x,t') \leq h_1(x)$ for all times $t' \in [a,b]$ for the always operator $G_{[a,b]}\mu_1$, while we only require that there is a time $t' \in [a,b]$ such that $\mathfrak{b}_1(x,t') \leq h_1(x)$ for the eventually operator $F_{[a,b]}\mu_1$. For the until operator $\mu_1 \, U_{[a,b]} \, \mu_2$, we require two candidate CBFs so that $p := 2$, since the until operator can be viewed as a combination of an always and an eventually operator. Using the property of the smooth approximation, stated in equation (4.4) in lemma 4.1, we can then ensure satisfaction of $\mu_1 \, U_{[a,b]} \, \mu_2$ if $\mathfrak{b}(x(t),t) \geq 0$ for all $t \in [0,b]$.

In step B, we generalize the conditions from step A to the case of a single temporal operator that may contain conjunctions of predicates. Let, therefore, $\psi_1 := \mu_1 \wedge \ldots \wedge \mu_{\tilde{p}_1}$ and $\psi_2 := \mu_{\tilde{p}_1+1} \wedge \ldots \wedge \mu_{\tilde{p}_1+\tilde{p}_2}$ with $\tilde{p}_1, \tilde{p}_2 \geq 1$ denote conjunctions of predicates. The conditions from step B are presented in table 4.2.

**Table 4.2** Step B conditions

| Step B Single Temporal Operators in Equation (4.3b) *with* Conjunctions | |
|---|---|
| $G_{[a,b]}\psi_1$ | $\forall t' \in [a,b]$, $\forall l \in \{1,\ldots,\tilde{p}_1\}$, $\mathfrak{b}_l(x,t') \leq h_l(x)$, $p := \tilde{p}_1$ |
| $F_{[a,b]}\psi_1$ | $\exists t' \in [a,b]$, $\forall l \in \{1,\ldots,\tilde{p}_1\}$ such that $\mathfrak{b}_l(x,t') \leq h_l(x)$, $p := \tilde{p}_1$ |
| $\psi_1 \, U_{[a,b]} \, \psi_2$ | $\exists t' \in [a,b]$, $\forall l' \in \{\tilde{p}_1+1,\ldots,\tilde{p}_1+\tilde{p}_2\}$ such that $\mathfrak{b}_{l'}(x,t') \leq h_{l'}(x)$ and $\forall t'' \in [0,t']$, $\forall l'' \in \{1,\ldots,\tilde{p}_1\}$, $\mathfrak{b}_{l''}(x,t'') \leq h_{l''}(x)$, $p := \tilde{p}_1 + \tilde{p}_2$ |

Note that, as in step A, $p$ is equal to the total number of predicates in the formula. The conditions for always, eventually, and until operators are now basically the same except for the added universal quantifier over all associated predicates.

In step C, conjunctions of single temporal operators are considered. The conditions on $\mathfrak{b}$ are based on the conditions from steps A and B. For instance, consider $G_{[a_1,b_1]}\psi_1 \wedge F_{[a_2,b_2]}\psi_2 \wedge \psi_3 \, U_{[a_3,b_3]} \, \psi_4$. Let $\mathfrak{b}_1$, $\mathfrak{b}_2$, and $\mathfrak{b}_3$ be the candidate time-varying CBFs associated with $G_{[a_1,b_1]}\psi_1$, $F_{[a_2,b_2]}\psi_2$, and $\psi_3 \, U_{[a_3,b_3]} \, \psi_4$, respectively, and constructed as in steps A and B. We can combine these simply as $\mathfrak{b}(x,t) := -\frac{1}{\eta}\ln(\sum_{l=1}^{3}\exp(-\eta\mathfrak{b}_l(x,t)))$.

**Deactivating Time-Varying CBFs to Remove Conservatism.** The function $\mathfrak{b}$ is now given as $\mathfrak{b}(x,t) := -\frac{1}{\eta}\ln(\sum_{l=1}^{p}\exp(-\eta\mathfrak{b}_l(x,t)))$ where $p$ is the number of functions $\mathfrak{b}_l$ that arise due to steps A, B, and C. Note particularly that each function $\mathfrak{b}_l$ corresponds to either an always, eventually, or until operator with a corresponding time interval $[a_l, b_l]$. Note also that several functions $\mathfrak{b}_l$ can correspond to the same temporal operator and time interval; for instance, $\phi := G_{[a,b]}(\mu_1 \wedge \mu_2)$ results in two functions $\mathfrak{b}_1$ and $\mathfrak{b}_2$ that correspond to $G_{[a,b]}$.

The function $\mathfrak{b}$ constructed in this way may, however, be overly conservative and not a candidate time-varying CBF when $p$ is too large or when the spatiotemporal constraints in $\phi$ are too constraining. For instance, consider the STL formula

$$\phi := F_{[5,15]}(\|x - \begin{bmatrix}10 & 0\end{bmatrix}^T\|^2 \leq 5^2) \wedge G_{[55,60]}(\|x - \begin{bmatrix}10 & 10\end{bmatrix}^T\|^2 \leq 1),$$

and let us construct candidate time-varying CBFs $\mathfrak{b}_1$ and $\mathfrak{b}_2$ for the eventually and always operators in $\phi$, respectively, according to step A. In particular, let us assume that $\mathfrak{b}_1$ is constructed as in example 4.1 with a decreasing function $\gamma_1$, which means that $\mathfrak{b}_1$ constrains the feasible set to be no larger than $\{x \in \mathbb{R}^n \| \|x - \begin{bmatrix}10 & 0\end{bmatrix}^T\|^2 \leq 5^2\}$ for all times $t \geq 15$. This will conflict with the function $\mathfrak{b}_2$, which constrains the feasible set to be no larger than $\{x \in \mathbb{R}^n | \|x - \begin{bmatrix} 10 & 10 \end{bmatrix}^T \|^2 \leq 1\}$ for all times $t \geq 55$. Consequently, the function $\mathfrak{b}(x,t) := -\ln(\exp(-\mathfrak{b}_1(x,t)) + \exp(-\mathfrak{b}_2(x,t)))$ cannot be a candidate time-varying CBF, and it holds that $\mathcal{C}(t) = \emptyset$ for all $t \geq 55$. However, if we instead construct the function $\mathfrak{b}$ as

$$\mathfrak{b}(x,t) := \begin{cases} -\frac{1}{\eta} \ln\left(\exp(-\eta\mathfrak{b}_1(x,t)) + \exp(-\eta\mathfrak{b}_2(x,t))\right) & \text{for all } t \in [0,15] \\ \mathfrak{b}_2(x,t) & \text{for all } t \in [15,60], \end{cases}$$

then we can show that the function $\mathfrak{b}$ is a candidate CBF over the intervals $[0,15]$ and $[15,60]$, respectively.

To make the encoding less conservative, we hence deactivate the function $\mathfrak{b}_l$ when the corresponding temporal specifications are satisfied. One could also consider activating the function $\mathfrak{b}_l$ before it becomes active, such as in the case of the always operator $G_{[55,60]}(\|x - \begin{bmatrix} 10 & 10 \end{bmatrix}^T \|^2 \leq 1)$ that only becomes active for $t \geq 55$, but we refrain from doing so for the sake of simplicity. For each temporal operator, the candidate time-varying CBF $\mathfrak{b}_l$ is deactivated at $t = b_l$. The function $\mathfrak{b}(x,t) := -\frac{1}{\eta} \ln\left(\sum_{l=1}^{p} \exp(-\eta\mathfrak{b}_l(x,t))\right)$ is now piecewise continuous in $t$, with switches at exactly those times where the function $\mathfrak{b}_l$ is deactivated. More formally, we integrate the switching function $\mathfrak{o}_l : \mathbb{R}_{\geq 0} \to \{0,1\}$ into $\mathfrak{b}$ as

$$\mathfrak{b}(x,t) := -\frac{1}{\eta} \ln\Big(\sum_{l=1}^{p} \mathfrak{o}_l(t) \exp(-\eta\mathfrak{b}_l(x,t))\Big). \tag{4.5}$$

Following our previous arguments, we define the switching functions as

$$\mathfrak{o}_l(t) := \begin{cases} 1 \text{ if } t < b_l \\ 0 \text{ if } t \geq b_l. \end{cases} \tag{4.6}$$

Finally, we denote the switching sequence as $\{s_0 := 0, s_1, \ldots, s_q\}$, where $q$ denotes the number of switches (i.e., the number of temporal operators in $\phi$) for which we know that $q \leq p$ holds. This sequence $\{s_0 := 0, s_1, \ldots, s_q\}$ is known in advance. Specifically, at time $t \geq s_j$, we know that the next switching time is

$$s_{j+1} := \text{argmin}_{b_l \in \{b_1, \ldots, b_p\}} \zeta(b_l, t),$$

where $\zeta$ is defined as

$$\zeta(b_l, t) := \begin{cases} b_l - t \text{ if } b_l - t > 0 \\ \infty \text{ otherwise.} \end{cases}$$

The switching sequence $\{s_0 := 0, s_1, \ldots, s_q\}$ will later define a switched (hybrid) time domain [118, chapter 2.2]. We next summarize our encoding of $\phi$ in $\mathfrak{b}$ and state its soundness.

**Theorem 4.2 (Soundness of $\mathfrak{b}$ in Encoding $\phi$)** *Let $\phi$ be an STL formula from the fragment in equation (4.3) and let $\mathfrak{b} : \mathbb{R}^n \times \mathbb{R}_{\geq 0} \to \mathbb{R}$ satisfy the conditions in steps A, B, and C for the formula $\phi$ and be according to equation (4.5) using the switching functions $\mathfrak{o}_l : \mathbb{R}_{\geq 0} \to \{0, 1\}$ as defined in equation (4.6). If a signal $x : \mathbb{R}_{\geq 0} \to \mathbb{R}^n$ is such that $\mathfrak{b}(x(t), t) \geq 0$ for all $t \geq 0$, then it holds that $(x, 0) \models \phi$.*

*Proof: The result follows by the construction of the function $\mathfrak{b}$ as described in steps A, B, and C, which adhere to the STL semantics in definition 3.3. Note that the functions $\mathfrak{o}_l$ are set to zero only when the corresponding temporal operators have been satisfied and do not affect soundness.*

Naturally, the next question is how we can design $u$ such that the solutions $x : \mathbb{R}_{\geq 0} \to \mathbb{R}^n$ to equation (4.1) are such that $\mathfrak{b}(x(t), t) \geq 0$ for all $t \geq 0$, and we show in the next section how we can achieve this when $\mathfrak{b}$ is a valid time-varying CBF on each domain $\mathbb{R}^n \times (s_j, s_{j+1})$. Note that we will consider open time intervals $(s_j, s_{j+1})$ since $\mathfrak{b}(x, t)$ is not differentiable at points where $t = s_j$ due to the switching function $\mathfrak{o}_l$.

## 4.3 Control Laws Based on Time-Varying Control Barrier Functions

In this section, we consider the dynamical control system in equation (4.1) with $t_0 := 0$ and an STL formula $\phi$ from the fragment in equation (4.3), and we want to construct a control law $u : \mathcal{D} \times \mathbb{R}_{\geq 0} \to \mathbb{R}^m$ so that each solution $x : \mathcal{I} \to \mathbb{R}^n$ to equation (4.1) with initial condition $x(0)$ is such that $(x, 0) \models \phi$. To enforce complete solutions under this control law $u$ ($\mathcal{I} = \mathbb{R}_{\geq 0}$), we need to ensure that the set $\mathcal{C}(t) := \{x \in \mathbb{R}^n | \mathfrak{b}(x, t) \geq 0\}$ is compact and contained within the workspace $\mathcal{D}$ of the system for all $t \in \mathbb{R}_{\geq 0}$ so that we can invoke lemma 2.2. If $x = 0$ is the center of the workspace, for instance, this can be done by adding the CBF

$$\mathfrak{b}_{p+1}(x, t) := D - \|x\|^2$$

to the function $\mathfrak{b}$, where $D \in \mathbb{R}$ is a large constant that is such that $\{x \in \mathbb{R}^n | \|x\|^2 \leq D\} \subseteq \mathcal{D}$.[5] In other words, we let

$$\mathfrak{b}(x,t) := -\frac{1}{\eta} \ln\Big(\sum_{l=1}^{p+1} \mathfrak{o}_l(t) \exp(-\eta \mathfrak{b}_l(x,t))\Big), \tag{4.7}$$

which now consists of $p+1$ functions $\mathfrak{b}_l$, where the $(p+1)$th switching function is such that

$$\mathfrak{o}_{p+1}(t) := 1 \text{ for all } t \in \mathbb{R}_{\geq 0}. \tag{4.8}$$

Moving forward in this section, let us assume that the function $\mathfrak{b}$

- satisfies the conditions of steps A, B, and C for the STL specification $\phi$; and
- is a valid time-varying CBF on each domain $\mathbb{R}^n \times (s_j, s_{j+1})$.

In general, the latter is a rather strong assumption that may constrain the class of systems that can be considered. In fact, in the next section, we will constrain the system in equation (4.1) further in order to be able to synthesize valid time-varying CBFs $\mathfrak{b}$ that encode more complex specifications $\phi$.

Before providing the main results, recall that the definitions of candidate and valid time-varying CBFs require $\mathfrak{b}$, and consequently each $\mathfrak{b}_l$, to be continuously differentiable. Fortunately, it holds that $\mathfrak{b}$ is continuously differentiable on each domain $\mathbb{R}^n \times (s_j, s_{j+1})$ if each function $\mathfrak{b}_l$ is continuously differentiable on this domain. The fact that $\mathfrak{b}$ is only continuously differentiable on the open interval $(s_j, s_{j+1})$, not on the closed interval $[s_j, s_{j+1}]$ does not pose any problems. This is because the Comparison Lemma, which we already used and will continue using, only requires the derivative condition to hold for almost all times $t \in [s_j, s_{j+1}]$. Since set $\mathcal{C}(t)$ is nondecreasing at the switching times $s_j$ where $\mathfrak{b}$ becomes nondifferentiable, we can thus sequentially reason over the time intervals $[s_j, s_{j+1}]$ and show that $\mathcal{C}(t)$ is forward invariant, by which it follows that $\phi$ is satisfied. We formally state this result next.

**Theorem 4.3 (Validity of $\mathfrak{b}$ Implies Satisfaction of $\phi$)** *Let the functions $f$, $g$, and $u$ that describe the system in equation (4.1) with $t_0 := 0$ be continuous in its first argument and piecewise continuous in its second argument. Let $\phi$ be an STL formula from the fragment in equation (4.3), and let $\mathfrak{b} : \mathbb{R}^n \times \mathbb{R}_{\geq 0} \to \mathbb{R}$ satisfy the conditions in steps A, B, and C for the formula $\phi$*

[5] Note that other choices of the function $\mathfrak{b}_{p+1}$ are possible so long as the set $\{x \in \mathbb{R}^n | \mathfrak{b}_{p+1}(x,t) \geq 0\}$ is compact and contained within the workspace $\mathcal{D}$.

*and be according to equation (4.7) using the switching functions* $\mathfrak{o}_l : \mathbb{R}_{\geq 0} \to \{0,1\}$ *defined in equations (4.6) and (4.8). Assume that* $\mathfrak{b}$ *is a valid CBF on each domain* $\mathcal{D} \times (s_j, s_{j+1})$*, and assume that* $u$ *is such that* $u(x,t) \in K_{CBF}(x,t)$ *for all* $(x,t) \in \mathcal{D} \times (s_j, s_{j+1})$*. Then it holds that every solution* $x : \mathcal{I} \to \mathcal{D}$ *to equation (4.1) with* $\mathfrak{b}(x(0),0) \geq 0$ *is such that* $(x,0) \models \phi$*.*

*Proof:* *The proof follows similar arguments as in theorems 2.1 and 4.1. However, we now need to account for the switching sequence* $\{s_0 := 0, s_1, \ldots, s_q\}$ *that results in a switched time domain for solutions* $x$*. Due to the assumptions on* $f$*,* $g$*, and* $u$*, there is at least one solution* $x : [0, \tau_{max}) \to \mathcal{D}$ *to equation (4.1) with* $\tau_{max} > 0$*. Note that* $u(x,t) \in K_{CBF}(x,t)$ *implies that, for all* $t \in (0, \min(\tau_{max}, s_1))$*,*

$$\dot{\mathfrak{b}}(x(t),t) \geq -\alpha(\mathfrak{b}(x(t),t)).$$

*By construction of* $\mathfrak{b}$*, it holds that* $\mathfrak{b}(x(0),0) \geq 0$*. Applying the Comparison Lemma to the IVP* $\dot{w}(t) = -\alpha(w(t))$ *with* $w_0 \leq \mathfrak{b}(x(0),0)$[6] *reveals that* $\mathfrak{b}(x(t),t) \geq 0$ *for all* $t \in [0, \min(\tau_{max}, s_1))$*. Assuming for a moment that the solution* $x$ *is defined on* $[s_0, s_1]$ *(i.e., that* $\tau_{max} \geq s_1$*), we know that* $x(s_1) \in \mathcal{C}(s_1)$ *due to continuity of* $x$ *and since* $\lim_{\tau \to s_1^-} \mathcal{C}(\tau) \subseteq \mathcal{C}(s_1)$ *where* $\lim_{\tau \to s_1^-}$ *denotes the limit from the left of* $s_1$*. It then holds that* $\mathfrak{b}(x(t),t) \geq 0$ *for all* $t \in [s_1, \min(\tau_{max}, s_2))$ *using similar arguments as before. Indeed, this argument can be repeated unless the solution* $x$ *ceases to exist (i.e., unless* $\tau_{max} < s_j$ *for some* $j$*). This, however, cannot happen since the set* $\mathcal{C}(t)$ *is compact and contained within* $\mathcal{D}$*. In particular, we know that* $\mathfrak{b}(x(t),t) \geq 0$ *implies* $\|x(t)\| \leq D$ *for all* $t \in [0, \tau_{max})$ *since* $\mathfrak{b}_{p+1}(x,t) := D - \|x\|$ *was added to* $\mathfrak{b}$*. Lemma 2.2 then guarantees complete solutions (i.e.,* $\tau_{max} = \infty$*), such that* $\mathfrak{b}(x(t),t) \geq 0$ *for all* $t \geq 0$*. By theorem 4.2, it follows that* $(x,0) \models \phi$*.*

We can now obtain a control law $u$ that satisfies $u(x,t) \in K_{\text{CBF}}(x,t)$ in the same way as in section 2.2.2 by solving either a convex optimization problem or even as a closed-form expression. For the former, let $u(x,t) := \hat{u}$, where $\hat{u}$ is given by

$$\underset{\hat{u}}{\operatorname{argmin}}\ \hat{u}^T \hat{u} \tag{4.9a}$$

$$\text{s.t.}\ \frac{\partial \mathfrak{b}(x,t)}{\partial x}(f(x) + g(x)\hat{u}) + \frac{\partial \mathfrak{b}(x,t)}{\partial t} \geq -\alpha(\mathfrak{b}(x,t)). \tag{4.9b}$$

[6] Recall from lemma 2.4 that the IVP $\dot{w}(t) = -\alpha(w(t))$ with $w_0 \geq 0$ has a complete solution such that $w(t) \geq 0$ for all $t \geq 0$.

Whenever $\mathfrak{b}$ is not differentiable with respect to $t$, which can happen only a finite number of times, we can technically set the control input to any value. In practice, however, we may simply choose $u(x(t),t)=\lim_{\tau\to s_j^-} u(x(\tau),\tau)$. To obtain a closed-form solution, we can instead define

$$B_0(x,t) := -\frac{\partial \mathfrak{b}(x,t)}{\partial x} f(x) - \frac{\partial \mathfrak{b}(x,t)}{\partial t} - \alpha(\mathfrak{b}(x,t)),$$
$$B_1(x,t) := \frac{\partial \mathfrak{b}(x,t)}{\partial x} g(x),$$

so that, following [106, chapter 4.2], the control law $u$ can also be defined as

$$u(x,t) := \begin{cases} \frac{B_1(x,t)^T B_0(x,t)}{B_1(x,t) B_1(x,t)^T} & \text{if } B_0(x,t) > 0 \\ 0 & \text{otherwise.} \end{cases}$$

To put what we learned into practice, consider example 4.3, in which a group of $N := 5$ omnidirectional robots is considered [188].

---

**Example 4.3** Let us denote each robot by $v_i$, with $i \in \{1,\dots,N\}$. The state of each robot is denoted by $x_i := \left[{p_i}^T \;\; \theta_i\right]^T \in \mathbb{R}^3$, where the vector $p_i := \left[p_{x,i} \;\; p_{y,i}\right]^T$ denotes the robot's two-dimensional positions $p_{x,i}$ and $p_{y,i}$ and $\theta_i$ denotes the robot's orientation. We then stack the states of all robots into the vector $x := \left[{x_1}^T \;\; \dots \;\; {x_N}^T\right]^T \in \mathbb{R}^{3N}$. The dynamics of each robot are now given as

$$\dot{x}_i = f_i(x) + \underbrace{\begin{bmatrix} \cos(\theta_i) & -\sin(\theta_i) & 0 \\ \sin(\theta_i) & \cos(\theta_i) & 0 \\ 0 & 0 & 1 \end{bmatrix} \left(B_i^T\right)^{-1} R_i}_{=:g_i(x_i)} u_i,$$

where $u_i$ is the angular velocity of the wheels that we can control, $R_i := 0.02$ is the radius of the wheels, and

$$B_i := \begin{bmatrix} 0 & \cos(\pi/6) & -\cos(\pi/6) \\ -1 & \sin(\pi/6) & \sin(\pi/6) \\ L_i & L_i & L_i \end{bmatrix}$$

describes geometric constraints, with $L_i := 0.2$ being the radius of the robot body [188]. We assume that each robot already uses a built-in collision avoidance mechanism, which is described by the function $f(x) := \left[f_1(x)^T \; \dots \; f_N(x)^T\right]^T \in \mathbb{R}^{3N}$, where each element $f_i(x) := [f_{x,i}(x) \;\; f_{y,i}(x) \;\; 0]^T$ is defined as

$$f_{x,i}(x) := \sum_{j=1, j\neq i}^{N} k_i \frac{p_{x,i} - p_{x,j}}{\|p_i - p_j\| + 0.00001}$$

$$f_{y,i}(x) := \sum_{j=1, j \neq i}^{N} k_i \frac{p_{y,i} - p_{y,j}}{\|p_i - p_j\| + 0.00001}$$

with sufficiently large gain $k_i > 0$. The function $f_i$ achieves a repelling effect between two agents when they come too close to each other. Note that the function $f$ is locally Lipschitz continuous to ensure the existence of solutions. By defining $g(x) := \operatorname{diag}(g_1(x_1), \ldots, g_N(x_N))$ and $u := \begin{bmatrix} u_1{}^T & \ldots & u_M{}^T \end{bmatrix}^T$, we recover the dynamics in equation (4.1).

As for the system specifications, we consider the STL formula

$$\phi := \phi_1 \wedge \phi_2 \wedge \phi_3 \wedge \phi_4,$$

where the first subspecification $\phi_1$ concerns robots $v_1$ and $v_2$ and is defined as

$$\begin{aligned}\phi_1 := &\underbrace{F_{[10,30]}(\|p_1 - p_2\|^2 \leq 10^2)}_{=:\phi_1'} \\ &\wedge \underbrace{F_{[25,50]}\left((\|p_1 - \begin{bmatrix} 40 & 75 \end{bmatrix}^T\|^2 \leq 10^2) \wedge (\|p_2 - \begin{bmatrix} 40 & 65 \end{bmatrix}^T\|^2 \leq 10^2)\right)}_{=:\phi_1''} \\ &\wedge \underbrace{F_{[40,90]}\left((\|p_1 - p_2\|^2 \leq 10^2) \wedge (\|p_1 - \begin{bmatrix} 70 & 75 \end{bmatrix}^T\|^2 \leq 10^2)\right)}_{=:\phi_1'''}.\end{aligned}$$

Stated in words, the subspecification $\phi_1$ consists of $\phi_1'$, $\phi_1''$, and $\phi_1'''$, which specify the following:

$\phi_1'$: robots $v_1$ and $v_2$ meet within the time interval $[10, 30]$, e.g., to perform a collaborative job;

$\phi_1''$: robots $v_1$ and $v_2$ reach individual locations within the time interval $[25, 50]$, e.g., to perform individual jobs; and

$\phi_1'''$: robots $v_1$ and $v_2$ meet within the time interval $[40, 90]$, while robot $v_1$ reaches an individual location.

The second subspecification $\phi_2$ concerns robots $v_2$ and $v_3$ and is defined as

$$\begin{aligned}\phi_2 := &\underbrace{(\|p_3 - \begin{bmatrix} 40 & 10 \end{bmatrix}^T\|^2 \leq 10^2)\, U_{[5,25]}\, (|\theta_3 - 90| \leq 10)}_{=:\phi_2'} \\ &\wedge \underbrace{F_{20,50}(\|p_2 - p_3\|^2 \leq 15^2)}_{=:\phi_2''} \wedge \underbrace{G_{[70,90]}(\|p_3 - \begin{bmatrix} 50 & 50 \end{bmatrix}^T\|^2 \leq 10^2)}_{=:\phi_2'''}.\end{aligned}$$

Here, $\phi_2$ specifies collaborative and individual jobs for robot $v_3$, including the robot's orientation. Specifically, specification $\phi_2$ consists of $\phi_2'$, $\phi_2''$, and $\phi_2'''$, which specify the following:

$\phi_2'$: robot $v_3$ stays within a location centered at $\begin{bmatrix}40 & 10\end{bmatrix}^T$ with radius 10 until, within the time interval $[5, 25]$, its orientation is approximately 90 degrees;

$\phi_2''$: robots $v_3$ and $v_2$ meet within the time interval $[20, 50]$; and

$\phi_2'''$: robot $v_3$ reaches and stays within an individual location between the time interval $[70, 90]$.

Finally, the third subspecification $\phi_3$ concerns robots $v_4$ and $v_5$ and is defined as

$$\phi_3 := \underbrace{G_{[10,50]}\left((\|p_4 - \begin{bmatrix}90 & 10\end{bmatrix}^T\|^2 \leq 10^2) \wedge (|\theta_4 - 135| \leq 10)\right)}_{=:\phi_3'}$$
$$\underbrace{F_{[50,50]}\left((\|p_5 - \begin{bmatrix}90 & 10\end{bmatrix}^T\|^2 \leq 10^2) \wedge (|\theta_5 - 135| \leq 10)\right)}_{=:\phi_3''}$$
$$\wedge \underbrace{F_{[50,90]}(\|p_4 - \begin{bmatrix}60 & 10\end{bmatrix}^T\|^2 \leq 10^2)}_{=:\phi_3'''}.$$

Specification $\phi_3$ encodes a handover job for robots $v_4$ and $v_5$ in which robot $v_4$ has to satisfy a job until robot $v_5$ can take over this job and robot $v_4$ can move to its next individual job. Specification $\phi_3$ consists of $\phi_3'$, $\phi_3''$, and $\phi_3'''$, which specify the following:

$\phi_3'$: robot $v_4$ stays within a location centered at $\begin{bmatrix}90 & 10\end{bmatrix}^T$ with radius 10 with an orientation of approximately 135 within the time interval $[10, 50]$;

$\phi_3''$: robot $v_5$ arrives at the same location as robot $v_4$ with the same orientation as robot $v_4$ at time 50; and

$\phi_3'''$: robot $v_4$ reaches another location within the time interval $[50, 90]$.

Before we show how the CBF $\mathfrak{b}$ encoding $\phi$ is constructed, we show the numerical results up front. Figure 4.2(a) shows the robot trajectories $x$ and it turns out that $\phi$ is satisfied (i.e., $(x, 0) \models \phi$). Figure 4.2(b) shows the time evolution of $\mathfrak{b}(x(t), t)$ where jumps occur due to the switching mechanism at the switching times $s_1 = 25$, $s_2 = 30$, and $s_3 = 50$. While the dimensionality of the multi-agent system is already $n = 15$ (5 agents with 3 states each), solving the quadratic program in equation (4.9) only took around 5 milliseconds on a standard laptop.[7]

[7]The computation times will naturally be larger on embedded hardware that has less computational resources or when the number of agents further increases. In chapter 5, we will discuss how to decentralize the computation of the control inputs of the agents.

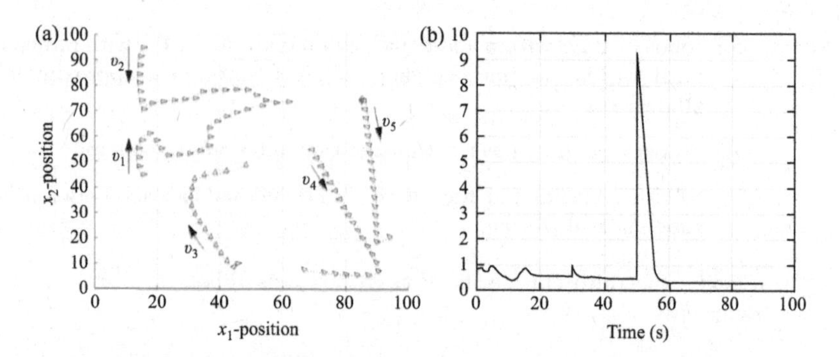


**Figure 4.2**
(a) Robot trajectories where triangles denote the robot orientation. (b) CBF evolution $\mathfrak{b}(x(t),t)$. Reprinted with permission from [172].

Next, we provide and discuss the MATLAB code that produces these results and showcases the construction of CBFs for STL specifications. In lines 1–18, we define initial conditions and goal locations and angles from specification $\phi$. In lines 20–96, the control loop is given, where we start with computing the system dynamics in line 21 (the definition of the "load_system" function for computing $f(x)$ and $g(x)$ is omitted for convenience). Lines 23–69 then construct the individual CBFs $\mathfrak{b}_l$ along with their partial derivatives $\frac{\mathfrak{b}_l(x,t)}{\partial x}$ and $\frac{\mathfrak{b}_l(x,t)}{\partial t}$. For example, for formula $\phi_1' := F_{[10,30]}(\|p_1 - p_2\|^2 \leq 10^2)$ we construct $\mathfrak{b}_1$ in line 24, while we compute $\frac{\mathfrak{b}_1(x,t)}{\partial x}$ and $\frac{\mathfrak{b}_1(x,t)}{\partial t}$ in lines 25 and 26, respectively. Note that, for $t=30$, it holds that $\mathfrak{b}_1$ encodes the predicate in $\phi_1'$ as desired. After having constructed each individual $\mathfrak{b}_l$, we construct the CBF $\mathfrak{b}$ in line 82. In the remainder of the code, we set up and solve the optimization problem (4.9).

```
1  clc; clear all; close all;
2
3  t=linspace(0,90,9001); % time
4  step=t(2)-t(1); % time increment
5  x(:,1)= transpose([15 95 0 15 45 0 45 10 deg2rad(10) 70 55
       deg2rad(180) 85 75 deg2rad(270)]); % initial condition
6
7  % Define locations and goal angles
8  g2        = [40; 75]; % phi_1''
9  g3        = [40; 65]; % phi_1''
10 g5        = [70; 75]; % phi_1'''
11 g6        = [40; 10]; % phi_2'
12 g7        = 90;        % phi_2'
13 g9        = [50; 50]; % phi_2'''
14 g10       = [90; 10]; % phi_3'
15 g11       = 135;       % phi_3'
16 g12       = [90; 10]; % phi_3''
17 g13       = 135;       % phi_3''
18 g14       = [60; 10]; % phi_3'''
19
```

```
20  for i=1:length(t)
21      [f,g] = load_system(x(:,i));
22
23      %phi1
24      b1 = 100-90/30*t(i)-norm(x(1:2,i)-x(4:5,i)); %phi_1'
25      db1 = -(x(1:2,i)-x(4:5,i))./(norm(x(1:2,i)-x(4:5,i)));
26      dt1 = -90/30;
27      b2 = 100-90/50*t(i)-norm(x(1:2,i)-g2);  % phi_1''
28      db2 = -(x(1:2,i)-g2)/norm(x(1:2,i)-g2);
29      dt2 = -90/50;
30      b3 = 100-90/50*t(i)-norm(x(4:5,i)-g3);  % phi_1''
31      db3 = -(x(4:5,i)-g3)/norm(x(4:5,i)-g3);
32      dt3 = -90/50;
33      b4 = 100-90/90*t(i)-norm(x(1:2,i)-x(4:5,i));  % phi_1'''
34      db4 = -(x(1:2,i)-x(4:5,i))/norm(x(1:2,i)-x(4:5,i));
35      dt4 = -90/90;
36      b5 = 100-90/90*t(i)-norm(x(1:2,i)-g5);  % phi_1'''
37      db5 = -(x(1:2,i)-g5)/norm(x(1:2,i)-g5);
38      dt5 = -90/90;
39
40      %phi2
41      b6 = 10-norm(x(7:8,i)-g6); % phi_2'
42      db6 = -(x(7:8,i)-g6)/norm(x(7:8,i)-g6);
43      dt6 = 0;
44      b7 = 150-140/25*t(i)-norm(rad2deg(x(9,i))-g7);  % phi_2'
45      db7 = deg2rad(-(rad2deg(x(9,i))-g7)/norm(rad2deg(x(9,i))-
            g7));
46      dt7 = -90/25;
47      b8 = 100-85/50*t(i)-norm(x(4:5,i)-x(7:8,i)); % phi_2''
48      db8 = -(x(4:5,i)-x(7:8,i))/norm(x(4:5,i)-x(7:8,i));
49      dt8 = -80/40;
50      b9= 995*exp(-0.07561*t(i))+5-norm(x(7:8,i)-g9);
51          % phi_2'''
52      db9= -(x(7:8,i)-g9)/norm(x(7:8,i)-g9);
53      dt9= -75.2319*exp(-0.07561*t(i));
54
55      %phi3
56      b10 = 45*exp(-0.2197*t(i))+5-norm(x(10:11,i)-g10);
57          % phi_3'
58      db10 = -(x(10:11,i)-g10)/norm(x(10:11,i)-g10);
59      dt10 = -9.8865*exp(-0.2197*t(i));
60      b11= 145*exp(-0.3367*t(i))+5-norm(rad2deg(x(12,i))-g11);
            % phi_3'
61      db11= deg2rad(-(rad2deg(x(12,i))-g11)/norm(rad2deg(x(12,
62          i))-g11));
63      dt11= -48.8215*exp(-0.3367*t(i));
64      b12= 100-90/50*t(i)-norm(x(13:14,i)-g12); % phi_3''
65      db12= -(x(13:14,i)-g12)/norm(x(13:14,i)-g12);
66      dt12= -90/50;
67      b13= 145*exp(-0.0673*t(i))+5-norm(rad2deg(x(15,i))-g13);
            % phi_3''
68      db13= -(rad2deg(x(15,i))-g13)/norm(rad2deg(x(15,i))-g13);
69      dt13= -9.7585*exp(-0.0673*t(i));
70      b14= 100-90/90*t(i)-norm(x(10:11,i)-g14); % phi_3'''
71      db14= -(x(10:11,i)-g14)/norm(x(10:11,i)-g14);
72      dt14= -90/90;
73
74      % Switching sequence
75      if t(i)<=25
76          i1=1;i2=1;i3=1;i4=1;i5=1;i6=1;i7=1;i8=1;i9=1;i10=1;
                i11=1;i12=1;i13=1;i14=1;
77      elseif t(i)<=30
78          i1=1;i2=1;i3=1;i4=1;i5=1;i6=0;i7=0;i8=1;i9=1;i10=1;
                i11=1;i12=1;i13=1;i14=1;
```

```
79        elseif t(i)<=50
80            i1=0;i2=1;i3=1;i4=1;i5=1;i6=0;i7=0;i8=1;i9=1;i10=1;
                  i11=1;i12=1;i13=1;i14=1;
81        elseif t(i)<=90
82            i1=0;i2=0;i3=0;i4=1;i5=1;i6=0;i7=0;i8=0;i9=1;i10=0;
                  i11=0;i12=0;i13=0;i14=1;
83        end
84
85        b    = -log(i1*exp(-b1)+i2*exp(-b2)+i3*exp(-b3)+i4*exp(-b4
              )+i5*exp(-b5)+i6*exp(-b6)+i7*exp(-b7)+i8*exp(-b8)+i9*
              exp(-b9)+i10*exp(-b10)+i11*exp(-b11)+i12*exp(-b12)+
              i13*exp(-b13)+i14*exp(-b14)); %min approximation
86        den = -1/(i1*exp(-b1)+i2*exp(-b2)+i3*exp(-b3)+i4*exp(-b4)
              +i5*exp(-b5)+i6*exp(-b6)+i7*exp(-b7)+i8*exp(-b8)+i9*
              exp(-b9)+i10*exp(-b10)+i11*exp(-b11)+i12*exp(-b12)+
              i13*exp(-b13)+i14*exp(-b14));
87
88        A   = den*[i1*exp(-b1)*(-db1)+i2*exp(-b2)*(-db2)+i4*exp
89            (-b4)*(-db4)+i5*exp(-b5)*(-db5);0; i1*exp(-b1)*(db1)+
                  i3*exp(-b3)*(-db3)+i4*exp(-b4)*(db4)+i8*exp(-b8)
                  *(-db8);0; i6*exp(-b6)*(-db6)+i8*exp(-b8)*(db8)+i9
                  *exp(-b9)*(-db9); i7*exp(-b7)*(-db7); i10*exp(-b10)
                  *(-db10)+i14*exp(-b14)*(-db14); i11*exp(-b11)*(-
                  db11); i12*exp(-b12)*(-db12); i13*exp(-b13)*(-db13)
                  ];
90        B   = den*(i1*exp(-b1)*(-dt1)+i2*exp(-b2)*(-dt2)+i3*exp
91            (-b3)*(-dt3)+i4*exp(-b4)*(-dt4)+i5*exp(-b5)*(-dt5)+i6
                  *exp(-b6)*(-dt6)+i7*exp(-b7)*(-dt7)+i8*exp(-b8)
                  *(-dt8)+i9*exp(-b9)*(-dt9)+i10*exp(-b10)*(-dt10)+
                  i11*exp(-b11)*(-dt11)+i12*exp(-b12)*(-dt12)+i13*
                  exp(-b13)*(-dt13)+i14*exp(-b14)*(-dt14))+b^2;
92
93        % Solve optimization program
94        Q = diag(ones(15,1));
95        opts = optimoptions('quadprog','Algorithm','interior-
              point-convex','Display','off');
96        u(:,i) = quadprog(Q,zeros(15,0),-transpose(A)*g,transpose
              (A)*f+B,[],[],[],[],[],opts);
97
98        % Euler method
99        x(:,i+1) = x(:,i)+step*(f+g*u(:,i));
100   end
```

## 4.4 Constructing Valid Time-Varying Control Barrier Functions

In this section, we show how to construct a valid time-varying CBF $\mathfrak{b}:\mathbb{R}^n \times \mathbb{R}_{\geq 0} \to \mathbb{R}$ that encodes the STL specification $\phi$ from the fragment in equation (4.3). As mentioned before, we need to restrict the class of systems that we can consider in order to be able to show the validity of the time-varying CBF $\mathfrak{b}$. For now, let us consider fully actuated systems in the following sense.

**Assumption 4.1** *Let the function $g$ that describes the input dynamics of the system in equation (4.1) be such that $g(x)g(x)^T$ is positive definite for each $x \in \mathcal{D}$.*

This assumption is a controllability assumption that implies that the system in equation (4.1) is feedback equivalent to a single integrator system. In fact, assumption 4.1 is equivalent to $g(x)$ being invertible for each $x \in \mathcal{D}$ if $n = m$.[8] We remark that we could consider relative degree systems such as double integrator or mechanical systems, using the same ideas that will be presented in this section by utilizing higher-order CBFs [288, 325, 328]. For the sake of simplicity, however, we omit this extension. Instead, we will show how we can extend the results from this section to nonholonomic unicycle dynamics in section 4.5.

In this section, we will formulate an optimization problem for constructing a valid time-varying CBF $\mathfrak{b}$ that encodes the semantics of an STL formula $\phi$ from the fragment in equation (4.3). For convenience, recall that the function $\mathfrak{b}$ takes the form

$$\mathfrak{b}(x,t) := -\frac{1}{\eta} \ln\Big(\sum_{l=1}^{p+1} \mathfrak{o}_l(t) \exp(-\eta \mathfrak{b}_l(x,t))\Big),$$

where each $\mathfrak{b}_l$ is associated either with an eventually formula $F_{[a_l,b_l]}\mu_l$ or an always formula $G_{[a_l,b_l]}\mu_l$. We will enforce the conditions from steps A, B, and C for the function $\mathfrak{b}$ as stated in section 4.2 by constructing a suitable time-varying function $\gamma_l : \mathbb{R}_{\geq 0} \to \mathbb{R}$ that is part of the definition of the function $\mathfrak{b}_l$. Therefore, we will design a template function $\gamma_l$ that contains several free variables, along with a constraint set over these variables that is in accordance with steps A, B, and C. We will then optimize these free variables within a constrained optimization problem. Finally, we note that we aim to satisfy $\phi$ with robustness $r \in \mathbb{R}_{\geq 0}$ by the construction of $\mathfrak{b}$; that is, we want to achieve $\rho^{\phi}(x,0) \geq r$. Let us illustrate this last point by continuing example 4.2.

---

**Example 4.4** Recall the STL formula $\phi := \phi_1 \wedge \phi_2$ from example 4.2 with

$$\phi_1 := F_{[5,15]}\big(\|x - \begin{bmatrix}10 & 0\end{bmatrix}^T\|^2 \leq 5^2\big)$$

$$\phi_2 := G_{[5,15]}\big(\|x - \begin{bmatrix}10 & 5\end{bmatrix}^T\|^2 \leq 10^2\big).$$

---

[8] Later in this book, we will also consider cases where parts of the dynamics are unknown so the system is not directly feedback equivalent to a single integrator.

We used $\eta := 1$ and defined the candidate time-varying CBF $\mathfrak{b}(x,t) := -\ln\big(\exp(-\mathfrak{b}_1(x,t)) + \exp(-\mathfrak{b}_2(x,t))\big)$, where

$$\mathfrak{b}_1(x,t) := \gamma_1(t) - \|x - \begin{bmatrix}10 & 0\end{bmatrix}^T\|^2,$$
$$\mathfrak{b}_2(x,t) := \gamma_2(t) - \|x - \begin{bmatrix}10 & 5\end{bmatrix}^T\|^2,$$

with $\gamma_1(t) := -6.3333t + 120$ and $\gamma_2(t) := 110\exp(-0.4796t) + 90$, for which $\gamma_1(15) = 5^2$ and $\gamma_2(5) = 10^2$. From here, it can be seen that $\mathfrak{b}(x(t),t) \geq 0$ for all times $t \in [0,15]$ implies that $\rho^\phi(x,0) \geq 0$. However, if we instead use $\gamma_1(t) := -6.3333t + 120 - r$ and $\gamma_2(t) := 110\exp(-0.4796t) + 90 - r$ for some $r \geq 0$ such that $\gamma_1(15) = 5^2 - r$ and $\gamma_2(5) = 10^2 - r$, then we can see that $\mathfrak{b}(x(t),t) \geq 0$ for all times $t \in [0,15]$ implies that $\rho^\phi(x,0) \geq r$.

---

We proceed in two steps. In step 1, we show how to construct $\mathfrak{b}$ when $\phi := F_{[a_l,b_l]}\mu_l$ or $\phi := G_{[a_l,b_l]}\mu_l$ where $\mu_l$ does not contain any conjunctions, i.e., $p = 1$. In step 2, we explain how to construct $\mathfrak{b}$ in the more general case when $\phi$ contains conjunctions (i.e., $p > 1$), and therefore present an optimization-based algorithm.

**Step 1.** Consider one of the two STL formulas $\phi := G_{[a_l,b_l]}\mu_l$ or $\phi := F_{[a_l,b_l]}\mu_l$, and define the predicate satisfaction time

$$t_l^* := \begin{cases} b_l & \text{if } F_{[a_l,b_l]}\mu_l \\ a_l & \text{if } G_{[a_l,b_l]}\mu_l, \end{cases} \tag{4.10}$$

which reflects the requirement that $\mu_l$ has to hold at least once between $[a_l, b_l]$ for the eventually operator $F_{[a_l,b_l]}\mu_l$ or at all times within $[a_l, b_l]$ for the always operator $G_{[a_l,b_l]}\mu_l$. For the eventually operator, we note that one could consider other values of $t_l^*$ within the range $[a_l, b_l]$. Besides the predicate satisfaction time, let us also define the maximum of the predicate function $h_l$ as

$$h_l^{\text{opt}} := \sup_{x \in \mathbb{R}^n} h_l(x),$$

for which it has to hold that $h_l^{\text{opt}} \geq 0$ because the predicate $\mu_l$ is not satisfiable otherwise. Note that computing $h_l^{\text{opt}}$ is a convex optimization problem if the function $h_l$ is concave. As the value of $h_l^{\text{opt}}$ constrains the quantitative semantics $\rho^\phi$ by which we can satisfy the formula $\phi$, we define a constraint set for the desired robustness $r$ as

$$r \in \begin{cases} (0, h_l^{\text{opt}}) & \text{if } t_l^* > 0 \\ (0, h_l(x(0))] & \text{if } t_l^* = 0. \end{cases} \tag{4.11}$$

While the first case in this definition is self-explanatory, the second case can be explained as follows: If $t_l^* = 0$, which happens only in the case of an always operator, then there is no signal $x : \mathbb{R}_{\geq 0} \to \mathbb{R}^n$ with initial condition $x(0)$ such that $\rho^\phi(x,0) > h_l(x(0))$.

Using the predicate satisfaction time $t_l^*$, the maximum predicate value $h_l^{\text{opt}}$, and the robustness $r$, we will now propose two different templates $\gamma_l$ that define the function $\mathfrak{b}_l$. In particular, we propose exponential and linear functions that we denote by $\gamma_l^{\text{exp}} : \mathbb{R}_{\geq 0} \to \mathbb{R}$ and $\gamma_l^{\text{lin}} : \mathbb{R}_{\geq 0} \to \mathbb{R}$, respectively. Both templates will inherit the same theoretical guarantees, but the choice of the template may affect the feasibility of the optimization problem, as well as the control performance in practice, as we will discuss later in this chapter. First, let us define the function $\mathfrak{b}_l$ as

$$\mathfrak{b}_l(x,t) := -\gamma_l(t) + h_l(x),$$

where $\gamma_l(t) \in \{\gamma_l^{\text{exp}}(t), \gamma_l^{\text{lin}}(t)\}$ can be either the exponential or the linear template.

Let us start with the exponential template, and let us define $\gamma_l^{\text{exp}}(t)$ as

$$\gamma_l^{\text{exp}}(t) := (\gamma_{l,0}^{\text{exp}} - \gamma_{l,\infty}^{\text{exp}}) \exp(-\mathfrak{l}_l^{\text{exp}} t) + \gamma_{l,\infty}^{\text{exp}},$$

where $\gamma_{l,0}^{\text{exp}}, \gamma_{l,\infty}^{\text{exp}} \in \mathbb{R}$ and $\mathfrak{l}_l^{\text{exp}} \in \mathbb{R}_{\geq 0}$ are free parameters that determine the initial value, the final value, and the steepness of $\gamma_l^{\text{exp}}$. To adhere to the conditions in step A from section 4.2, we constrain these three parameters as follows:

$$\gamma_{l,0}^{\text{exp}} \in \begin{cases} \big(-\infty, h_l(x(0))\big] & \text{if } t_l^* > 0 \\ \big[r, h_l(x(0))\big] & \text{otherwise;} \end{cases} \tag{4.12a}$$

$$\gamma_{l,\infty}^{\text{exp}} \in (\max(r, \gamma_{l,0}^{\text{exp}}), h_l^{\text{opt}}); \tag{4.12b}$$

$$\mathfrak{l}_l^{\text{exp}} \in \begin{cases} -\ln\big(\frac{r-\gamma_{l,\infty}^{\text{exp}}}{\gamma_{l,0}^{\text{exp}}-\gamma_{l,\infty}^{\text{exp}}}\big)/t_l^* & \text{if } \gamma_{l,0}^{\text{exp}} < r \\ 0 & \text{otherwise.} \end{cases} \tag{4.12c}$$

Let us now parse these constraints carefully and explain their meaning. The choice of $\gamma_{l,0}^{\text{exp}}$ as in equation (4.12a) ensures that $\mathfrak{b}_l(x(0),0) \geq 0$, as required in theorem 4.2; and that $\mathfrak{b}_l(x(0),0) \leq h_l(x(0)) - r$ if $t_l^* = 0$, such that we can achieve $\rho^\phi(x,0) \geq 0$. The choices of $\gamma_{l,\infty}^{\text{exp}}$ and $\mathfrak{l}_l^{\text{exp}}$ as in equations (4.12b) and (4.12c) ensure that $\mathfrak{b}_l(x,t') \leq h_l(x) - r$ for all $t' \geq t_l^*$ and $x \in \mathbb{R}^n$. If a signal $x : \mathbb{R}_{\geq 0} \to \mathbb{R}^n$ is such that $\mathfrak{b}_l(x(t'),t') \geq 0$ for all $t' \geq t_l^*$, then it follows that $h_l(x(t')) \geq r$, which implies $\rho^\phi(x,0) \geq r$ by the choices of $t_l^*$ and $r$. We emphasize that the function $\gamma_l^{\text{exp}}$ explicitly depends on the initial condition $x(0)$.

From a practical point of view, an exponential template function $\gamma_l^{\exp}(t)$ may result in large control inputs, as the partial derivatives of the function $\mathfrak{b}$ may get large. This can be avoided with a piecewise linear template function $\gamma_l^{\text{lin}}$, which we define as a piecewise-continuous function:

$$\gamma_l^{\text{lin}}(t) := \begin{cases} \frac{\gamma_{l,\infty}^{\text{lin}} - \gamma_{l,0}^{\text{lin}}}{t_l^*} t + \gamma_{l,0}^{\text{lin}} & \text{if } t < t_l^* \\ \gamma_{l,\infty}^{\text{lin}} & \text{otherwise,} \end{cases}$$

where the two free parameters $\gamma_{l,0}^{\text{lin}}, \gamma_{l,\infty}^{\text{lin}} \in \mathbb{R}$ determine the initial and final values of $\gamma_l^{\text{lin}}$. Similar to the exponential case, we constrain these two parameters to adhere to the conditions in step A from section 4.2 as follows:

$$\gamma_{l,0}^{\text{lin}} \in \big(-\infty, h_l(x(0))\big] \tag{4.13a}$$

$$\gamma_{l,\infty}^{\text{lin}} \in \big(\max(r, \gamma_{l,0}^{\text{lin}}), h_l^{\text{opt}}\big). \tag{4.13b}$$

With the same motivation as for the exponential template function, equations (4.13a) and (4.13b) ensure that $0 \le \mathfrak{b}_l(x(0), 0)$, and that $\mathfrak{b}_l(x(0), 0) \le h_l(x(0)) - r$ if $t_l^* = 0$. By the choice of $\gamma_{l,\infty}^{\text{lin}}$, we have that $\mathfrak{b}_l(x, t') \le h_l(x) - r$ for all $x \in \mathbb{R}^n$ and $t' \ge t_l^*$. For a signal $x : \mathbb{R}_{\ge 0} \to \mathbb{R}^n$, it hence again holds that $\mathfrak{b}_l(x(t'), t') \ge 0$ for all $t' \ge t_l^*$ implies $\rho^\phi(x, 0) \ge r$. Note that $\gamma_l^{\text{lin}}$ is piecewise-continuously differentiable on $(s_0 := 0, s_1)$, as $\gamma_l^{\text{lin}}$ is only continuously differentiable on $(s_0, t_l^*)$ and $(t_l^*, s_1)$, where $s_1$ is the switching time. This, however, does not affect our guarantees from theorem 4.2, as we can simply consider the modified switching sequence $\{s_0 = 0, t_l^*, s_1 = b_l\}$ instead of $\{s_0 = 0, s_1 = b_l\}$.

Finally, assume now that $\mathfrak{b}(x, t) := \mathfrak{b}_l(x, t)$. Note that $\gamma_l^{\exp}$ and $\gamma_l^{\text{lin}}$ ensure that step A from section 4.2 holds for $\mathfrak{b}$ and the STL formulas $\phi := G_{[a_l, b_l]}\mu_l$ or $\phi := F_{[a_l, b_l]}\mu_l$. We also note that $\gamma_l^{\exp}$ and $\gamma_l^{\text{lin}}$ are nondecreasing functions. By these construction rules, it is straightforward to conclude that $\mathfrak{b}$ is a candidate time-varying CBF. We present an illustrative example next, and refer the reader back to example 4.3 for more examples.

---

**Example 4.5** Consider the STL formula

$$\phi := G_{[7.5, 10]}\big(\|x\|^2 < 5^2\big)$$

that yields $h_l(x) := 5^2 - \|x\|^2$ and $h_l^{\text{opt}} = 5^2$ so we can choose $r := 0.5$. Assume the initial condition $x(0) := \begin{bmatrix} 3.68 & 3.68 \end{bmatrix}^T$ such that $h_l(x(0)) = -2.07$. We select $t_l^* := 7.5$, $\gamma_{l,0}^{\text{lin}} := -2.5$, and $\gamma_{l,\infty}^{\text{lin}} := 0.5$ in accordance with equations (4.10) and (4.13). Recall that $\mathfrak{b}_l(x, t) := -\gamma_l^{\text{lin}}(t) + h_l(x)$ and note that $\mathfrak{b}_l(x(t), t) \ge 0$ for all $t \ge 0$ is equivalent to $h_l(x(t)) \ge \gamma_l^{\text{lin}}(t)$ for

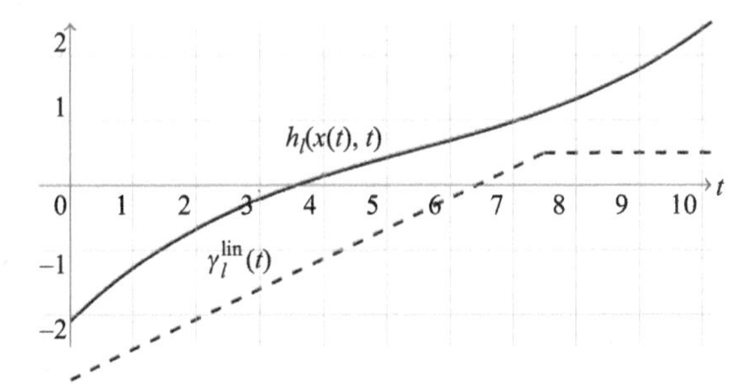


**Figure 4.3**
The functions $\gamma_l^{\text{lin}}(t)$ (dashed line) and $h_l(x(t),t)$ (solid line) for $\phi := G_{[7.5,10]}(\|x\|^2 < 5^2)$ with $r := 0.5$ and a candidate trajectory $x: \mathbb{R}_{\geq 0} \to \mathbb{R}^n$ satisfying $\phi$. Reprinted with permission from [175].

all $t \geq 0$. This leads to $\rho^{\phi}(x,0) > r$; that is, $(x,0) \models \phi$, by the construction of $\gamma_l^{\text{lin}}$ as illustrated in figure 4.3.

---

**Step 2.** For cases where the STL formula $\phi$ contains conjunctions (i.e., when $p > 1$), a more elaborate procedure is needed as individually constructing function $\mathfrak{b}_l$ may not result in a candidate time-varying CBF $\mathfrak{b}(x,t) := -\frac{1}{\eta}\ln\left(\sum_{l=1}^{p+1} \mathfrak{o}_l(t)\exp(-\eta\mathfrak{b}_l(x,t))\right)$. In the remainder let us, as in step 1, consider $\mathfrak{b}_l(x,t) := -\gamma_l(t) + h_l(x)$ for $l \in \{1,\ldots,p\}$ with $\gamma_l(t) \in \{\gamma_l^{\exp}(t), \gamma_l^{\text{lin}}(t)\}$ where $\gamma_l^{\exp}(t)$ and $\gamma_l^{\text{lin}}(t)$ are constrained according to equations (4.12) and (4.13), respectively. To later ensure that $\mathfrak{b}$ is a candidate time-varying CBF, we further pose assumption 4.2.

**Assumption 4.2** *Let the STL formula $\phi$ from the fragment in equation (4.3) be such that each predicate function $h_l : \mathbb{R}^n \to \mathbb{R}$ contained in $\phi$ is concave.*

Concave predicate functions $h_l$ include linear functions, as well as functions that express, for instance, reachability objectives using predicates such as $\|x - r\|^2 \leq \epsilon$ for $r \in \mathbb{R}^n$ and $\epsilon \in \mathbb{R}_{\geq 0}$. See also example 4.3 for further examples of concave predicate functions. Assumption 4.2, however, does not permit formulas like

$$\phi := G_{[0,15]}(\|x - \begin{bmatrix}50 & 50\end{bmatrix}^T\|^2 \geq 15^2) \wedge F_{[5,15]}(\|x - \begin{bmatrix}90 & 90\end{bmatrix}^T\|^2 \leq 10^2).$$

Formally, the reason is that the two predicate functions $h_1(x) := \|x - \begin{bmatrix}50 & 50\end{bmatrix}^T\|^2 - 15^2$ and $h_2(x) := 10^2 - \|x - \begin{bmatrix}90 & 90\end{bmatrix}^T\|^2$ contained in $\phi$ are convex and concave, respectively, resulting in local optima in the resulting candidate CBF $\mathfrak{b}$. As a consequence, it may happen that the partial derivative

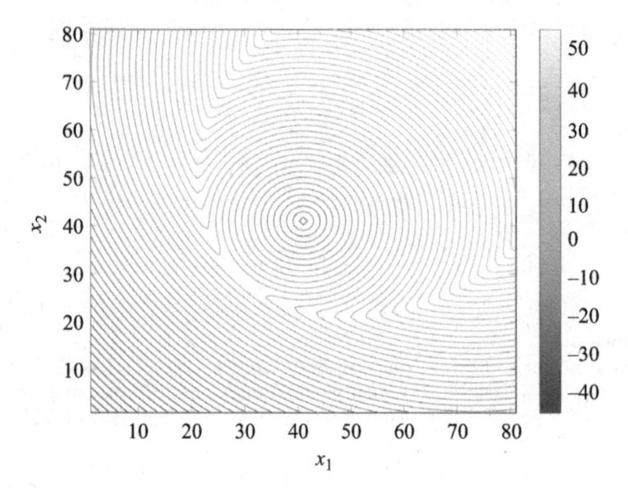


**Figure 4.4**
Level curves of $\mathfrak{b}(x,t)$ at $t = 1.2$ showing a critical point at $x_1 = x_2 \approx 38$. Reprinted with permission from [172].

of $\mathfrak{b}$ with respect to $x$ vanishes (i.e., that $\frac{\partial \mathfrak{b}(x,t)}{\partial x} = 0$ for some $(x,t) \in \mathcal{D} \times \mathbb{R}_{\geq 0}$). As we know from equation (4.2), this may result in a loss of controllability and imply that $\mathfrak{b}$ may not be a valid time-varying CBF. For the formula $\phi$, there is a time $t \in [0, 15]$ and state $x \in \mathcal{D}$, which lies on a line connecting the points $\begin{bmatrix} 50 & 50 \end{bmatrix}^T$ and $\begin{bmatrix} 90 & 90 \end{bmatrix}^T$, where $\frac{\partial \mathfrak{b}(x,t)}{\partial x} = 0$. The intuition here is that the gradients of the two predicate functions $h_1$ and $h_2$ point in opposite directions, indicating conflicting objectives, and lead to $\frac{\partial \mathfrak{b}_1(x,t)}{\partial x} = -\frac{\partial \mathfrak{b}_2(x,t)}{\partial x}$. The level curves for this specific example and an appropriately designed candidate time-varying CBF are shown in figure 4.4. Note that, in this example, the set of points $(x,t) \in \mathcal{D} \times \mathbb{R}_{\geq 0}$ where $\frac{\partial \mathfrak{b}(x,t)}{\partial x} = 0$ is a set of measure zero (i.e., a set without a "volume") that is not attractive. While this may hence not be a problem in practice, more complex formulas $\phi$ may lead to larger and attracting sets where the partial derivative of $\mathfrak{b}$ vanishes, which is prevented by assumption 4.2. Note that such reach-avoid specifications, as considered here, can be dealt with when the proposed CBF–based control laws are combined with automata-based planning later in part IV of the book. Besides automata-based approaches, there are other ways to to deal with certain types of local optima at the expense of a more complex control design (see e.g., [67, 320]).

Assumption 4.2 will allow us later to show that $\mathfrak{b}$ is a candidate and a valid time-varying CBF (theorems 4.4 and 4.5) by using the fact that the resulting function $\mathfrak{b}$ is concave in $x$, as shown next.

**Lemma 4.2 (Concavity of $\mathfrak{b}$)** *Let $\mathfrak{b}:\mathbb{R}^n\times\mathbb{R}_{\geq 0}\to\mathbb{R}$ be according to equation (4.7) using the switching functions $\mathfrak{o}_l:\mathbb{R}_{\geq 0}\to\{0,1\}$ defined in equations (4.6) and (4.8) and with $\mathfrak{b}_l(x,t):=-\gamma_l(t)+h_l(x)$ for $l\in\{1,\dots,p\}$ and $\mathfrak{b}_{p+1}(x,t):=D-\|x\|^2$. Further, let assumption 4.2 be satisfied. Then, for each fixed $t'\geq 0$, the function $\mathfrak{b}(x,t')$ is concave in $x$.*

*Proof:* *For each fixed $t'\geq 0$, note that $\eta\mathfrak{b}_l(x,t')$ is concave in $x$ so that $-\eta\mathfrak{b}_l(x,t')$ is convex in $x$. It is easy to see that* $\exp(-\eta\mathfrak{b}_l(x,t'))$ *is log-convex in $x$, i.e.,* $\ln(\exp(-\eta\mathfrak{b}_l(x,t')))=-\eta\mathfrak{b}_l(x,t')$ *is convex in $x$. Due to [54, section 3.5] we know that a sum of log-convex functions is log-convex again such that*

$$\mathfrak{b}(x,t')=-\frac{1}{\eta}\ln\Big(\sum_{l=1}^{p+1}\mathfrak{o}_l(t)\exp(-\eta\mathfrak{b}_l(x,t'))\Big)$$

*is concave in $x$.*

Compared to step 1, it is now not enough to select $\gamma_{l,0}^{\text{exp}}$ and $\gamma_{l,0}^{\text{exp}}$ as in equations (4.12a) and (4.13a), respectively, to ensure $\mathfrak{b}(x(0),0)\geq 0$, as evident by equation (4.4). To illustrate this, consider $\mathfrak{b}(x,t):=-\frac{1}{\eta}\ln\big(\exp(-\eta\mathfrak{b}_1(x,t))+\exp(-\eta\mathfrak{b}_2(x,t))\big)$. If $\mathfrak{b}_1(x(0),0)\geq 0$ and $\mathfrak{b}_2(x(0),0)\geq 0$, which is ensured by both equations (4.12a) and (4.13a), then it does not necessarily hold that $\mathfrak{b}(x(0),0)\geq 0$ depending on the value of $\eta$. Therefore, $\eta$ now needs to be selected to be sufficiently large, as we recall that increasing $\eta$ increases the accuracy of the approximation used for conjunctions. More importantly, $\gamma_{l,\infty}^{\text{exp}}$ and $\gamma_{l,\infty}^{\text{lin}}$, which have to be selected according to equations (4.12b) and (4.13b), respectively, and the desired robustness $r$ need to be selected such that for all $t\in[s_0,s_q]$ there is $x\in\mathbb{R}^n$ so that $\mathfrak{b}(x,t)\geq 0$. This is a necessary condition for $\mathfrak{b}$ to be a candidate and hence valid time-varying CBF.

To address the aforementioned issue, we now propose an optimization problem to find the free parameters of the exponential and linear template functions. For exponential template functions $\gamma_l^{\text{exp}}$, let us first define

$$\gamma_0^{\text{exp}}:=\begin{bmatrix}\gamma_{1,0}^{\text{exp}} & \dots & \gamma_{p,0}^{\text{exp}}\end{bmatrix}^T,$$
$$\gamma_\infty^{\text{exp}}:=\begin{bmatrix}\gamma_{1,\infty}^{\text{exp}} & \dots & \gamma_{p,\infty}^{\text{exp}}\end{bmatrix}^T,$$
$$\mathfrak{l}^{\text{exp}}:=\begin{bmatrix}\mathfrak{l}_1^{\text{exp}} & \dots & \mathfrak{l}_p^{\text{exp}}\end{bmatrix}^T,$$

which contain the free parameters $\gamma_{l,0}^{\text{exp}}$, $\gamma_{l,\infty}^{\text{exp}}$, and $\mathfrak{l}_l$ for each eventually and always operator encoded in $\mathfrak{b}_l$. For linear template functions $\gamma_l^{\text{lin}}$, similarly define

$$\gamma_0^{\text{lin}} := \begin{bmatrix}\gamma_{1,0}^{\text{lin}} & \dots & \gamma_{p,0}^{\text{lin}}\end{bmatrix}^T,$$
$$\gamma_\infty^{\text{lin}} := \begin{bmatrix}\gamma_{1,\infty}^{\text{lin}} & \dots & \gamma_{p,\infty}^{\text{lin}}\end{bmatrix}^T.$$

Now let $\xi_1, \dots, \xi_q \in \mathbb{R}^n$ with $\xi := \begin{bmatrix}\xi_1{}^T & \dots & \xi_q{}^T\end{bmatrix}^T$ be another set of free parameters that will ensure that for all $t \in [s_0, s_q]$, there exists $x \in \mathbb{R}^n$ such that $\mathfrak{b}(x,t) \geq 0$. For exponential templates $\gamma_l^{\text{exp}}$, compute now $\gamma_0^{\text{exp}}$, $\gamma_\infty^{\text{exp}}$, and $\mathfrak{l}^{\text{exp}}$ as the solution of the following optimization problem:

$$\underset{\eta, r, \gamma_0^{\text{exp}}, \gamma_\infty^{\text{exp}}, \mathfrak{l}^{\text{exp}}, \xi}{\text{argmax}} \quad r, \tag{4.14a}$$

$$\text{such that } \mathfrak{b}(x(0), 0) \geq \chi, \tag{4.14b}$$

$$\lim_{\tau \to s_j^-} \mathfrak{b}(\xi_j, \tau) \geq \chi \text{ for each } j \in \{1, \dots, q\}, \tag{4.14c}$$

$$\gamma_{l,0}^{\text{exp}} \text{ as in (4.12}a\text{) for each } l \in \{1, \dots, p\}, \tag{4.14d}$$

$$\gamma_{l,\infty}^{\text{exp}} \text{ as in (4.12}b\text{) for each } l \in \{1, \dots, p\}, \tag{4.14e}$$

$$\mathfrak{l}_l^{\text{exp}} \text{ as in (4.12}c\text{) for each } l \in \{1, \dots, p\}, \tag{4.14f}$$

$$\eta > 0 \text{ and } r \text{ as in equation (4.11) and}$$

$$t_l^* \text{ as in equation (4.10)}, \tag{4.14g}$$

where $\chi \geq 0$ is a design parameter that we can ignore for now, or simply assume that $\chi := 0$. In equation (4.14a), we aim to maximize the desired robustness $r$; in equations (4.14b) and (4.14c), we guarantee that $\mathfrak{b}$ is a candidate time-varying CBF (as shown later in this discussion), and in equations (4.14d), (4.14e), and (4.14f), we ensure that $\mathfrak{b}$ encodes $\phi$. Some additional comments are in order. First, note that the limit $\lim_{\tau \to s_j^-} \mathfrak{b}(\xi_j, \tau)$ from the left of $s_j$ can in practice easily be evaluated, at least approximately, by using $\mathfrak{b}(\xi_j, \tau')$ instead of $\lim_{\tau \to s_j^-} \mathfrak{b}(\xi_j, \tau)$, where we fix a value $\tau'$ such that $s_j - \tau'$ is positive but close to zero. Second, while our ultimate goal is to compute $\gamma_0^{\text{exp}}$, $\gamma_\infty^{\text{exp}}$, and $\mathfrak{l}^{\text{exp}}$ we remark that $\eta$, $r$, and $\xi$ are additional decision variables.

For linear templates $\gamma_l^{\text{lin}}$, we proceed similarly and compute $\gamma_0^{\text{lin}}$ and $\gamma_\infty^{\text{lin}}$ as the solution of the following optimization problem:

$$\underset{\eta, r, \gamma_0^{\text{lin}}, \gamma_\infty^{\text{lin}}, \xi}{\text{argmax}} \quad r, \tag{4.15a}$$

$$\text{such that } \mathfrak{b}(x(0), 0) \geq \chi, \tag{4.15b}$$

$$\lim_{\tau \to s_j^-} \mathfrak{b}(\xi_j, \tau) \geq \chi \text{ for each } j \in \{1, \dots, q\}, \tag{4.15c}$$

$$\gamma_{l,0}^{\text{lin}} \text{ as in equation (4.13a) for each } l \in \{1, \ldots, p\}, \tag{4.15d}$$

$$\gamma_{l,\infty}^{\text{lin}} \text{ as in equation (4.13b) for each } l \in \{1, \ldots, p\} \tag{4.15e}$$

$$\eta > 0 \text{ and } r \text{ as in equation (4.11), and } t_l^* \text{ as in equation (4.10).} \tag{4.15f}$$

Some comments regarding the complexity of the optimization problems in equations (4.14) and (4.15) are in order. These optimization problems are generally nonconvex. However, one can introduce relaxations to solve these optimization problems efficiently in practice. In the following discussion, we assume that the limit $\lim_{\tau \to s_j^-} \mathfrak{b}(\xi_j, \tau)$ is approximated by $\mathfrak{b}(\xi_j, \tau')$, with $\tau' < s_j$ and $s_j - \tau'$ close to zero, as previously discussed. First, simply setting $\eta$ to a large value (e.g., $\eta := 5$ works well in practice) removes nonconvexities from equations (4.14b) and (4.14c), as well as equations (4.15b) and (4.15c). For linear templates, we note that the constraints in equations (4.15b), (4.15c), (4.15d), and (4.15e) then become jointly convex in $\gamma_0^{\text{lin}}$, $\gamma_\infty^{\text{lin}}$, $r$, and $\xi$, as discussed next. Specifically for the constraints in equations (4.15b) and (4.15c), recall that $\mathfrak{b}(x,t) := -\frac{1}{\eta} \ln\left(\sum_{l=1}^{p+1} \mathfrak{o}_l(t) \exp(-\eta \mathfrak{b}_l(x,t))\right)$. As $\mathfrak{b}_l(x,t) = -\gamma_l(t) + h_l(x)$, we can note that $-\eta \mathfrak{b}_l(x,t)$ is linear in $\gamma_0^{\text{lin}}$, and $\gamma_\infty^{\text{lin}}$ and convex in $x$ and consequently jointly convex in $\gamma_0^{\text{lin}}$, $\gamma_\infty^{\text{lin}}$, and $x$. Again using the property that the sum of log-convex functions is log-convex [54, section 3.5], we note that $\mathfrak{b}(x,t)$ is jointly concave in $\gamma_0^{\text{lin}}$, $\gamma_\infty^{\text{lin}}$, and $x$. In addition, the constraint $\gamma_{l,\infty}^{\text{lin}} \in \left(\max(r, \gamma_{l,0}^{\text{lin}}), h_l^{\text{opt}}\right)$ in equation (4.15e) can be rewritten as the set of constraints $\gamma_{l,\infty}^{\text{lin}} > r$, $\gamma_{l,\infty}^{\text{lin}} > \gamma_{l,0}^{\text{lin}}$, and $\gamma_{l,\infty}^{\text{lin}} < h_l^{\text{opt}}$. Consequently, under these modifications, equation (4.15) becomes a convex optimization problem. For exponential templates, the problem stays generally nonconvex as $\gamma_l$ is nonconvex in the decision parameters. If, however, maximization of $r$ is not of interest, then we can fix $r$ and instead solve a feasibility program where we do not optimize over $r$. This results in removing nonconvexities from the constraint in equation (4.14f). Not surprisingly, we remark that we observed faster computation times for linear functions $\gamma_l^{\text{lin}}$ in equation (4.15) compared with exponential functions $\gamma_l^{\text{exp}}$ in equation (4.14). We also remark that equation (4.15) has fewer decision variables than equation (4.14). Despite these computational advantages, exponential templates may still be beneficial over linear templates in case one wants to shape the closed-loop response of the system. Finally, we emphasize that equations (4.14) and (4.15) can be solved offline before system deployment.

In lemma 4.3, we state the soundness of the function $\mathfrak{b}$ that we obtain as the solutions of equations (4.14) and (4.15) with respect to the STL formula $\phi$.

**Lemma 4.3 (Soundness of $\mathfrak{b}$ in Encoding $\phi$)** *Let $\phi$ be an STL formula from the fragment in equation (4.3), and let assumption 4.2 hold. Further, let $\mathfrak{b}:\mathbb{R}^n \times \mathbb{R}_{\geq 0} \to \mathbb{R}$ be according to equation (4.7) using the switching functions $\mathfrak{o}_l:\mathbb{R}_{\geq 0} \to \{0,1\}$ defined in equations (4.6) and (4.8) and with $\mathfrak{b}_l(x,t) := -\gamma_l(t) + h_l(x)$ for $l \in \{1,\dots,p\}$ and $\mathfrak{b}_{p+1}(x,t) := D - \|x\|^2$ where $\gamma_l$ is the solution of equation (4.14) or equation (4.15). If a signal $x:\mathbb{R}_{\geq 0} \to \mathbb{R}^n$ is such that $\mathfrak{b}(x(t),t) \geq 0$ for all $t \in \mathbb{R}_{\geq 0}$, then it holds that $(x,0) \models \phi$. It also holds that $\rho^\phi(x,0) \geq r$ where $r$ is also obtained from equation (4.14) or equation (4.15).*

*Proof:* *The result simply follows from the construction of $\gamma_l^{exp}$ and $\gamma_l^{lin}$ according to equations (4.12) and (4.13), respectively, which ensure that the conditions from steps A, B, and C hold for $\mathfrak{b}$. By theorem 4.2, it then simply follows that $(x,0) \models \phi$ if $\mathfrak{b}(x(t),t) \geq 0$ for all $t \in \mathbb{R}_{\geq 0}$. It can further be shown that $\rho^\phi(x,0) \geq r$. The detailed proof for the latter result is not provided but follows from the definition of the quantitative semantics $\rho^\phi$ and as the formula $\phi$ does not contain any negations.*

In lemma 4.3, we have assumed that $\mathfrak{b}(x(t),t) \geq 0$ for all $t \in \mathbb{R}_{\geq 0}$. We next show that $\mathfrak{b}$ is a candidate and a valid time-varying CBF for each time interval $[s_j, s_{j+1})$ so we can synthesize a control law $u$ such that the solution $x:\mathcal{I} \to \mathbb{R}^n$ to equation (4.1) is such that $\mathfrak{b}(x(t),t) \geq 0$. We first show that $\mathfrak{b}$ is a candidate time-varying CBF.[9]

**Theorem 4.4 (Candidacy of $\mathfrak{b}$)** *Let the conditions from lemma 4.3 hold. Then, it holds that the function $\mathfrak{b}$ is a candidate time-varying CBF on each domain $\mathcal{D} \times [s_j, s_{j+1})$.*

*Proof:* *To show that $\mathfrak{b}$ is a candidate CBF on the domain $\mathcal{D} \times [s_j, s_{j+1})$, we first have to show that feasibility of equation (4.14) or equation (4.15) implies that $\mathcal{C}(t)$ is nonempty for all $t \in [s_j, s_{j+1})$. Note, therefore, that equation (4.14c) or equation (4.15c) ensures that there is a state $\xi_j \in \mathbb{R}^n$ such that $\lim_{\tau \to s_j^-} \mathfrak{b}(\xi_j, \tau) \geq 0$; that is, the set $\mathcal{C}(t)$ is nonempty at the end of each interval $[s_j, s_{j+1})$. The function $\gamma$ is nondecreasing by construction in*

[9] Intuitively, we assume that domain $\mathcal{D}$ contains the set $\mathcal{C}(t)$ for all times $t \geq 0$.

*equations (4.14d)–(4.14f) or equation (4.15d)–(4.15e) so that* $\mathcal{C}(t_1) \supseteq \mathcal{C}(t_2)$ *for* $s_j \le t_1 < t_2 < s_{j+1}$*, which in turn implies that the state* $\xi_j$ *also ensures that* $\mathfrak{b}(\xi_j, \tau) \ge 0$ *for all* $\tau \in [s_j, s_{j+1})$*. In addition, there needs to be an absolutely continuous function* $x : [s_j, s_{j+1}) \to \mathbb{R}^n$ *for each* $x(s_j) \in \mathcal{C}(s_j)$ *such that* $x(t) \in \mathcal{C}(t)$ *for all* $t \in [s_j, s_{j+1})$*. Since* $\mathfrak{b}(x,t)$ *is concave in* $x$ *for each fixed* $t$*, it holds that all superlevel sets of* $\mathfrak{b}(x,t)$ *are convex [54, section 3.1.6] so that the set* $\mathcal{C}(t)$ *is connected. Since* $\frac{\partial \mathfrak{b}(x,t)}{\partial t}$ *is finite, the existence of an absolutely continuous function* $x : [s_j, s_{j+1}) \to \mathbb{R}^n$ *such that* $\mathfrak{b}(x(t), t) \ge 0$ *for all* $t \in [s_j, s_{j+1})$ *follows.*

We have shown that $\mathfrak{b}$ defined via function $\gamma$, which is the solution of equation (4.14) or equation (4.15), is a candidate time-varying CBF on each domain $\mathcal{D} \times [s_j, s_{j+1})$. We now derive conditions under which $\mathfrak{b}$ is a valid time-varying CBF. Recall that being a valid time-varying CBF requires that equation (4.2) in definition 4.2 holds. Satisfying equation (4.2) is always possible if $\frac{\partial \mathfrak{b}(x,t)}{\partial x} g(x) \ne 0$; that is, if $\mathfrak{b}$ has a relative degree of 1. However, this may not be the case, and it may hold that $\frac{\partial \mathfrak{b}(x,t)}{\partial x} g(x) = 0$ for some $(x,t) \in \mathcal{D} \times (s_j, s_{j+1})$. At such points, the constraint in equation (4.2) is satisfied if and only if

$$\frac{\partial \mathfrak{b}(x,t)}{\partial x} f(x) + \frac{\partial \mathfrak{b}(x,t)}{\partial t} \ge -\alpha(\mathfrak{b}(x,t)),$$

which may generally not hold. Note in this respect that $\frac{\partial \mathfrak{b}(x,t)}{\partial x} g(x) = 0$ iff $g(x)^T \frac{\partial \mathfrak{b}(x,t)}{\partial x}^T = 0$. Due to assumption 4.1, it holds that $\text{rank}(g(x)^T) = \text{rank}(g(x)) = n$ so that the nullspace of $g(x)^T$ is empty due to the rank-nullity theorem. Assumption 4.1 hence implies that

$$\frac{\partial \mathfrak{b}(x,t)}{\partial x} g(x) = 0 \text{ iff } \frac{\partial \mathfrak{b}(x,t)}{\partial x} = 0.$$

In this way, assumption 4.1 enables us to decouple the construction of CBFs from the dynamics of the system dynamics. In theorem 4.5, we use this fact and show that $\alpha$ can be selected such that $\mathfrak{b}$ is a valid time-varying CBF on each $\mathcal{D} \times (s_j, s_{j+1})$.

**Theorem 4.5 (Validity of $\mathfrak{b}$)** *Let the functions* $f$ *and* $g$ *that describe the system in equation (4.1) with* $t_0 := 0$ *be continuous, and let* $g$ *satisfy assumption 4.1. Let* $\phi$ *be an STL formula from the fragment in equation (4.3), and let assumption 4.2 hold. Further, let* $\mathfrak{b} : \mathbb{R}^n \times \mathbb{R}_{\ge 0} \to \mathbb{R}$ *be according to equation (4.7) using the switching functions* $\mathfrak{o}_l : \mathbb{R}_{\ge 0} \to \{0, 1\}$*, defined in equations (4.6) and (4.8) and with* $\mathfrak{b}_l(x,t) := -\gamma_l(t) + h_l(x)$ *for* $l \in \{1, \dots, p\}$ *and* $\mathfrak{b}_{p+1}(x,t) := D - \|x\|^2$*, where* $\gamma_l$ *is the solution of equation (4.14) or equation (4.15), which* *is solved for* $\chi > 0$. *Then the function* $\alpha$ *can be selected such that* $\mathfrak{b}$ *is a valid time-varying CBF on each domain* $\mathcal{D} \times (s_j, s_{j+1})$.

We provide the proof of this theorem in section 4.7. The intuition is that $\mathfrak{b}(x,t)$ is concave in $x$ by lemma 4.2 so that $\frac{\partial \mathfrak{b}(x,t)}{\partial x} = 0$ only happens for points $(x,t) \in \mathcal{D} \times (s_j, s_{j+1})$ where $\mathfrak{b}(x,t)$ is strictly positive (this follows as (4.14) or (4.15) is solved for $\chi > 0$). Consequently, a "large enough" $\alpha$ leads to the constraint $\frac{\partial \mathfrak{b}(x,t)}{\partial t} \geq -\alpha(\mathfrak{b}(x,t))$ being satisfied if $\frac{\partial \mathfrak{b}(x,t)}{\partial x} = 0$.

Finally, we put everything together and show that the control law $u$ defined by equation (4.9) is continuous. This means that theorem 4.3 can be invoked so that solutions $x : \mathcal{I} \to \mathcal{D}$ to the system in equation (4.1) satisfy $\phi$. The proof is again provided in section 4.7.

**Theorem 4.6 (Continuity of $u$ as per Equation (4.9))** *Let the conditions from theorem 4.5 hold and let function* $\alpha$ *be such that* $\mathfrak{b}$ *is a valid time-varying CBF on each domain* $\mathcal{D} \times (s_j, s_{j+1})$ *(as instructed in the proof of theorem 4.5). Let the control law* $u : \mathcal{D} \times [s_0, s_q]$ *be defined according to equation (4.9). Then it holds that* $u$ *is continuous in* $x$ *and piecewise continuous in* $t$*, and also that every solution* $x : \mathcal{I} \to \mathcal{D}$ *to equation (4.1) with* $t_0 := 0$ *is such that* $(x, 0) \models \phi$. *In addition, it holds that* $r \leq \rho^\phi(x, 0)$.

## 4.5 Unicycle Models and Unknown Dynamics

As previously mentioned, assumption 4.1 implies that the system in equation (4.1) is feedback equivalent to a single integrator. While the ideas from section 4.4 can also be applied to higher relative degree systems (e.g., double integrator or mechanical systems) we will instead focus on unicycle dynamics and show how to deal with additive unknown dynamics. We now focus on a single unicycle agent, but note that we can consider $N$ unicycle agents simultaneously without much change by stacking the individual agent dynamics. Recall example 2.2, and let

$$z(t) := \begin{bmatrix} x(t)^T & \theta(t) \end{bmatrix}^T \in \mathbb{R}^3$$

be the unicycle's state at time $t$, where $x(t) := \begin{bmatrix} p_x(t) & p_y(t) \end{bmatrix}^T$ and $\theta(t)$ are the two-dimensional position and orientation, respectively, with the dynamics

$$\dot{z}(t) = g(z(t))u + c(z(t), t), \tag{4.16}$$

where the function $c:\mathbb{R}^3 \times \mathbb{R}_{\geq 0} \to \mathbb{R}^3$ models unknown internal dynamics, and where again $u := \begin{bmatrix} v & \omega \end{bmatrix}^T \in \mathbb{R}^2$ is the control input. In particular, we let

$$g(z) := \begin{bmatrix} \cos(\theta) & 0 \\ \sin(\theta) & 0 \\ 0 & 1 \end{bmatrix} \quad c(z,t) := \begin{bmatrix} c_x(z,t) \\ c_\theta(z,t) \end{bmatrix}, \tag{4.17}$$

where $c_x$ and $c_\theta$ are locally Lipschitz continuous in the first argument and piecewise continuous in the second argument. We also assume that the function $c$ is bounded. Assumption 4.3 formalizes these requirements.

**Assumption 4.3** *Let the function $c:\mathbb{R}^3 \times \mathbb{R}_{\geq 0} \to \mathbb{R}^3$ be locally Lipschitz continuous in the first argument and piecewise continuous in the second argument. Further, let $c$ be bounded with a known constant $C$ (i.e., $\|c(z,t)\| \leq C$ for all $(z,t) \in \mathbb{R}^3 \times \mathbb{R}_{\geq 0}$).*

Assume again an STL specification $\phi$ from the STL fragment in equation (4.3), where now, however, the satisfaction of $\phi$ depends only on the position $x$ and not on the orientation $\theta$. The reason for this restriction is that the time-varying CBF $\mathfrak{b}$ encoding $\phi$ will only depend on $x$ and not on $\theta$, which will enable us to show that $\mathfrak{b}$ is valid. Consequently, let us explicitly define the safe set as

$$\mathcal{C}(t) := \{(x,\theta) \in \mathbb{R}^3 | \mathfrak{b}(x,t) \geq 0\},$$

where $\mathfrak{b}$ encodes $\phi$ as discussed in section 4.2 of this chapter. We first account for the fact that we do not know the internal dynamics encoded in the function $c$ by using a robust notion of time-varying CBFs as

$$\sup_{u \in \mathbb{R}^2} \frac{\partial \mathfrak{b}(x,t)}{\partial z}^T g(z)u + \frac{\partial \mathfrak{b}(x,t)}{\partial t} \geq -\alpha(\mathfrak{b}(x,t)) + \left\| \frac{\partial \mathfrak{b}(x,t)}{\partial z} \right\| C. \tag{4.18}$$

If the constraint in equation (4.18) holds for all $(z,t) \in \mathbb{R}^3 \times \mathbb{R}_{\geq 0}$, it is easy to see that

$$\sup_{u \in \mathbb{R}^2} \frac{\partial \mathfrak{b}(x,t)}{\partial z}^T (g(z)u + c(z,t)) + \frac{\partial \mathfrak{b}(x,t)}{\partial t} \geq -\alpha(\mathfrak{b}(x,t))$$

holds for all $(z,t) \in \mathbb{R}^3 \times \mathbb{R}_{\geq 0}$ as $\|\frac{\partial \mathfrak{b}(x,t)}{\partial z}\| C \geq -\frac{\partial \mathfrak{b}(x,t)}{\partial z} c(z,t)$. Following similar arguments as in theorem 4.6, we could then construct a control law $u$ so $\phi$ is satisfied. The constraint in equation (4.18), however, may not be feasible when

$$\frac{\partial \mathfrak{b}(x,t)}{\partial z}^T g(z) = \begin{bmatrix} \frac{\partial \mathfrak{b}(x,t)}{\partial p_x}^T \cos(\theta) + \frac{\partial \mathfrak{b}(x,t)}{\partial p_y}^T \sin(\theta) & 0 \end{bmatrix}$$

is equivalent to the zero vector (i.e., when $\frac{\partial \mathfrak{b}(x,t)}{\partial x}$ and $\begin{bmatrix}\cos(\theta) & \sin(\theta)\end{bmatrix}^T$ are perpendicular). Instead, we use a near-identity diffeomorphism to obtain a condition similar to equation (4.18), but in a transformed domain where these singular points are avoided [218, 85]. In particular, we introduce the coordinate transformation

$$k := x + l_t R(\theta) e_1, \tag{4.19}$$

where $e_1 := \begin{bmatrix}1 & 0\end{bmatrix}^T$ is a unit vector, $R(\theta) := \begin{bmatrix}\cos(\theta) & -\sin(\theta) \\ \sin(\theta) & \cos(\theta)\end{bmatrix}$ is a rotation matrix, and $l_t > 0$ can be an arbitrarily small constant (we provide conditions on $l_t$ next). Note that the original coordinates are recovered as $l_t \to 0$. In original $x$-coordinates, we can write the dynamics as

$$\dot{x} = R(\theta) e_1 u_1 + c_x(z,t)$$

such that we can derive the dynamics in transformed $k$ coordinates as

$$\dot{k} = g_k(z) u + c_x(z,t),$$

where the transformation matrix

$$g_k(z) := \begin{bmatrix}\cos(\theta) & -\sin(\theta) l_t \\ \sin(\theta) & \cos(\theta) l_t\end{bmatrix}$$

is known to have full rank [218, lemma 1].

We can now construct a time-varying CBF $\mathfrak{b}(k,t)$ in the $k$-coordinates to satisfy $\phi$ in the original $x$-coordinates. Recall that $\phi$ consists of $p$ predicates $\mu_l$ and predicate functions $h_l$ for which we now also assume that $h_l^{\text{opt}} := \sup_{x \in \mathbb{R}^2} h_l(x)$ is strictly positive (i.e., that $h_l^{\text{opt}} > 0$ as opposed to $h_l^{\text{opt}} \geq 0$). This is needed, as we will need to tighten the predicates to account for the change of coordinates. The STL formula $\phi$ is now transformed into the tightened STL formula $\bar{\phi}$ by replacing each predicate $\mu_l$ by the tightened predicate

$$\bar{\mu}_l(x) := \begin{cases} \top & \text{if } h_l(x) - \chi_l \geq 0 \\ \bot & \text{otherwise} \end{cases} \tag{4.20}$$

for a constant $\chi_l > 0$, which ensures that the tightened predicate $\bar{\mu}_l$ is satisfiable (i.e., $\chi_l$ is chosen so that $h_l^{\text{opt}} - \chi_l > 0$ for each $l \in \{1, \ldots, p+1\}$). We can then construct a candidate time-varying CBF $\mathfrak{b}$ for $\bar{\phi}$, as detailed in the previous sections. Finally, let $L_h^l$ be the Lipschitz constant of $h_l$ (i.e., so that $|h_l(x) - h_l(k)| \leq L_h^l \|x - k\| \leq L_h^l l_t$), and let $l_t$ be such that $l_t \leq \chi_l \,/\, L_h^l$ for each $l \in \{1, \ldots, p+1\}$. This choice implies that $k$ approximates $x$ accurate enough, in the sense that $h_l(k) - \chi_l \geq 0$ implies $h_l(x) \geq 0$. We can now derive a control law $u$ based on the time-varying CBF $\mathfrak{b}$ such that $\phi$ is satisfied.

**Theorem 4.7** *Let the functions $g$ and $c$ defined in equation (4.17) describe the system in equation (4.16), and let the function $c$ satisfy assumption 4.3. Let $\phi$ be an STL formula from the fragment in equation (4.3) defined over the $x$ component of equation (4.16), and let assumption 4.2 hold. Let $\mathfrak{b}:\mathbb{R}^2 \times \mathbb{R}_{\geq 0} \to \mathbb{R}$ and $\alpha:\mathbb{R} \to \mathbb{R}$ be constructed as in lemma 4.3 and the proof of theorem 4.5, respectively, but for the formula $\bar{\phi}$, which is obtained from $\phi$ by replacing all predicates $\mu_l$ with predicates $\bar{\mu}_l$ as defined in equation (4.20), where $\chi_l > 0$ is such that $h_l^{opt} - \chi_l > 0$. If the control law $u(z,t) := \hat{u}$ is the solution of*

$$\operatorname*{argmin}_{\hat{u}} \hat{u}^T \hat{u} \tag{4.21a}$$

$$s.t.\ \frac{\partial \mathfrak{b}(k,t)}{\partial k} g_k(z)\hat{u} + \frac{\partial \mathfrak{b}(k,t)}{\partial t} \geq -\alpha(\mathfrak{b}(k,t)) + \left\| \frac{\partial \mathfrak{b}(k,t)}{\partial k} \right\| C, \tag{4.21b}$$

*where $k$ is the coordinate transformation in equation (4.19) with $l_t$ being such that $l_t \leq \chi_l \,/\, L_h^l$, then every solution $x:\mathcal{I} \to \mathcal{D}$ to equation (4.16) is such that $(x,0) \models \phi$.*

*Proof:* *The near-identity diffeomorphism results in equation (4.21) being always feasible since $g_k$ has full rank. Similar to the proof of theorem 4.6, we can then show that $u$ is continuous in the first argument and piecewise continuous in the second argument so that we know that $\mathfrak{b}(k(t),t) \geq 0$ for all $t \geq 0$. By the construction of $\mathfrak{b}$, we know that this implies $(k,0) \models \bar{\phi}$. By the choice of $l_t$ and $\chi_l$ and Lipschitz continuity of $h_l$, it holds that $h_l(k(t)) - \chi_l \geq 0$ which implies $h_l(x(t)) \geq 0$. It thus follows that $(x,0) \models \phi$ as the syntax of $\phi$ and $\bar{\phi}$ is the same except for the use of tightened predicates.*[10]

## 4.6 Notes and References

The results from this chapter are mainly taken from [172]. Section 4.4 presents ideas from [175], while the results from section 4.5 appeared in [181].

In section 4.1, we introduced a general notion of time-varying CBFs. We presented this notion originally in [172], while similar notions have also

[10] We note that solutions $z:\mathcal{I} \to \mathbb{R}^3$ are complete (i.e., $\mathcal{I} = \mathbb{R}_{\geq 0}$), as it is easy to see that $\theta(t)$ will be bounded under the given assumptions on $c_\theta$.

appeared in [328, 327]. Conceptually different but related are finite time CBFs [283, 284], which were extended to stochastic systems in [262]. In addition, fixed time control Lyapunov functions (CLFs) were proposed [108, 48].

In sections 4.2–4.4, we presented barrier function–based control laws for STL fragments that exclude disjunction operators. Control laws for STL fragments that include disjunction operators have recently been presented in [128, 67, 320, 335]. CBF for systems with actuation constraints under STL specifications were presented in [59]. Similar to the ideas presented in sections 4.2–4.4, finite-time CBFs have been used to enforce linear temporal logic (LTL) specifications [283, 282] by using CBF–based control laws to enforce traversing edges within the automaton that represents the specifications. Barrier functions have also been more tightly coupled with the automata representation of LTL specifications in [50, 51]. For stochastic systems and LTL specifications, barrier function certificates were proposed in [136]. Fixed-time CLFs have also been used to enforce spatiotemporal constraints in [109, 110, 108]. While we have presented the log-exp approximation of the min operator in equation (4.4), concepts of nonsmooth CBFs for the composition of min and max operators were presented in [117, 116] as well as multiobjective CBFs in [315, 328, 327, 329]. Multiple CBFs for input constrained systems appeared in [82, 55]. Input constraints have further been considered by integral CBFs and control-dependent CBFs, see [17] and [132], respectively.

In several places throughout this chapter, we discussed the continuity of control laws derived from a CBF. Continuity properties of CBF–based control laws have been analyzed in [212]. Finally, the robust CBF condition presented in section 4.5 shares similarities with material presented in [330, 254], while related notions of input-to-state safety have appeared in [156, 253].

## 4.7 Additional Proofs

*Proof of theorem 4.5:* For each point $(x,t)\in\mathcal{D}\times(s_j,s_{j+1})$ where $\frac{\partial\mathfrak{b}(x,t)}{\partial x}\neq 0$, a suitable control input $u\in\mathbb{R}^m$ can be found such that the constraint in equation (4.2) is satisfied. In the remainder, we hence focus on the points $(x,t)\in\mathcal{D}\times(s_j,s_{j+1})$, where $\frac{\partial\mathfrak{b}(x,t)}{\partial x}=0$. Concavity of $\mathfrak{b}(x,t)$ in $x$ implies that for each fixed time $t'\in(s_j,s_{j+1})$, the points $x\in\mathcal{D}$ where $\frac{\partial\mathfrak{b}(x,t')}{\partial x}=0$ are maxima of the function $\mathfrak{b}(x,t')$.[11] Each such maxima, which we denote as $x^*_{t'}$, is such that $x^*_{t'} \in \text{int}(\mathcal{C}(t'))$ since we have solved the optimization problem in equation (4.14) or equation (4.15) with $\chi > 0$. It hence holds that $\mathfrak{b}(x^*_{t'},t') \geq \chi > 0$ for each $t' \in (s_j, s_{j+1})$ such that

$$\mathfrak{b}_l(x^*_{t'},t') \geq \chi > 0$$

for each $l \in \{1,\ldots,p+1\}$ with $\mathfrak{o}_l(t')=1$. Next, note that there is a constant $\mathfrak{b}_l^{\max}$ for each $l \in \{1,\ldots,p+1\}$ such that $\mathfrak{b}_l(x^*_{t'},t') \leq \mathfrak{b}_l^{\max}$ for each $t' \in (s_j, s_{j+1})$ due to piecewise continuity of $\mathfrak{b}_l(x,t)$ in $t$. Let $\mathfrak{b}^{\max} := \max(\mathfrak{b}_1^{\max},\ldots,\mathfrak{b}_{p+1}^{\max})$ such that

$$\max(\mathfrak{b}_1(x^*_{t'},t'),\ldots,\mathfrak{b}_{p+1}(x^*_{t'},t')) \leq \mathfrak{b}^{\max}$$

for all $t' \in (s_j, s_{j+1})$ and, in case of exponential templates $\gamma_l^{\exp}$, let

$$\Delta_l := \sup_{t\geq 0} \left|\frac{\partial \mathfrak{b}_l(x,t)}{\partial t}\right| = |\mathfrak{l}_l^{\exp}(\gamma_{l,0}^{\exp} - \gamma_{l,\infty}^{\exp})|$$

be the maximum partial derivative of $\mathfrak{b}$ with respect to $t$. It follows that

$$\begin{aligned}\frac{\partial \mathfrak{b}(x^*_{t'},t')}{\partial t} &= \frac{\sum_{l=1}^{p+1} \mathfrak{o}_l(t')\exp(-\eta \mathfrak{b}_l(x^*_{t'},t'))\frac{\partial \mathfrak{b}_l(x^*_{t'},t')}{\partial t}}{\sum_{l=1}^{p+1} \mathfrak{o}_l(t')\exp(-\eta \mathfrak{b}_l(x^*_{t'},t'))} \\ &= \frac{-\sum_{l=1}^{p+1} \mathfrak{o}_l(t')\exp(-\eta \mathfrak{b}_l(x^*_{t'},t'))|\frac{\partial \mathfrak{b}_l(x^*_{t'},t')}{\partial t}|}{\sum_{l=1}^{p+1} \mathfrak{o}_l(t')\exp(-\eta \mathfrak{b}_l(x^*_{t'},t'))} \\ &\geq \frac{-\exp(-\eta\chi)\max_l \Delta_l}{\exp(-\eta \mathfrak{b}^{\max})} =: \zeta\end{aligned}$$

where $\zeta$ is negative. If we now guarantee, by the choice of function $\alpha$, that $\zeta \geq -\alpha(\chi) + \epsilon$ for some small $\epsilon \geq 0$, then it holds that $\frac{\partial \mathfrak{b}(x^*_{t'},t')}{\partial t} \geq -\alpha(\mathfrak{b}(x^*_{t'},t') + \epsilon$ for all $t' \in (s_j, s_{j+1})$ such that the constraint in equation (4.2) holds, and we can conclude that $\mathfrak{b}$ is a valid time-varying CBF on each domain $\mathcal{D} \times (s_j, s_{j+1})$.[12] We can achieve this by the specific choice of $\alpha(\chi) := \kappa\chi$ where we select $\kappa \geq \frac{\epsilon - \zeta}{\chi}$. A similar analysis holds for linear template functions $\gamma_l^{\text{lin}}$ where $\Delta_l := \sup_{t\geq 0}|\frac{\partial \mathfrak{b}_l(x,t)}{\partial t}| = \frac{\gamma_{l,\infty}^{\text{lin}} - \gamma_{l,0}^{\text{lin}}}{t_l^*}$ if $t_l^* > 0$ and $\Delta_l := \sup_{t\geq 0}|\frac{\partial \mathfrak{b}_l(x,t)}{\partial t}| = 0$ if $t_l^* = 0$.

[11] In fact, the inclusion of the quadratic function $\mathfrak{b}_{p+1}(x,t) := D - \|x\|^2$ into the time-varying CBF $\mathfrak{b}$ as in equation (4.7) guarantees that this maximum is bounded.

[12] The observant reader may wonder why we have introduced the positive constant $\epsilon$. This constant is technically not needed here, but it is introduced as it will ensure in theorem 4.6 that the control law $u(x,t)$ obtained by solving equation (4.9) is continuous in $x$ and piecewise continuous in $t$.

*Proof of theorem 4.6:* We first show that $u(x,t)$ is continuous in $x$ and piecewise continuous in $t$. For points $(x,t) \in \mathcal{D} \times (s_j, s_{j+1})$ where $\frac{\partial \mathfrak{b}(x,t)}{\partial x} \neq 0$, the optimization problem in equation (4.9b) is feasible and $u(x,t)$ is locally Lipschitz continuous around $(x,t)$ due to [330, theorem 8]. For points $(x,t) \in \mathcal{D} \times (s_j, s_{j+1})$ where $\frac{\partial \mathfrak{b}(x,t)}{\partial x} = 0$, the optimization problem in equation (4.9b) is satisfied due to the previously discussed choice of $\alpha$, which makes $\mathfrak{b}$ a valid time-varying CBF on each domain $\mathcal{D} \times (s_j, s_{j+1})$ such that $u(x,t) := 0$. In particular, it holds that $\frac{\partial \mathfrak{b}(x,t)}{\partial t} \geq -\alpha(\mathfrak{b}(x,t)) + \epsilon$ as shown in theorem 4.5. Due to the continuity of $\frac{\partial \mathfrak{b}(x,t)}{\partial t}$ and $\alpha(\mathfrak{b}(x,t))$, there is a neighborhood $\mathcal{U}$ around $(x,t)$ such that, for each $(x',t') \in \mathcal{U}$, $\frac{\partial \mathfrak{b}(x',t')}{\partial t} \geq -\alpha(\mathfrak{b}(x',t'))$ and consequently, it holds that $u(x',t') = 0$. We can hence conclude that $u$ is continuous on each domain $\mathcal{D} \times (s_j, s_{j+1})$. By theorem 4.3, it follows that every solution $x : \mathcal{I} \to \mathcal{D}$ to the system in equation (4.1) satisfies $\phi$ (i.e., $(x,0) \models \phi$). By lemma 4.3, it further follows that $r \leq \rho^{\phi}(x,0)$.

# Chapter 5

# Decentralized Time-Varying Control Barrier Functions

In this chapter, we will design decentralized control laws for multi-agent control systems under global and local signal temporal logic (STL) specifications using time-varying control barrier functions (CBF). Specifically, consider $N$ agents where agent $i \in \{1,\dots,N\}$ is now modeled as

$$\dot{x}_i = f_i(x_i) + g_i(x_i)u_i + c_i(x,t), \tag{5.1}$$

where $x_i \in \mathbb{R}^{n_i}$, $u_i \in \mathbb{R}^{m_i}$, $f_i : \mathbb{R}^{n_i} \to \mathbb{R}^{n_i}$, $g_i : \mathbb{R}^{n_i} \to \mathbb{R}^{n_i \times m_i}$, and $c_i : \mathbb{R}^n \times \mathbb{R}_{\geq 0} \to \mathbb{R}^{n_i}$ denote the state, control input, internal dynamics, input dynamics, and dynamical couplings of agent $i$, respectively, while $x := \begin{bmatrix} x_1^T & \dots & x_N^T \end{bmatrix}^T$ denotes the multi-agent state. For simplicity and without loss of generality, the function $f_i$ and $g_i$ do not depend on time. The functions $f_i$, $g_i$, and $c_i$ are locally Lipschitz continuous and, in this section, the functions $f_i$ and $g_i$ are only known to agent $i$, while the coupling term $c_i$ is generally unknown but bounded. Similar to chapter 4, we make the following assumptions.

**Assumption 5.1** *Let the function $g_i : \mathbb{R}^{n_i} \to \mathbb{R}^{n_i \times m_i}$ be such that $g_i(x_i)g_i(x_i)^T$ is positive definite for each $x_i \in \mathbb{R}^{n_i}$. Further, let the function $c_i : \mathbb{R}^n \times \mathbb{R}_{\geq 0} \to \mathbb{R}^{n_i}$ be bounded with known constant $C$ (i.e., $\|c_i(x,t)\| \leq C$ for all $(x,t) \in \mathbb{R}^n \times \mathbb{R}_{\geq 0}$).*

Global specifications are collaborative specifications in which all agents share the same objective. The satisfaction of a global STL specification $\phi$

depends on the multi-agent state $x$. In principle, we could apply the time-varying CBF method described in chapter 4 for global specifications in a centralized manner. In other words, each agent $i$ could compute the stacked control input $u := \begin{bmatrix} u_1^T & \dots & u_N^T \end{bmatrix}^T$ following the CBF condition in equation (4.9) for the multi-agent dynamics $\dot{x} = f(x) + g(x)u + c(x,t)$ where $f$, $g$, and $c$ are the stacked versions of all $f_i$, $g_i$, and $c_i$ as defined in section 2.3, and then apply the individual control $u_i$. However, such a centralized approach has two main limitations: (1) it would require that each agent knows the dynamics of all other agents, and (2) this approach would not be efficient, as each agent solves a control problem with $m := \sum_i^N m_i$ decision variables. An alternative approach could be that a central control unit (e.g., a designated agent) computes $u$ and communicates the control portions $u_i$ to agent $i$. However, this approach is not robust to failure of the central control unit and requires excessive communication.

In section 5.1, we therefore present a decentralized control approach for global STL specifications based on time-varying CBFs where each agent directly computes $u_i$ without knowing the dynamics of other agents. Effectively, a set of smaller control problems will be solved in a decentralized manner by decomposing the CBF condition in equation (4.9) so that we can parallelize the computation of the stacked control input $u$.

Assigning global and collaborative specifications to a multi-agent system represents a top-down approach, while in practice a bottom-up approach may be desirable. In section 5.2, we consider local STL specifications $\phi_i$ that are individually assigned to each agent $i$. Local specifications may not be known to other agents, but the satisfaction of a local specification $\phi_i$ can still depend on the behavior of other agents $j \neq i$. Consequently, the set of local specifications may be conflicting (or adversarial) in the sense that it may not be possible to satisfy all of them (i.e., satisfiability of each local task does not imply satisfiability of the conjunction of all local tasks). We present a decentralized control approach based on time-varying CBFs that finds locally least-violating solutions that locally maximize the quantitative semantics associated with $\phi_i$.

## 5.1 Decentralized Collaborative Control

We will first discuss challenges related to deriving a decentralized version of the CBF condition in equation (4.2). It will turn out that a decentralized time-varying CBF condition can lead to discontinuous control laws. The use of discontinuous control laws will naturally require heavier technical machinery, as the existence of classical solutions to an initial value problem (IVP) as per lemma 2.1 is not guaranteed anymore. We thus consider Filippov solutions and present results for CBFs with discontinuous feedback control laws. Based on these results, we will present a decentralized feedback control law for global STL specifications.

We again consider the STL fragment from chapter 4 as per equation (4.3). Specifically, we consider the global STL specification

$$\phi := \phi_1 \wedge \ldots \wedge \phi_K,$$

where each STL formula $\phi_k$ for $k \in \{1, \ldots, K\}$ is from the STL fragment in equation (4.3). The satisfaction of $\phi_k$ depends on the behavior of the set of agents in $\mathcal{V}_k \subseteq \{1, \ldots, N\}$. By "behavior of agent $i$," we mean its solution $x_i : \mathbb{R}_{\geq 0} \to \mathbb{R}^{n_i}$ to equation (5.1). In other words, the satisfaction of $\phi_k$ depends on the solutions $x_i$ to equation (5.1) for $i \in \mathcal{V}_k$. The reason why we write $\phi$ explicitly as a conjunction of $K$ formulas $\phi_k$ is that not all agents may be coupled by a specification, while all agents may be coupled dynamically via $c_i$. We assume that the sets of agents $\mathcal{V}_1, \ldots, \mathcal{V}_K \subseteq \{1, \ldots, N\}$ are disjoint (i.e., that $\mathcal{V}_{k_1} \cap \mathcal{V}_{k_2} = \emptyset$ for all $k_1, k_2 \in \{1, \ldots, K\}$ with $k_1 \neq k_2$). There are hence no formula dependencies between agents in $\mathcal{V}_{k_1}$ and agents in $\mathcal{V}_{k_2}$. For the set $\mathcal{V}_k$, we define the stacked state vector of all agents within this set $j_1, \ldots, j_{|\mathcal{V}_k|} \in \mathcal{V}_k$ as

$$\bar{x}_k := \begin{bmatrix} x_{j_1}{}^T & \ldots & x_{j_{|\mathcal{V}_k|}}{}^T \end{bmatrix}^T \in \mathbb{R}^{\bar{n}_k},$$

with state dimension $\bar{n}_k := n_{j_1} + \ldots + n_{j_{|\mathcal{V}_k|}}$. The stacked agent dynamics are thus

$$\dot{\bar{x}}_k = \bar{f}_k(\bar{x}_k) + \bar{g}_k(\bar{x}_k)\bar{u}_k(\bar{x}_k, t) + \bar{c}_k(x, t), \tag{5.2}$$

with $\bar{u}_k(\bar{x}_k, t)$ explicitly depending on $\bar{x}_k$ and $t$ and where

$$\begin{aligned}
\bar{f}_k(\bar{x}_k) &:= \begin{bmatrix} f_{j_1}(x_{j_1})^T & \ldots & f_{j_{|\mathcal{V}_k|}}(x_{j_{|\mathcal{V}_k|}})^T \end{bmatrix}^T, \\
\bar{g}_k(\bar{x}_k) &:= \operatorname{diag}\big(g_{j_1}(x_{j_1}), \ldots, g_{j_{|\mathcal{V}_k|}}(x_{j_{|\mathcal{V}_k|}})\big), \\
\bar{u}_k(\bar{x}_k, t) &:= \begin{bmatrix} u_{j_1}(\bar{x}_k, t)^T & \ldots & u_{j_{|\mathcal{V}_k|}}(\bar{x}_k, t)^T \end{bmatrix}^T, \\
\bar{c}_k(x, t) &:= \begin{bmatrix} c_{j_1}(x, t)^T & \ldots & c_{j_{|\mathcal{V}_k|}}(x, t)^T \end{bmatrix}^T.
\end{aligned}$$

Finally, let $\mathfrak{b}_k : \mathbb{R}^{\bar{n}_k} \times \mathbb{R}_{\geq 0} \to \mathbb{R}$ be a time-varying CBF that encodes the STL formula $\phi_k$, as instructed in chapter 4.

### 5.1.1 Challenges in Decentralizing Time-Varying CBFs

For the stacked agent dynamics in equation (5.2), the centralized CBF condition from equation (4.2) is

$$\frac{\partial \mathfrak{b}_k(\bar{x}_k,t)}{\partial \bar{x}_k}(\bar{f}_k(\bar{x}_k)+\bar{g}_k(\bar{x}_k)\bar{u}_k(\bar{x}_k,t)+\bar{c}_k(x,t))+\frac{\partial \mathfrak{b}_k(\bar{x}_k,t)}{\partial t}\geq -\alpha_k(\mathfrak{b}_k(\bar{x}_k,t)), \tag{5.3}$$

where $\alpha_k:\mathbb{R}\to\mathbb{R}$ is again an extended class $\mathcal{K}$ function. However, the function $\bar{c}_k$ is unknown, so we cannot use equation (5.3). Instead, we consider the condition

$$\begin{aligned}\frac{\partial \mathfrak{b}_k(\bar{x}_k,t)}{\partial \bar{x}_k}(\bar{f}_k(\bar{x}_k)+\bar{g}_k(\bar{x}_k)\bar{u}_k(\bar{x}_k,t))+\frac{\partial \mathfrak{b}_k(\bar{x}_k,t)}{\partial t}&\geq -\alpha_k(\mathfrak{b}_k(\bar{x}_k,t))\\ &+\left\|\frac{\partial \mathfrak{b}_k(\bar{x}_k,t)}{\partial \bar{x}_k}\right\|_1 C,\end{aligned} \tag{5.4}$$

where the constant $C$ is known from assumption 5.1. Specifically, note that equation (5.4) implies equation (5.3) since

$$\left\|\frac{\partial \mathfrak{b}_k(\bar{x}_k,t)}{\partial \bar{x}_k}\right\|_1 C\geq\left\|\frac{\partial \mathfrak{b}_k(\bar{x}_k,t)}{\partial \bar{x}_k}\right\|_1\|\bar{c}_k(x,t)\|_\infty\geq -\frac{\partial \mathfrak{b}_k(\bar{x}_k,t)}{\partial \bar{x}_k}\bar{c}_k(x,t),$$

where the first inequality follows by the definition of the $\infty$-norm, the fact that the Euclidean norm upper bounds the $\infty$-norm, and assumption 5.1 so that

$$\|\bar{c}_k(x,t)\|_\infty=\max_{i\in\mathcal{V}_k}\|c_i(x,t)\|_\infty\leq\max_{i\in\mathcal{V}_k}\|c_i(x,t)\|\leq C$$

and where the second inequality follows by Hölder's inequality. While using the Euclidean norm would suffice here, we remark that the use of the 1-norm and the $\infty$-norm will be needed later.

Following section 4.4 and theorem 4.5, we know—under the assumptions made in chapter 4—that the functions $\mathfrak{b}_k$ and $\alpha_k$ can be designed so that

$$\frac{\partial \mathfrak{b}_k(\bar{x}_k,t)}{\partial t}\geq -\alpha_k(\mathfrak{b}_k(\bar{x}_k,t)) \text{ if } \frac{\partial \mathfrak{b}_k(\bar{x}_k,t)}{\partial \bar{x}_k}=0.$$

This implies that we can satisfy the centralized CBF condition in equation (5.4) by an appropriate choice of $\bar{u}_k(\bar{x}_k,t)$ for all $(\bar{x}_k,t)\in\mathbb{R}^{\bar{n}_k}\times\mathbb{R}_{\geq 0}$. We can thus use the condition in equation (5.4) to obtain a centralized control law $\bar{u}_k$ that is continuous following similar arguments as in theorem 4.6.

As motivated before, we would like to obtain a decentralized control law and would thus like to define a decentralized version of the CBF condition in equation (5.4). A straightforward idea is to let each agent $i\in\mathcal{V}_k$ solve

$$\frac{\partial \mathfrak{b}_k(\bar{x}_k,t)}{\partial x_i}(f_i(x_i)+g_i(x_i)u_i(\bar{x}_k,t)) \geq \\ -D_i\Big(\frac{\partial \mathfrak{b}_k(\bar{x}_k,t)}{\partial t}+\alpha_k(\mathfrak{b}_k(\bar{x}_k,t))\Big)+\Big\|\frac{\partial \mathfrak{b}_k(\bar{x}_k,t)}{\partial x_i}\Big\|_1 C, \tag{5.5}$$

where the weight $D_i := \frac{1}{|\mathcal{V}_k|}$ distributes the inequality in equation (5.4) equally among all agents. Note that other weights $D_i$ could be possible so long as $\sum_{i\in\mathcal{V}_k} D_i = 1$. This is since summing over both sides of equation (5.5) for all agents $i \in \mathcal{V}_k$ will imply that equation (5.4) holds. Specifically, if each agent $i$ solves equation (5.5), it follows that

$$\sum_{i\in\mathcal{V}_k}\frac{\partial \mathfrak{b}_k(\bar{x}_k,t)}{\partial x_i}(f_i(x_i)+g_i(x_i)u_i(\bar{x}_k,t)) \geq \\ \sum_{i\in\mathcal{V}_k} -D_i\Big(\frac{\partial \mathfrak{b}_k(\bar{x}_k,t)}{\partial t}+\alpha_k(\mathfrak{b}_k(\bar{x}_k,t))\Big)+\sum_{i\in\mathcal{V}_k}\Big\|\frac{\partial \mathfrak{b}_k(\bar{x}_k,t)}{\partial x_i}\Big\|_1 C$$

$$\Leftrightarrow \frac{\partial \mathfrak{b}_k(\bar{x}_k,t)}{\partial \bar{x}_k}(\bar{f}_k(\bar{x}_k)+\bar{g}_k(\bar{x}_k)\bar{u}_k(\bar{x}_k,t)) \geq \\ -\frac{\partial \mathfrak{b}_k(\bar{x}_k,t)}{\partial t}-\alpha_k(\mathfrak{b}_k(\bar{x}_k,t))+\Big\|\frac{\partial \mathfrak{b}_k(\bar{x}_k,t)}{\partial \bar{x}_k}\Big\|_1 C, \tag{5.6}$$

where we used the property of the 1-norm such that $\sum_{i\in\mathcal{V}_k}\big\|\frac{\partial \mathfrak{b}_k(\bar{x}_k,t)}{\partial x_i}\big\|_1 = \big\|\frac{\partial \mathfrak{b}_k(\bar{x}_k,t)}{\partial \bar{x}_k}\big\|_1$. From here, we immediately see that equation (5.6) is equivalent to equation (5.4).

This analysis is based on the assumption that each agent $i \in \mathcal{V}_k$ can solve equation (5.5). With the choice of $D_i := \frac{1}{|\mathcal{V}_k|}$, however, this may not always be feasible. In particular, in cases where $\frac{\partial \mathfrak{b}_k(\bar{x}_k,t)}{\partial x_i} = 0$ while $\exists j \in \mathcal{V}_k \setminus \{i\}$ such that $\frac{\partial \mathfrak{b}_k(\bar{x}_k,t)}{\partial x_j} \neq 0$, it turns out that equation (5.5) is infeasible for agent $i$ if the right side of equation (5.5) is positive. This may easily happen due to the nature of time-varying CBFs via the term $\frac{\partial \mathfrak{b}_k(\bar{x}_k,t)}{\partial t}$. We remark that when $\frac{\partial \mathfrak{b}_k(\bar{x}_k,t)}{\partial \bar{x}_k} \neq 0$, $\bar{x}_k$ is not the global optimum of $\mathfrak{b}_k$ (recall that $\mathfrak{b}_k$ is concave). Furthermore, a control law based on equation (5.5) may be unbounded when the gradient $\frac{\partial \mathfrak{b}_k(\bar{x}_k,t)}{\partial x_i}$ approaches zero. Specifically, it may happen that $\|u_i(x_i,t)\| \to \infty$ as $\frac{\partial \mathfrak{b}_k(\bar{x}_k,t)}{\partial x_i} \to 0$ if $\exists j \in \mathcal{V}_k \setminus \{i\}$ such that $\frac{\partial \mathfrak{b}_k(\bar{x}_k,t)}{\partial x_j} \to d$ for some vector $d \neq 0$. Consequently, a weight function $D_i : \mathbb{R}^{\bar{n}_k} \times \mathbb{R}_{\geq 0} \to \mathbb{R}_{\geq 0}$ is needed instead of a constant $D_i$. As it will turn out this weight function will be discontinuous, which may lead to discontinuous control laws requiring a nonsmooth version of time-varying CBFs.

### 5.1.2 CBFs for Discontinuous Systems

To present a CBF condition for discontinuous systems, we will introduce Filippov solutions [104] since the existence of classical solutions as in lemma 2.1 is no longer guaranteed. The technical machinery in this section will be more sophisticated. For the reader that is more interested in the main results, we recommend going directly to section 5.1.3 and revisit this section later. When doing so, one can think of Filippov solutions simply as suitably defined solutions for discontinuous dynamical systems. Consider now the dynamical system

$$\dot{x}(t) = H(x(t), t), \ x(0) \in \mathbb{R}^n, \tag{5.7}$$

where $H : \mathbb{R}^n \times \mathbb{R}_{\geq 0} \to \mathbb{R}^n$ is locally bounded and measurable, but not necessarily continuous in $x$. In general, Filippov solutions relax the requirement on the continuity of the system; that is, the function $H$ does not need to be continuous everywhere in its first argument. This includes discontinuous functions $H$ as a special case, which we will encounter later in this chapter. In this way, the function $H$ can represent a dynamical system $\dot{x} = f(x, t) + g(x, t)u$ with a discontinuous feedback control law $u$. We will consider Filippov solutions that are defined for discontinuous systems such as in equation (5.7), and we define the Filippov set–valued map for $H$ as

$$F[H](x, t) := \overline{\text{co}}\{\lim_{j \to \infty} H(x_j, t) | x_j \to x, x_j \notin N \cup N_f\},$$

where $\overline{\text{co}}$ denotes the convex closure, $x_j \to x$ is a sequence of states $x_j$ that converge to $x$, $N_f$ denotes the set of the Lebesgue measure zero where $H(x, t)$ is discontinuous, while $N$ denotes an arbitrary set of Lebesgue measure zero. At points where $H$ is discontinuous, the Filippov set–valued map consequently consists of the convex combination of the vector fields in a small neighborhood around these points. Filippov solutions are then defined as functions $x : \mathcal{I} \to \mathbb{R}^n$ that satisfy the differential inclusion described by the Filippov set–valued map for almost all $t \in \mathcal{I}$.

**Definition 5.1 (Filippov Solution to the IVP in Equation (5.7))** *A Filippov solution to the IVP in equation (5.7) over a time interval $\mathcal{I} \subseteq \mathbb{R}_{\geq 0}$ is an absolutely continuous function $x : \mathcal{I} \to \mathbb{R}^n$ that satisfies $\dot{x}(t) \in F[H](x(t), t)$ for almost all $t \in \mathcal{I}$.*

The existence of Filippov solutions is guaranteed under much weaker assumptions than classical solutions. That is, only local boundedness and measurability of the function $H$ are required [81, proposition 3].

**Lemma 5.1 (Existence of Filippov Solutions)** *Assume that the function $H:\mathbb{R}^n\times\mathbb{R}_{\geq 0}\to\mathbb{R}^n$ is locally bounded and measurable. Then, there is a Filippov solution $x:\mathcal{I}\to\mathbb{R}^n$ to the IVP in equation (5.7) over the time interval $\mathcal{I}\subseteq\mathbb{R}_{\geq 0}$.*

Consider again a continuously differentiable function $\mathfrak{b}:\mathbb{R}^n\times\mathbb{R}_{\geq 0}$ so that its gradient is well defined.[1] We can now define the set-valued Lie derivative of $\mathfrak{b}$ with respect to the Filippov set-valued map $F[H]$ as

$$\mathcal{L}_{F[H]}\mathfrak{b}(x,t):=\Big\{\frac{\partial\mathfrak{b}(x,t)}{\partial x}\zeta^T+\frac{\partial\mathfrak{b}(x,t)}{\partial t}\Big|\zeta\in F[H](x,t)\Big\}.$$

For the case where $H$ is continuous, it is easy to see that the set-valued Lie derivative reduces to the Lie derivative as $F[H](x,t)$ reduces to a singleton. According to [273, theorem 2.2], a Filippov solution $x:\mathcal{I}\to\mathbb{R}^n$ satisfies

$$\dot{\mathfrak{b}}(x(t),t)\in\mathcal{L}_{F[H]}\mathfrak{b}(x(t),t)$$

for almost all $t\in\mathcal{I}$. For convenience, we also define

$$\hat{\mathcal{L}}_{F[H]}\mathfrak{b}(x,t):=\Big\{\frac{\partial\mathfrak{b}(x,t)}{\partial x}^T\zeta\Big|\zeta\in F[H](x,t)\Big\}$$

such that we can rewrite the set-valued Lie derivative as

$$\mathcal{L}_{F[H]}\mathfrak{b}(x,t)=\hat{\mathcal{L}}_{F[H]}\mathfrak{b}(x,t)\oplus\Big\{\frac{\partial\mathfrak{b}(x,t)}{\partial t}\Big\}.$$

Let us now derive a discontinuous CBF condition for the discontinuous control system in equation (5.7).

**Theorem 5.1 (Forward Invariance of Discontinuous Systems)** *Let the function $H$ that defines the system in equation (5.7) be locally bounded and measurable. Assume that $\mathfrak{b}:\mathbb{R}^n\times[t_0,t_1]$ is a candidate time-varying CBF on $\mathcal{D}\times[t_0,t_1]$, and that there is a locally Lipschitz-continuous extended class $\mathcal{K}$ function $\alpha:\mathbb{R}\to\mathbb{R}$ such that*

$$\min\mathcal{L}_{F[H]}\mathfrak{b}(x,t)\geq-\alpha(\mathfrak{b}(x,t))$$

*for all $(x,t)\in\mathcal{D}\times[t_0,t_1]$. Then $x(t_0)\in\mathcal{C}(t_0)$ implies $x(t)\in\mathcal{C}(t)$ for all $t\in\mathcal{I}\cap[t_0,t_1]$ for all Filippov solutions to the IVP in equation (5.7).*

*Proof:* *Each Filippov solution $x:\mathcal{I}\to\mathbb{R}^n$ satisfies $\dot{\mathfrak{b}}(x(t),t)\in\mathcal{L}_{F[H]}\mathfrak{b}(x(t),t)$ for almost all $t\in\mathcal{I}$ [273, theorem 2.2]. We can now see that $\dot{\mathfrak{b}}(x(t),t)\geq-\alpha(\mathfrak{b}(x(t),t)$ for almost all $t\in\mathcal{I}$ since we have assumed that*

[1] Nonsmooth analysis generally allows $\mathfrak{b}$ to be discontinuous, which is not needed in our case.

$\min \mathcal{L}_{F[H]}\mathfrak{b}(x,t) \geq -\alpha(\mathfrak{b}(x,t))$ *for all* $(x,t) \in \mathcal{D} \times [t_0, t_1]$. *From here, the proof is the same as in theorem 4.1.*

If $\mathcal{I} \supset [t_0, t_1]$, then set $\mathcal{C}$ is forward invariant over the time interval $[t_0, t_1]$. We can again achieve this by assuming that $\mathcal{C}(t)$ is compact for all times $t \in [t_0, t_1]$.

We would now like to extend the previous result to obtain a centralized control barrier condition for the function $\mathfrak{b}_k$ that encodes the STL specifications $\phi_k$ and the stacked agent dynamics of $\bar{x}_k$ in equation (5.2). In doing so, we need the following lemma that concerns the set-valued Lie derivative of the sum of two functions.

**Lemma 5.2** *Let the function* $\mathfrak{b} : \mathbb{R}^n \times \mathbb{R}_{\geq 0} \to \mathbb{R}$ *be continuously differentiable. Consider* $\dot{x} = H_1(x,t) + H_2(x,t)$, *where* $H_1 : \mathbb{R}^n \times \mathbb{R}_{\geq 0} \to \mathbb{R}^n$ *and* $H_2 : \mathbb{R}^n \times \mathbb{R}_{\geq 0} \to \mathbb{R}^n$ *are locally bounded and measurable. It then holds that*

$$\mathcal{L}_{F[H_1+H_2]}\mathfrak{b}(x,t) \subseteq \hat{\mathcal{L}}_{F[H_1]}\mathfrak{b}(x,t) \oplus \hat{\mathcal{L}}_{F[H_2]}\mathfrak{b}(x,t) \oplus \Big\{\frac{\partial \mathfrak{b}(x,t)}{\partial t}\Big\}.$$

*Proof: Applying the definition of* $\mathcal{L}_{F[H]}\mathfrak{b}$ *results in*

$$\begin{aligned}
\mathcal{L}_{F[H_1+H_2]}\mathfrak{b}(x,t) &:= \Big\{\frac{\partial \mathfrak{b}(x,t)}{\partial x}\zeta^T + \frac{\partial \mathfrak{b}(x,t)}{\partial t}\Big| \zeta \in F[H_1+H_2](x,t)\Big\} \\
&\subseteq \Big\{\frac{\partial \mathfrak{b}(x,t)}{\partial x}\zeta^T + \frac{\partial \mathfrak{b}(x,t)}{\partial t}\Big| \zeta \in F[H_1](x,t) \oplus F[H_2](x,t)\Big\} \\
&= \Big\{\frac{\partial \mathfrak{b}(x,t)}{\partial x}\zeta \Big| \zeta \in F[H_1](x,t) \oplus F[H_2](x,t)\Big\} \oplus \Big\{\frac{\partial \mathfrak{b}(x,t)}{\partial t}\Big\} \\
&= \hat{\mathcal{L}}_{F[H_1]}\mathfrak{b}(x,t) \oplus \hat{\mathcal{L}}_{F[H_2]}\mathfrak{b}(x,t) \oplus \Big\{\frac{\partial \mathfrak{b}(x,t)}{\partial t}\Big\},
\end{aligned}$$

*where we used the fact that* $F[H_1+H_2](x,t) \subseteq F[H_1](x,t) \oplus F[H_2](x,t)$ *due to [222, theorem 1].*

We are now ready to present a centralized control barrier condition for the function $\mathfrak{b}_k$ that encodes the STL specification $\phi_k$ and the stacked agent dynamics of $\bar{x}_k$ in equation (5.2). We remark that we will use this centralized CBF condition in the next section to present a decentralized control law.

**Theorem 5.2 (CBFs for $\phi_k$ with Discontinuous Control Law)** *Let the functions* $f_i$, $g_i$, *and* $c_i$ *that define the system in equation (5.1) be locally Lipschitz continuous and satisfy assumption 5.1, while the function* $u_i$ *is locally bounded and measurable. Let the function* $\mathfrak{b}_k : \mathbb{R}^{\bar{n}_k} \times \mathbb{R}_{\geq 0} \to \mathbb{R}$ *encode the STL* *specification $\phi_k$ from the fragment in equation (4.3) and be constructed as in lemma 4.3. If, for each $k \in \{1, \dots, K\}$, there is a locally Lipschitz-continuous extended class $\mathcal{K}$ function $\alpha_k : \mathbb{R} \to \mathbb{R}$ such that*

$$\min \mathcal{L}_{F[\bar{f}_k + \bar{g}_k \bar{u}_k]} \mathfrak{b}_k(\bar{x}_k, t) \geq -\alpha_k(\mathfrak{b}_k(\bar{x}_k, t)) + \left\| \frac{\partial \mathfrak{b}_k(\bar{x}_k, t)}{\partial \bar{x}_k} \right\|_1 C \tag{5.8}$$

*for all $(\bar{x}_k, t) \in \mathbb{R}^{\bar{n}_k} \times (s_j^k, s_{j+1}^k)$,[2] then each Filippov solution $x : \mathcal{I} \to \mathbb{R}^n$ to equation (5.1) is such that $(x, 0) \models \phi_1 \wedge \ldots \wedge \phi_K$.*

*Proof:* Following the same arguments as in section 5.1.1, we note that the condition in equation (5.8) implies that

$$\min \mathcal{L}_{F[\bar{f}_k + \bar{g}_k \bar{u}_k]} \mathfrak{b}_k(\bar{x}_k, t) \oplus \frac{\partial \mathfrak{b}_k(\bar{x}_k, t)}{\partial \bar{x}_k} \bar{c}_k(x, t) \geq -\alpha_k(\mathfrak{b}_k(\bar{x}_k, t)). \tag{5.9}$$

We further know that $\hat{\mathcal{L}}_{F[\bar{c}_k]} \mathfrak{b}_k(\bar{x}_k, t) = \frac{\partial \mathfrak{b}_k(\bar{x}_k, t)}{\partial \bar{x}_k} \bar{c}_k(x, t)$ since $\mathfrak{b}_k$ is continuously differentiable and $\bar{c}_k$ is locally Lipschitz continuous. Using lemma 5.2, we find that

$$\begin{aligned} \min \mathcal{L}_{F[\bar{f}_k + \bar{g}_k \bar{u}_k + \bar{c}_k]} \mathfrak{b}_k(\bar{x}_k, t) &\geq \min\{\mathcal{L}_{F[\bar{f}_k + \bar{g}_k \bar{u}_k]} \mathfrak{b}_k(\bar{x}_k, t) \oplus \hat{\mathcal{L}}_{F[\bar{c}_k]} \mathfrak{b}_k(\bar{x}_k, t)\} \\ &= \min \mathcal{L}_{F[\bar{f}_k + \bar{g}_k \bar{u}_k]} \mathfrak{b}_k(\bar{x}_k, t) \oplus \frac{\partial \mathfrak{b}_k(\bar{x}_k, t)}{\partial \bar{x}_k} \bar{c}_k(x, t) \\ &\geq -\alpha_k(\mathfrak{b}_k(\bar{x}_k, t)). \end{aligned}$$

For the initial time, we know that there is a Filippov solution $x : [0, \tau_{max}) \to \mathbb{R}^n$ to equation (5.1). For all $k \in \{1, \dots, K\}$, we remark that

$$\dot{\mathfrak{b}}_k(\bar{x}_k(t), t) \in \mathcal{L}_{F[\bar{f}_k + \bar{g}_k \bar{u}_k + \bar{c}_k]} \mathfrak{b}_k(\bar{x}_k(t), t)$$

has to hold for almost all $t \in (0, \tau_{max})$ by [273, theorem 2.2]. We can hence conclude that

$$\dot{\mathfrak{b}}_k(\bar{x}_k(t), t) \geq \min \mathcal{L}_{F[\bar{f}_k + \bar{g}_k \bar{u}_k + \bar{c}_k]} \mathfrak{b}_k(\bar{x}_k(t), t) \geq -\alpha_k(\mathfrak{b}_k(\bar{x}_k(t), t)).$$

for almost all $t \in (0, \tau_{max})$. If now the initial state $\bar{x}_k(0)$ is such that $\mathfrak{b}_k(\bar{x}_k(0), 0) \geq 0$, we can once again apply the Comparison Lemma, by which it follows that $\mathfrak{b}_k(\bar{x}_k(t), t) \geq 0$ for all $t \in [0, \tau_{max})$. The rest of the proof continues as in theorem 4.3 by sequentially showing that the solution $x$ is defined until the next switching time $s_j^k$.

[2] By construction, the function $\mathfrak{b}_k : \mathbb{R}^{\bar{n}_k} \times \mathbb{R}_{\geq 0} \to \mathbb{R}$ is continuously differentiable on $\mathbb{R}^{\bar{n}_k} \times (s_j^k, s_{j+1}^k)$, where $\{s_0^k := 0, s_1^k, \dots, s_{p_k}^k\}$ is the associated switching sequence, as discussed in sections 4.2 and 4.4.

### 5.1.3 Decentralized Collaborative Control Laws

Let us first recall the decentralized CBF condition from equation (5.5) as

$$\begin{aligned}&\frac{\partial \mathfrak{b}_k(\bar{x}_k,t)}{\partial x_i}(f_i(x_i)+g_i(x_i)u_i(\bar{x}_k,t))\geq\\&-D_i(\bar{x}_k,t)\Big(\frac{\partial \mathfrak{b}_k(\bar{x}_k,t)}{\partial t}+\alpha_k(\mathfrak{b}_k(\bar{x}_k,t))\Big)+\Big\|\frac{\partial \mathfrak{b}_k(\bar{x}_k,t)}{\partial x_i}\Big\|_1 C,\end{aligned}\tag{5.10}$$

where we, as motivated before, now consider a function $D_i:\mathbb{R}^{\bar{n}_k}\times\mathbb{R}_{\geq 0}\to\mathbb{R}$ to share the centralized CBF condition in equation (5.8) among all agents $i\in\mathcal{V}_k$. Before proving the main result, we state the choice of $D_i$ up front as

$$D_i(\bar{x}_k,t):=\begin{cases}\frac{\|\frac{\partial \mathfrak{b}_k(\bar{x}_k,t)}{\partial x_i}\|_1}{\sum_{v\in\mathcal{V}_k}\|\frac{\partial \mathfrak{b}_k(\bar{x}_k,t)}{\partial x_v}\|_1} & \text{if } \sum_{v\in\mathcal{V}_k}\|\frac{\partial \mathfrak{b}_k(\bar{x}_k,t)}{\partial x_v}\|_1\neq 0\\ 1/|\mathcal{V}_k| & \text{otherwise,}\end{cases}\tag{5.11}$$

for which it can be confirmed that $\sum_{i\in\mathcal{V}_k}D_i(\bar{x}_k,t)=1$. At the same time, the function $D_i$ is discontinuous for points $(\bar{x}_k,t)\in\mathbb{R}^{\bar{n}_k}\times\mathbb{R}_{\geq 0}$, where $\sum_{v\in\mathcal{V}_k}\|\frac{\partial \mathfrak{b}_k(\bar{x}_k,t)}{\partial x_v}\|_1=0$ (i.e., for points where $\frac{\mathfrak{b}_k(\bar{x}_k,t)}{\partial \bar{x}_k}=0$). Using equation (5.10) for control design may hence result in a control law $u_i$ that is discontinuous, which motivated us to present the centralized CBF condition in equation (5.8) for discontinuous control laws. For convenience, we next make the assumption that we can construct a valid time-varying CBF $\mathfrak{b}_k$ that encodes $\phi_k$ along with an extended class $\mathcal{K}$ function $\alpha_k$, as discussed in chapter 4.

**Assumption 5.2** *The function $\mathfrak{b}_k:\mathbb{R}^{\bar{n}_k}\times\mathbb{R}_{\geq 0}\to\mathbb{R}$ encodes the STL specification $\phi_k$ from the fragment in equation (4.3) and is constructed as in lemma 4.3. The function $\alpha_k:\mathbb{R}\to\mathbb{R}$ is a locally Lipschitz-continuous extended class $\mathcal{K}$ function constructed as in the proof of theorem 4.5.*

Based on the decentralized CBF condition in equation (5.10) with the choice of the function $D_i$ in equation (5.11), we can now obtain a decentralized control law so that $\phi_1\wedge\ldots\wedge\phi_K$ is satisfied.

**Theorem 5.3 (Decentralized CBFs for $\phi_k$)** *Let the functions $f_i$, $g_i$, and $c_i$ that define the system in equation (5.1) be locally Lipschitz continuous and satisfy assumption 5.1. Let the functions $\mathfrak{b}_k:\mathbb{R}^{\bar{n}_k}\times\mathbb{R}_{\geq 0}\to\mathbb{R}$ and $\alpha_k:\mathbb{R}\to\mathbb{R}$ satisfy assumption 5.2. If, for each $k\in\{1,\ldots,K\}$, each agent $i\in\mathcal{V}_k$ applies the control law $u_i(\bar{x}_k,t):=\hat{u}_i$, where $\hat{u}_i$ is the solution of*

$$\underset{\hat{u}_i \in \mathbb{R}^{m_i}}{\operatorname{argmin}}\ \hat{u}_i^T \hat{u}_i \tag{5.12a}$$

$$\begin{aligned} \textit{such that } & \frac{\partial \mathfrak{b}_k(\bar{x}_k,t)}{\partial x_i}(f_i(x_i)+g_i(x_i)\hat{u}_i) \geq \\ & -D_i(\bar{x}_k,t)\Big(\frac{\partial \mathfrak{b}_k(\bar{x}_k,t)}{\partial t}+\alpha_k(\mathfrak{b}_k(\bar{x}_k,t))\Big)+\Big\|\frac{\partial \mathfrak{b}_k(\bar{x}_k,t)}{\partial x_i}\Big\|_1 C, \end{aligned} \tag{5.12b}$$

*with* $D_i$ *defined in equation (5.11), then each Filippov solution* $x:\mathcal{I}\to\mathbb{R}^n$ *to equation (5.1) is such that* $(x,0)\models\phi_1\wedge\ldots\wedge\phi_K$.

*Proof: The formal proof of the result is lengthy and provided in section 5.4. The proof essentially consists of showing that the optimization problem in equation (5.12) is always feasible, the control law* $u_i$ *is locally bounded and measurable, and satisfaction of equation (5.12b) allows us to apply theorem (5.2).*

The optimization problem in equation (5.12) is a computationally tractable convex quadratic program with $m_i$ decision variables. It can be observed that the control inputs $u_i$ are computed locally by each agent, as opposed to computing $\bar{u}_k$, which centrally requires knowledge of the stacked agent dynamics $\bar{f}_k$ and $\bar{g}_k$. As $\mathfrak{b}_k$ is obtained by solving equation (4.14) or equation (4.15), it holds in fact that $\rho^{\phi_1\wedge\ldots\wedge\phi_K}(x,0)\geq r$ where $r$ is maximized.

In the following two examples, we provide a simulation case study and an experiment where we use exponential functions $\gamma_l^{\exp}(t)$ and linear functions $\gamma_l^{\text{lin}}(t)$, respectively, to construct $\mathfrak{b}_k$.

---

**Example 5.1** Consider $N:=4$ agents with positions $x_i:=\begin{bmatrix}p_{x,i} & p_{y,i}\end{bmatrix}^T\in\mathbb{R}^2$. The dynamics are $\dot{x}_i=u_i+c_i(x,t)$, where $c_i(x,t):=f_i^{\text{c}}(x)+w(t)$ consist of noise $w_i(t)$ from the set $\{w_i\in\mathbb{R}^2 \mid \|w_i\|\leq 0.1\}$ and dynamical couplings

$$f_i^{\text{c}}(x):=\begin{cases} 0.5\text{sat}_1(x_4-x_i) \text{ if } i\in\{1,2,3\} \\ 0.25(\text{sat}_1(x_1-x_4)+\text{sat}_1(x_2-x_4)) \text{ if } i\in\{4\} \end{cases}$$

that are defined by the saturation function $\text{sat}_1(\zeta):=\begin{bmatrix}\bar{\zeta}_1 & \bar{\zeta}_2\end{bmatrix}^T$ with $\bar{\zeta}_d:=\zeta_d$ if $|\zeta_d|\leq 1$, $\bar{\zeta}_d:=1$ if $\zeta_d>1$, and $\bar{\zeta}_d:=-1$ if $\zeta_d<-1$ for $d\in\{1,2\}$. We set the control input as $u_i:=f_i^{\text{u}}(x)+v_i$, where $f_i^{\text{u}}$ is used for collision avoidance and defined as

$$f_i^{\text{u}}(x):=\sum_{j=1}^{4}\frac{x_i-x_j}{\|x_i-x_j\|+0.01}.$$

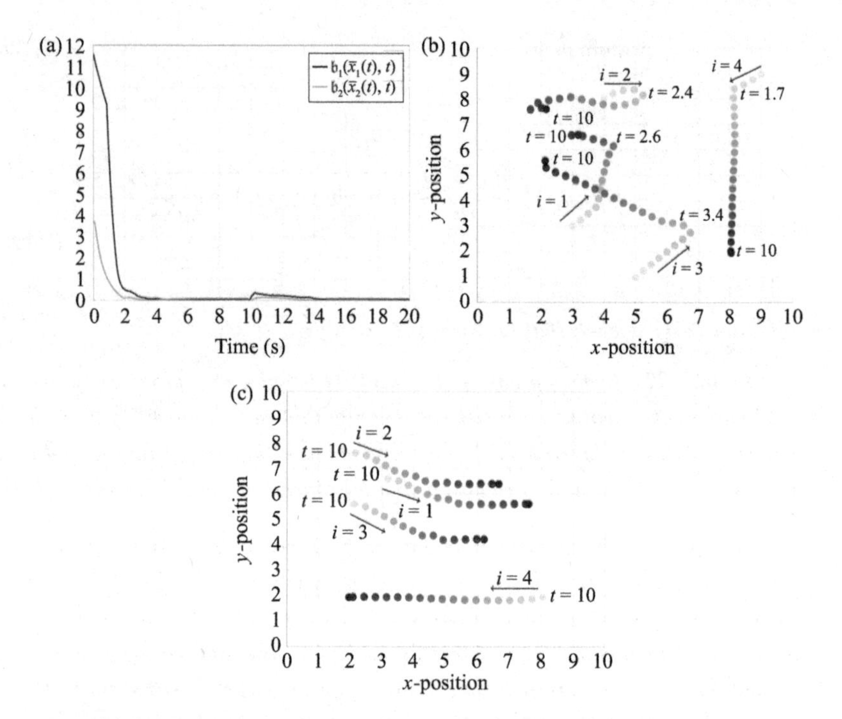


**Figure 5.1**
Barrier function evolution and agent trajectories for example 5.1. (a) Evolution of $\mathfrak{b}_1(\bar{x}_1(t), t)$ and $\mathfrak{b}_2(\bar{x}_2(t), t)$. (b) Agent trajectories from 0–10 seconds. (c) Agent trajectories from 10–20 seconds. Reprinted with permission from [173].

In the remainder, we use $(\|x_i\|_\infty \leq 1)$ to express $(p_{x,i} \leq 1) \wedge (-p_{x,i} \leq 1) \wedge (p_{y,i} \leq 1) \wedge (-p_{y,i} \leq 1)$. For agents 1, 2, and 3, assume the task

$$\begin{aligned}\phi_1 := &G_{[5,10]}(\|x_1 - p_A\|_\infty \leq 0.5)\\ &\wedge G_{[10,20]}\left((\|x_2 - \begin{bmatrix} p_{x,1}-1 & p_{y,1}+1\end{bmatrix}^T\|_\infty \leq 0.5)\right.\\ &\left.\wedge (\|x_3 - \begin{bmatrix} p_{x,1}-1 & p_{y,1}-1\end{bmatrix}^T\|_\infty \leq 0.5)\right)\\ &\wedge F_{[10,20]}(\|x_1 - p_B\|_\infty \leq 0.5),\end{aligned}$$

where $p_A := \begin{bmatrix} 2.5 & 7\end{bmatrix}^T$ and $p_B := \begin{bmatrix} 8 & 6\end{bmatrix}^T$; that is, agent 1 should always be in region $p_A$ within 5–10 time units, such as to prepare an object for transportation, while agents 2 and 3 should form a formation with agent 1 for all times within 10–20 time units, and eventually agent 1 should reach region $p_B$ within 10–20 time units, such as for collaborative transportation of this object. Furthermore, for agent 4, consider

$$\begin{aligned}\phi_2 := &F_{[5,10]}(\|x_4 - p_C\|_\infty \leq 1) \wedge G_{[0,10]}(p_{x,4} \geq 8)\\ &\wedge F_{[15,20]}(\|x_4 - p_D\|_\infty \leq 1) \wedge G_{[10,20]}(p_{y,4} \leq 2),\end{aligned}$$

where $p_C := \begin{bmatrix} 9 & 1 \end{bmatrix}^T$ and $p_D := \begin{bmatrix} 1 & 1 \end{bmatrix}^T$ (e.g., a surveillance task). We obtain $\mathfrak{b}_1$ and $\mathfrak{b}_2$ for $\phi_1$ and $\phi_2$, respectively, by solving equation (4.14) as a feasibility problem with computation times of 50 and 13 seconds, respectively, on a two-core, 1.8-GHz central processing unit with 4 gigabytes (GB) of random access memory (RAM). Computing equation (5.12) took, on average, 0.006 seconds. The CBFs are plotted in figure 5.1(a), while figures 5.1(b) and 5.1(c) show the agent trajectories from 0–10 and from 10–20 time units, respectively. In figure 5.1(b), agents 1, 2, 3, and 4 first approach each other due to dynamical couplings $f_i^c$. At 2.6, 2.4, 3.4, and 1.7 time units, respectively, the barrier functions force the agents not to approach each other any further and instead work toward satisfying $\phi_1$ and $\phi_2$. At 10 time units, agent 1 reaches region $p_A$, while agents 2 and 3 create a formation. Figure 5.1(c) shows how this formation is maintained as agent 1 approaches $p_B$. Agents do not get too close to each other due to $f_i^u$. It holds that $\rho^{\phi_1 \wedge \phi_2}(x, 0) \geq 0.005$.

**Example 5.2** We consider three Nexus 4WD Mecanum Robotic Cars (see figure 5.2), which are equipped with low-level PID controllers that track translational and rotational velocities. The state of robot $i$ is $x_i := \begin{bmatrix} p_i^T & \theta_i \end{bmatrix}^T$, where $p_i := \begin{bmatrix} p_{x,i} & p_{y,i} \end{bmatrix}^T$ denotes the two-dimensional position, while $\theta_i$ denotes the orientation. The agent dynamics are given by $\dot{x}_i = u_i + f_i^u(x) + c_i(x, t)$ where the coupling term $c_i$ models disturbances such as those induced by the digital implementation of the continuous-time control law or inaccuracies in the low-level PID controllers for which we set $C := 2$. The function $f_i^u$ again describes induced dynamical couplings used for collision avoidance, as in example 5.1. The robots are subject to the specification

$$\phi := \phi' \wedge \phi'' \wedge \phi''' \wedge \phi'''',$$

with

$$\phi' := G_{[15,90]}(\|p_1 + p_x - p_2\|^2 \leq \epsilon^2)$$

$$\phi'' := G_{[25,35]}(\|p_1 + p_y - p_3\|^2 \leq \epsilon^2) \wedge F_{[30,35]}(\|p_1 - p_B\|^2 \leq \epsilon^2)$$

**Figure 5.2**
The Nexus 4WD Mecanum Robotic Cars used in this experiment.

$$\phi''' := F_{[40,60]}(\|p_3 - p_C\|^2 \leq \epsilon^2)$$

$$\phi'''' := F_{[50,90]}((\|p_1 - p_A\|^2 \leq \epsilon^2) \wedge (\|p_2 + p_x - p_3\|^2 \leq \epsilon^2)),$$

where $\epsilon := 0.33$, $p_A := \begin{bmatrix}-1.2 & 1.2\end{bmatrix}^T$, $p_B := \begin{bmatrix}1.2 & 1.2\end{bmatrix}^T$, $p_C := \begin{bmatrix}1.2 & -1.2\end{bmatrix}^T$, $p_x := \begin{bmatrix}0.8 & 0\end{bmatrix}^T$, and $p_y := \begin{bmatrix}0 & -0.8\end{bmatrix}^T$. The implementation within the *Robot Operating System* framework [239] is available in [2]. The quadratic program in equation (5.12) is solved using *CVXGEN* [203] at a frequency of 50 Hz. The CBF $\mathfrak{b}$ encoding $\phi$ is obtained offline and in *MATLAB* by solving equation (4.15) as a feasibility problem using *YALMIP* [191] with the "fmincon option." The synthesis of $\mathfrak{b}$ took 4.2 seconds on an Intel Core i7-6600U with 16 GB of RAM.

The experimental result is shown in figure 5.3 as well as in [3], which provides a video of the experiment. We have intentionally chosen an initial condition $x(0)$ of the robots that does not coincide with the initial condition $x(0) := 0$ used in equation (4.15) to construct $\mathfrak{b}$. In figure 5.3(a), it is hence visible that initially $\mathfrak{b}(x(0), 0) \approx -0.62$. We have done this to

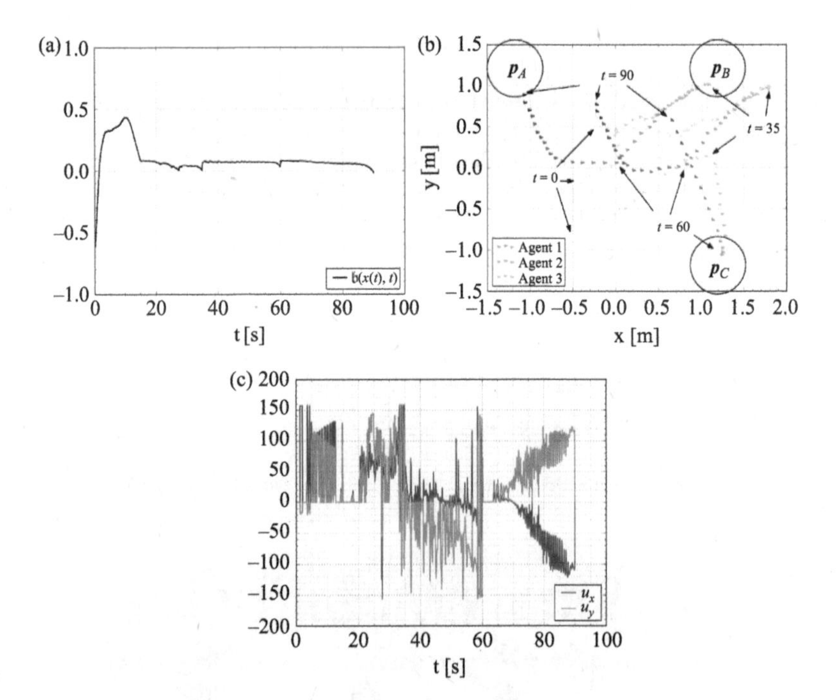


**Figure 5.3**
Barrier function evolution, agent trajectories, and control inputs for example 5.2. (a) Evolution of $\mathfrak{b}(x(t), t)$. (b) Robot trajectories. (c) Evolution of $u_3(x(t), t)$. Reprinted with permission from [175].

illustrate the attractivity property of set $\mathcal{C}$, similar to [330]. In particular, note that $\dot{\mathfrak{b}}(x(t),t) \geq -\alpha(\mathfrak{b}(x(t),t)) > 0$ if $\mathfrak{b}(x(t),t) < 0$, which pushes $x(t)$ toward $\mathcal{C}$. In our experiment, indeed, after approximately $t \approx 2$ seconds, it holds that $\mathfrak{b}(x(t),t) \geq 0$ and the robots have recovered from this misaligned initial configuration. Furthermore, note that $\mathfrak{b}(x(t),t) \geq 0$ for the rest of the experiment, from which we can see that $(x,0) \models \phi$ with $\rho^{\phi}(x,0) \geq 0.05$. Figure 5.3(b) shows the corresponding robot trajectories. As emphasized previously, the control law $u_i$ is discontinuous. This is shown in figure 5.3(c) where we plot the $x$- and $y$-components of $u_3$. We avoided to use an additional filter on $u_i$ to smoothen the control input in order to show the nature of the discontinuous control law. The low-level PID controllers, however, filter $u_i$ when applied to the motors of the robots. We further observe that using linear functions $\gamma_l^{\text{lin}}$ to construct $\mathfrak{b}$ according to equation (4.15) compared to using exponential ones according to equation (4.14) is beneficial since input saturations are less likely to occur. An exponential function $\gamma_l^{\text{exp}}$ would sometimes induce high control inputs, while at other times, almost no control action would be needed. A linear function $\gamma_l^{\text{lin}}$ distributes the needed control action more uniformly over time and is hence better suited for experiments. Finally, note that collisions are avoided by using $f_i^{\text{u}}$, especially in the first 5 seconds when a collision would occur between robots 1 and 2 without induced dynamical couplings.

## 5.2 Decentralized Control under Conflicting Local Specifications

We consider now a bottom-up setting in which local STL specifications $\phi_i$ are individually assigned to agents. The satisfaction of a local specification $\phi_i$ depends on agent $i$, but it also can depend on other agents $j \neq i$. The set of local specifications may be conflicting (or adversarial), in the sense that it may not be possible to satisfy all of them, as in section 5.1. We present a decentralized control approach based on time-varying CBFs, which finds locally least-violating solutions with and without collaboration between agents.

We proceed is as follows. We first provide a CBF–based control law that guarantees the satisfaction of a local task despite dynamical couplings when the task does not depend on other agents, similar to section 5.1. In this setting, contrary to section 5.1, we can obtain a continuous control law. We then propose a control law that finds locally least-violating solutions for the case when the local tasks are conflicting and when a collaboration between agents is not possible or desired. Finally, we introduce a local detection mechanism to detect critical events that may lead to a violation of the local task and that may be resolved or benefit from online collaboration with other agents.

Let the dynamics of agent $i$ again be described by equation (5.1), subject to assumption 5.1. As before, we consider the STL fragment from chapter 4 as per equation (4.3). However, now each agent $i$ is assigned a local task $\phi_i$ from the STL fragment in equation (4.3). The satisfaction of $\phi_i$ depends on the behavior of agent $i$ and possibly of other agents $j \neq i$. This set of agents is denoted by $\mathcal{V}_i \subseteq \{1, \ldots, N\}$, for which it holds that $|\mathcal{V}_i| \geq 1$ since $i \in \mathcal{V}_i$. We emphasize that, as opposed to section 5.1, the sets of agents $\mathcal{V}_1, \ldots, \mathcal{V}_N \subseteq \{1, \ldots, N\}$ do not need to be disjoint; that is, we may have that $\mathcal{V}_i \cap \mathcal{V}_j \neq \emptyset$ for some $i, j \in \{1, \ldots, N\}$. Similar to section 5.1, we now define the stacked state vector of all agents within this set $j_1, \ldots, j_{|\mathcal{V}_i|} \in \mathcal{V}_i$ as

$$\bar{x}_i := \begin{bmatrix} x_{j_1}{}^T & \ldots & x_{j_{|\mathcal{V}_i|}}{}^T \end{bmatrix}^T \in \mathbb{R}^{\bar{n}_i},$$

with state dimension $\bar{n}_i := n_{j_1} + \ldots + n_{j_{|\mathcal{V}_i|}}$. Without loss of generality, we here assume that $x_{j_1} = x_i$; that is, that the first element within $\bar{x}_i$ is the state $x_i$ of agent $i$. The stacked agent dynamics are thus

$$\dot{\bar{x}}_i = \bar{f}_i(\bar{x}_i) + \bar{g}_i(\bar{x}_k)\bar{u}_i(\bar{x}_i, t) + \bar{c}_i(x, t), \tag{5.13}$$

where the stacked dynamics $\bar{f}_i : \mathbb{R}^{\bar{n}_i} \to \mathbb{R}^{\bar{n}_i}$, $\bar{g}_i : \mathbb{R}^{\bar{n}_i \times \bar{m}_i}$, $\bar{u}_i : \mathbb{R}^{\bar{n}_i} \to \mathbb{R}^{\bar{m}_i}$, and $\bar{c}_i : \mathbb{R}^n \times \mathbb{R}_{\geq 0} \to \mathbb{R}^{\bar{n}_i}$ with $\bar{m}_i := m_{j_1} + \ldots + m_{j_{|\mathcal{V}_i|}}$ are defined analogous to equation (5.2). Finally, let $\mathfrak{b}_i : \mathbb{R}^{\bar{n}_i} \times \mathbb{R}_{\geq 0} \to \mathbb{R}$ be a time-varying CBF that encodes the STL formula $\phi_i$ as usual.

**Assumption 5.3** *The function $\mathfrak{b}_i : \mathbb{R}^{\bar{n}_i} \times \mathbb{R}_{\geq 0} \to \mathbb{R}$ encodes the STL specification $\phi_i$ from the fragment in equation (4.3) and is constructed according to theorem 4.4. The function $\alpha_i : \mathbb{R} \to \mathbb{R}$ is a locally Lipschitz-continuous, extended class $\mathcal{K}$ function constructed according to theorem 4.5.*

For the simple case where there are no task dependencies (i.e., when $|\mathcal{V}_i| = 1$), we can directly derive a continuous control law $u_i$ that results in satisfaction of $\phi_i$.

**Theorem 5.4 (Satisfaction of $\phi_i$ without Task Dependency)** *Let the functions $f_i$, $g_i$, and $c_i$ that define the system in equation (5.1) be locally Lipschitz continuous and satisfy assumption 5.1. Let $|\mathcal{V}_i| = 1$, and let the functions $\mathfrak{b}_i : \mathbb{R}^{\bar{n}_i} \times \mathbb{R}_{\geq 0} \to \mathbb{R}$ and $\alpha_i : \mathbb{R} \to \mathbb{R}$ satisfy assumption 5.3. If each agent*

$i \in \{1,\ldots,N\}$ *applies the control law* $u_i(x_i,t) = \hat{u}_i$*, where* $\hat{u}_i$ *is given by*

$$\underset{\hat{u}_i}{\text{argmin}}\, \hat{u}_i^T \hat{u}_i \tag{5.14a}$$

$$\text{such that } \frac{\partial \mathfrak{b}_i(x_i,t)}{\partial x_i}(f_i(x_i)+g_i(x_i)\hat{u}_i) + \frac{\partial \mathfrak{b}_i(x_i,t)}{\partial t} \geq -\alpha_i(\mathfrak{b}_i(x_i,t)) + \left\|\frac{\partial \mathfrak{b}_i(x_i,t)}{\partial x_i}\right\| C, \tag{5.14b}$$

*then each solution* $x : \mathcal{I} \to \mathbb{R}^n$ *to equation (5.1) is such that* $(x,0) \models \phi_1 \wedge \ldots \wedge \phi_N$.

*Proof: The proof is similar to previous ones. By assumptions 5.1 and 5.3, we know that the optimization problem in equation (5.14) is always feasible. The continuity of* $u_i$ *can be shown in the same way as in the proof of theorem 4.6 such that there are solutions* $x : [0,\tau_{max}) \to \mathbb{R}^n$ *to equation (5.1). Next, note that equation (5.14b) implies*

$$\frac{\partial \mathfrak{b}_i(x_i,t)}{\partial x_i}(f_i(x_i)+g_i(x_i)u_i+c_i(x,t)) + \frac{\partial \mathfrak{b}_i(x_i,t)}{\partial t} \geq -\alpha_i(\mathfrak{b}_i(x_i,t)),$$

*such that* $\dot{\mathfrak{b}}_i(x_i(t),t) \geq -\alpha_i(\mathfrak{b}_i(x_i,t))$ *for all* $t \in (0,\tau_{max})$*. If now the initial state* $x_i(0)$ *is such that* $\mathfrak{b}_i(x_i(0),0) \geq 0$*, we can once again apply the Comparison Lemma, by which it follows that* $\mathfrak{b}_i(x_i(t),t) \geq 0$ *for all* $t \in [0,\tau_{max})$*. As in theorem 4.3, we can then sequentially show that* $\tau_{max} = \infty$*. Finally, we can conclude that* $(x,0) \models \phi_1 \wedge \ldots \wedge \phi_N$ *since* $\mathfrak{b}_i$ *encodes* $\phi_i$*.*

However, if for one agent $i$ the task $\phi_i$ depends on other agents $j \neq i$ (i.e., if $|\mathcal{V}_i| > 1$), then satisfiability of each $\phi_i$ separately does not ensure satisfiability of $\phi_1 \wedge \ldots \wedge \phi_M$. This is exactly the case when local STL formulas are conflicting.

---

**Example 5.3** Consider $N := 6$ agents. We consider agents 1, 2, and 3 to be autonomous; that is, we cannot manipulate their behavior via the control input $u_i$. Agents 1, 2, and 3 are then known to follow the periodic trajectories

$$x_i(t) := \begin{bmatrix}\sin(t+\frac{2(i-1)}{3}\pi) & \cos(t+\frac{2(i-1)}{3}\pi)\end{bmatrix}^T.$$

Agents 4, 5, and 6 are supposed to track agents 1, 2, and 3, respectively, while staying close to each other (e.g., to satisfy connectivity constraints). In STL language, this can be expressed as follows:

$$\phi_4 := G_{[10,\infty)} \bigwedge_{j=1,5,6} (\|x_4 - x_j\| \leq 0.3),$$

$$\phi_5 := G_{[10,\infty)} \bigwedge_{j=2,4,6} (\|x_5 - x_j\| \leq 0.3),$$

$$\phi_6 := G_{[10,\infty)} \bigwedge_{j=3,4,5} (\|x_6 - x_j\| \leq 0.3).$$

Each of $\phi_4$, $\phi_5$, or $\phi_6$ is satisfiable on its own; however, $\phi_4 \wedge \phi_5 \wedge \phi_6$ is not satisfiable.

---

In the general case where $|\mathcal{V}_i| > 1$, we remark that the CBF condition changes to

$$\frac{\partial \mathfrak{b}_i(\bar{x}_i, t)}{\partial \bar{x}_i}(\bar{f}_i(\bar{x}_i) + \bar{g}_i(\bar{x}_i)\bar{u}_i + \bar{c}_i(x,t)) + \frac{\partial \mathfrak{b}_i(\bar{x}_i, t)}{\partial t} \geq -\alpha_i(\mathfrak{b}_i(\bar{x}_i, t)). \quad (5.15)$$

Due to the construction of $\mathfrak{b}_i$ and $\alpha_i$ as per assumption 5.3, the specification $\phi_i$ can be satisfied if all agents in $\mathcal{V}_i$ collaborate, such as by designing a centralized control law $\bar{u}_i$ or by designing a decentralized control law as previously presented in section 5.1. Since each agent $\phi_i$, however, is subject to its own task $\phi_i$, such collaboration is not feasible. This means that infeasibilities of equation (5.15) are in fact expected and caused by conflicting local objectives.

### 5.2.1 Locally Least Violating Solutions without Collaboration

We first consider cases where collaboration among agents is not desired (e.g., agents are not willing to collaborate) or possible (e.g., there are communication limitations between agents) and investigate the behavior of agent $i$ in isolation. Therefore, let us rewrite the stacked agent dynamics of $\bar{x}_i$ from equation (5.13) as

$$\begin{aligned}\dot{\bar{x}}_i &= \bar{f}_i(\bar{x}_i) + \bar{g}_i(\bar{x}_i)\bar{u}_i + \bar{c}_i(x,t)\\ &= \tilde{f}_i(x_i) + \tilde{g}_i(x_i)u_i + \tilde{c}_i(x,t),\end{aligned}$$

where the functions $\tilde{f}_i$, $\tilde{g}_i$, and $\tilde{c}_i$ are defined as

$$\begin{aligned}\tilde{f}_i(x_i) &:= \left[f_i(x_i)^T \ 0^T \ \ldots \ 0^T\right]^T,\\ \tilde{g}_i(x_i) &:= \left[g_i(x_i)^T \ 0 \ \ldots \ 0\right]^T,\\ \tilde{c}_i(x,t) &:= \bar{c}_i(x,t) + \left[0^T \ d_{j_2}(x,t)^T \ \ldots \ d_{j_{|\mathcal{V}_i|}}(x,t)^T\right]^T,\end{aligned}$$

with $d_j(x,t) := f_j(x_j) + g_j(x_j)u_j(x,t)$ describing the nominal dynamics of an agent $j \in \mathcal{V}_i \setminus \{i\}$. From the point of view of agent $i$, the nominal dynamics $d_j$ of agent $j$ are treated as an additional disturbance. As we look at agent $i$ in an isolated manner, we will assume that other agents $j \in \mathcal{V}_i \setminus \{i\}$ evolve within a bounded set.

**Assumption 5.4** *Each agent $j \in \{1, \ldots, N\} \setminus \{i\}$ applies a bounded and continuous control law $u_j$ that results in $x_j(t) \in \mathfrak{B}_j$ for all $t \geq 0$ where $\mathfrak{B}_j \subseteq \mathbb{R}^{n_j}$ is a compact set.*

Using the rewritten dynamics, we can now see that the CBF condition from equation (5.15) is equivalent to

$$\frac{\partial \mathfrak{b}_i(\bar{x}_i, t)}{\partial x_i}(f_i(x_i) + g_i(x_i)u_i) + \frac{\partial \mathfrak{b}_i(\bar{x}_i, t)}{\partial \bar{x}_i}\tilde{c}_i(x, t) + \frac{\partial \mathfrak{b}_i(\bar{x}_i, t)}{\partial t}$$
$$\geq -\alpha_i(\mathfrak{b}_i(\bar{x}_i, t)). \tag{5.16}$$

The satisfaction of the constraint in equation (5.16) depends on the term $\frac{\partial \mathfrak{b}_i(\bar{x}_i, t)}{\partial \bar{x}_i}\tilde{c}_i(x, t)$, and hence on the behavior of the agents $j \in \mathcal{V}_i \setminus \{i\}$ according to $\tilde{c}_i$. However, the function $\tilde{c}_i(x, t)$ is unknown to agent $i$ and may either help in or act against satisfying equation (5.16). As a consequence, the constraint in equation (5.16) may not be feasible if

$$\frac{\partial \mathfrak{b}_i(\bar{x}_i, t)}{\partial x_i} g_i(x_i) = 0 \text{ and } \frac{\partial \mathfrak{b}_i(\bar{x}_i, t)}{\partial \bar{x}_i}\tilde{c}_i(x, t) \neq 0.$$

In some sense, we encountered this problem before in section 5.1.1, when we realized that the decentralized CBF condition in equation (5.5) is not always feasible unless the weight $D_i$ is chosen properly.

Since we treat the function $\tilde{c}_i$ as a disturbance that is completely unknown, we again assume that $\tilde{c}_i$ is bounded.

**Assumption 5.5** *Let $\tilde{C}_i$ be a positive constant such that $\|\tilde{c}_i(x, t)\| \leq \tilde{C}_i$ for all $(x, t) \in \mathcal{D} \times \mathbb{R}_{\geq 0}$, where the set $\mathcal{D} \in \mathbb{R}^n$ denotes the workspace of all agents. We assume that $\mathcal{D}$ is an open set for which it holds that $P_i(\mathcal{D}) \supseteq \mathcal{C}_i(t) := \{\bar{x}_i \in \mathbb{R}^{\bar{n}_i} | \mathfrak{b}(\bar{x}_i, t) \geq 0\}$ for all times $t \geq 0$, as well as $P_j(\mathcal{D}) \supseteq \mathfrak{B}_j$ for all agents $j \neq \{1, \ldots, N\} \setminus \{i\}$ where $P_i$ projects set $\mathcal{D}$ from $\mathbb{R}^n$ to its components in $\mathbb{R}^{\bar{n}_i}$.*[3]

It is clear that such a constant $\tilde{C}_i$ exists if $f_j$ and $g_j$ are bounded functions or if $\mathcal{D}$ is a bounded set, implying that $f_j$ and $g_j$ are bounded on $\mathcal{D}$, since the functions $c_j$ and $u_j$ are assumed to be bounded. We will comment more on the existence and choice of $\tilde{C}_i$ at the end of this section.

---

[3]We can define the projection map $p_i : \mathbb{R}^n \to \mathbb{R}^{\bar{n}_i}$ as $p_i(x) := \bar{x}_i$ and let the projection from a set $\mathcal{D} \in \mathbb{R}^n$ to $\mathbb{R}^{\bar{n}_i}$ be $P_i(\mathcal{D}) := \{\bar{x}_i \in \mathbb{R}^{\bar{n}_i} | \exists x \in \mathcal{D}, p_i(x) = \bar{x}_i\}$.

Our goal is now to define a control law that finds locally least-violating solutions. By that, we mean that each agent $i \in \{1, \ldots, N\}$ aims to locally—without involving agents $j \in \mathcal{V}_i \setminus \{i\}$—maximize $\rho^{\phi_i}(\bar{x}_i, 0)$ when $\rho^{\phi_i}(\bar{x}_i, 0) < 0$ (i.e., when $\phi_i$ is violated). For that purpose, consider the control law $u_i(\bar{x}_i, t) := \hat{u}_i$, where $\hat{u}_i$ is given by

$$\underset{\hat{u}_i, \hat{\epsilon}_i}{\operatorname{argmin}} K_{i,1} \hat{u}_i^T \hat{u}_i + K_{i,2} \hat{\epsilon}_i^2 \tag{5.17a}$$

$$\begin{aligned} \text{such that } & \frac{\partial \mathfrak{b}_i(\bar{x}_i, t)}{\partial x_i}(f_i(x_i) + g_i(x_i)\hat{u}_i) + \frac{\partial \mathfrak{b}_i(\bar{x}_i, t)}{\partial t} \geq -\alpha_i(\mathfrak{b}_i(\bar{x}_i, t)) \\ & + \left\| \frac{\partial \mathfrak{b}_i(\bar{x}_i, t)}{\partial \bar{x}_i} \right\| \tilde{C}_i - \hat{\epsilon}_i, \end{aligned} \tag{5.17b}$$

with slack variable $\hat{\epsilon}_i$ and design parameters $K_{i,1}, K_{i,2} \in [0, 1]$ such that $K_{i,1} + K_{i,2} = 1$. The constants $K_{i,1}$ and $K_{i,2}$ trade off between minimizing the input energy $\hat{u}_i^T \hat{u}_i$ and the squared slack variable $\hat{\epsilon}_i^2$. Importantly, the slack variable $\hat{\epsilon}_i$ in the constraint in equation (5.17b) ensures that the optimization problem in equation (5.17) is always feasible. Further, note that the satisfaction of the constraint in equation (5.17b) with $\hat{\epsilon}_i = 0$ implies that the CBF condition in equation (5.16) is satisfied.

Let us next define $\epsilon_i(\bar{x}_i, t) := \hat{\epsilon}_i$ as the slack variable obtained as the solution of equation (5.17) for a point $(\bar{x}_i, t) \in \mathbb{R}^{\bar{n}_i} \times \mathbb{R}_{\geq 0}$. Let us also define the worst-case slack variable of equation (5.17) as

$$\epsilon_{i,\mathrm{wc}} := \sup_{(\bar{x}_i, t) \in P_i(\mathcal{D}) \times \mathbb{R}_{\geq 0}} \epsilon_i(\bar{x}_i, t).$$

We now state our main result, which quantifies the worst-case violation with respect to set $C_i$ and the quantitative semantics $\rho^{\phi_i}$ in terms of the worst-case slack $\epsilon_{i,\mathrm{wc}}$.

**Theorem 5.5 (Locally Least-Violating Solutions for $\phi_i$ without Collaboration)** *Let the functions $f_i$, $g_i$, and $c_i$ that define the system in equation (5.1) be locally Lipschitz continuous, and let assumptions 5.4 and 5.5 hold. Let the functions $\mathfrak{b}_i : \mathbb{R}^{\bar{n}_i} \times \mathbb{R}_{\geq 0} \to \mathbb{R}$ and $\alpha_i : \mathbb{R} \to \mathbb{R}$ satisfy assumption 5.3, where now $\alpha_i$ is assumed to be a linear function. If agent $i$ applies the control law $u_i(x_i, t) = \hat{u}_i$ where $\hat{u}_i$ is given by equation (5.17), then it holds that*

$$C_{i,wc}(t) := \{\bar{x}_i \in \mathbb{R}^{\bar{n}_i} \mid \mathfrak{b}_i(\bar{x}_i, t) \geq \alpha_i^{-1}(-\epsilon_{i,wc})\}$$

*is forward invariant if $P_i(\mathcal{D}) \supseteq C_{i,wc}(t)$ for all $t \geq 0$. In this case, it holds that $\rho^{\phi_i}(\bar{x}_i, 0) \geq \alpha_i^{-1}(-\epsilon_{i,wc})$.*

*Proof:* *Note that $u_i$ is locally Lipschitz continuous due to [330, theorem 8]. This, together with assumption 5.4, implies that there is a solution $x:[0,\tau_{max})\rightarrow\mathcal{D}$ to equation (5.1). Due to equation (5.17b) and assumption 5.5, it holds that $\dot{\mathfrak{b}}_i(\bar{x}_i(t),t)\geq -\alpha_i(\mathfrak{b}_i(\bar{x}_i(t),t))-\epsilon_{i,wc}$ for all $t\in[0,\tau_{max})$. Now by lemma 2.4 and the Comparison Lemma, we know that $\mathfrak{b}_i(\bar{x}_i(t),t)\geq \beta(|\mathfrak{b}_i(\bar{x}_i(0),0)|,t)+\alpha_i^{-1}(-\epsilon_{i,wc})\geq\alpha_i^{-1}(-\epsilon_{i,wc})$ for all $t\in[0,\tau_{max})$. As in theorem 4.3, we can then show that $\tau_{max}=\infty$ as $\mathcal{C}_{i,wc}(t)$ is compact since $P_i(\mathcal{D})\supseteq\mathcal{C}_{i,wc}(t)$. The fact that $\rho^{\phi_i}(\bar{x}_i)\geq\alpha_i^{-1}(-\epsilon_{i,wc})$ follows by construction of $\mathfrak{b}_i$ following assumption 5.3.*

In this case, we do not need to assume that $\mathfrak{b}_i$ is a valid time-varying CBF, and in particular, we do not need to assume that $g_i$ has full row rank. The estimate of $\rho^{\phi_i}(\bar{x}_i)\geq\alpha_i^{-1}(-\epsilon_{i,\text{wc}})$ may be conservative. For a given initial condition $x(0)$ and the solution $x:[0,\tau)\rightarrow\mathcal{D}$ to equation (5.1) until time $\tau>0$, it does not necessarily hold that $\epsilon_i(\bar{x}_i(t),t)=\epsilon_{i,\text{wc}}$ for some $t\in[0,\tau)$. We can instead compute the maximum slack variable of $\epsilon_i(\bar{x}_i(t),t)$ obtained after observing (or computing) the solution $x$ by defining

$$\epsilon_{i,\max}(\tau):=\sup_{t\in[0,\tau)}\epsilon_i(\bar{x}_i(t),t).$$

Note that it necessarily has to hold that $\epsilon_{i,\max}(\tau)\leq\epsilon_{i,\text{wc}}$.

**Corollary 5.1** *Let the conditions of theorem 5.5 hold. Given the solution $x:[0,\tau)\rightarrow\mathcal{D}$ to equation (5.1) until time $\tau>0$. If $P_i(\mathcal{D})\supset\mathcal{C}_{i,max}(t)$ for all $t\in[0,\tau)$, where*

$$\mathcal{C}_{i,max}(t):=\{\bar{x}_i\in\mathbb{R}^{\bar{n}_i}\,|\,\mathfrak{b}_i(\bar{x}_i,t)\geq\alpha_i^{-1}(-\epsilon_{i,max}(\tau))\},$$

*then it holds that $\bar{x}_i(t)\in\mathcal{C}_{i,max}(t)$ for all $t\in[0,\tau)$. If $\tau=\infty$, then it holds that $\rho^{\phi_i}(\bar{x}_i,0)\geq\alpha_i^{-1}(-\epsilon_{i,max}(\tau))$.*

*Proof:* *It holds that $\dot{\mathfrak{b}}_i(\bar{x}_i(t),t)\geq-\alpha_i(\mathfrak{b}_i(\bar{x}_i(t),t))-\epsilon_{i,max}(\tau)$ for all $t\in[0,\tau)$. By lemma 2.4 and the Comparison Lemma, $\bar{x}_i(t)\in\mathcal{C}_{i,max}(t)$ for all $t\in[0,\tau)$. If $\tau=\infty$, the same conclusion as in theorem 5.5 can be drawn.*

Corollary 5.1 tells us that $\bar{x}_i(t)\in\mathcal{C}_{i,\max}(t)$ for all $t\in[0,\tau)$, but it does not tell us whether $\bar{x}_i(t)\in\mathcal{C}_{i,\max}(t)$ for all $t\geq\tau$. However, the corollary motivates that minimizing $\hat{\epsilon}_i$ results in a locally least-violating solution; that is, achieving $\bar{x}_i(t)\in\mathcal{C}_{i,\max}(t)$ for all $t\geq0$ depends on ensuring that $\epsilon_i(\bar{x}_i(t),t)\leq\epsilon_{i,\max}(\tau)$ for $t\geq\tau$. By a locally least-violating solution, we accordingly refer to a solution $\bar{x}_i(t)$ such that $\bar{x}_i(t)\in\mathcal{C}_{i,\max}(t)$, where $\epsilon_{i,\max}(\infty)$ is minimized locally. This observation will be used in the online collaboration part presented in subsection 5.2.2.

The previous results rely on assumption 5.4. If each agent $i$ solves equation (5.17), then one can show that assumption 5.4 becomes obsolete and each agent evolves within a compact set $\mathfrak{B}$, no matter which $\tilde{C}_i$ has been chosen. Then the question arises of how an estimate of $\tilde{C}_i$ can be obtained to satisfy assumption 5.5. This is possible if the functions $f_i$ and $g_i$ are bounded and subject to input constraints (i.e., $u_i \in \mathcal{U}_i$ for some compact set $\mathcal{U}_i$). This will be assumed in section 5.2.2. Before proceeding further, we will present an example to illustrate the previous results.

---

**Example 5.4** Consider $N := 6$ agents with $n_i := m_i := 2$. Agents 1, 2, and 3 are as in example 5.3. Agents 4, 5, and 6 are subject to

$$\dot{x}_i = c_i(x,t) + u_i,$$

where the function $c_i$ is defined as

$$c_i(x,t) := \sum_{j \in \{4,5,6\} \setminus \{i\}} \text{sat}_1(x_j - x_i),$$

with $\text{sat}_1$ again being the saturation function. We impose $u_i \in \mathcal{U}_i := [-2,2]^2$ so that $\tilde{C}_i = 4$. The function $\mathfrak{b}_i$ for $i \in \{4,5,6\}$ is constructed as in section 4.4 with exponential function $\gamma_l^{\text{exp}}$, while we set $\alpha_i(r) := 10r$. Agents 4, 5, and 6 are subject to $\phi_4$, $\phi_5$, and $\phi_6$ as in example 5.3; that is,

$$\phi_4 := G_{[10,\infty)} \bigwedge_{j=1,5,6} (\|x_4 - x_j\| \le 0.3),$$

$$\phi_5 := G_{[10,\infty)} \bigwedge_{j=2,4,6} (\|x_5 - x_j\| \le 0.3),$$

$$\phi_6 := G_{[10,\infty)} \bigwedge_{j=3,4,5} (\|x_6 - x_j\| \le 0.3).$$

The simulation results are shown in figure 5.4 with design parameters $K_{i,1} := 0.1$ and $K_{i,2} := 0.9$. Figure 5.4a shows that $\kappa_4 = -0.395$, $\kappa_5 = -0.382$, and $\kappa_6 = -0.39$, where $\kappa_i := \inf_{t \ge 0} \mathfrak{b}_i(\bar{x}_i(t),t)$, while figure 5.4b shows that $\epsilon_{4,\max}(\infty) = 7.006$, $\epsilon_{5,\max}(\infty) = 6.808$, and $\epsilon_{6,\max}(\infty) = 7.271$ with $\epsilon_{i,\max}(\infty) = \sup_{t \ge 0} \epsilon_i(\hat{x}_i(t),t)$. Since we selected $\alpha_i(r) := 10r$, theorem 5.6 predicts that $\mathfrak{b}_4(\hat{x}_4(t),t) \ge -0.7006$, $\mathfrak{b}_5(\hat{x}_5(t),t) \ge -0.6808$, and $\mathfrak{b}_6(\hat{x}_6(t),t) \ge -0.7271$, which gives a more conservative estimate than what is actually obtained when comparing with $\kappa_4$, $\kappa_5$, and $\kappa_6$. The trajectories from 0–10 s and from 10–35 s are shown in figures 5.4c and 5.4d, respectively.

---

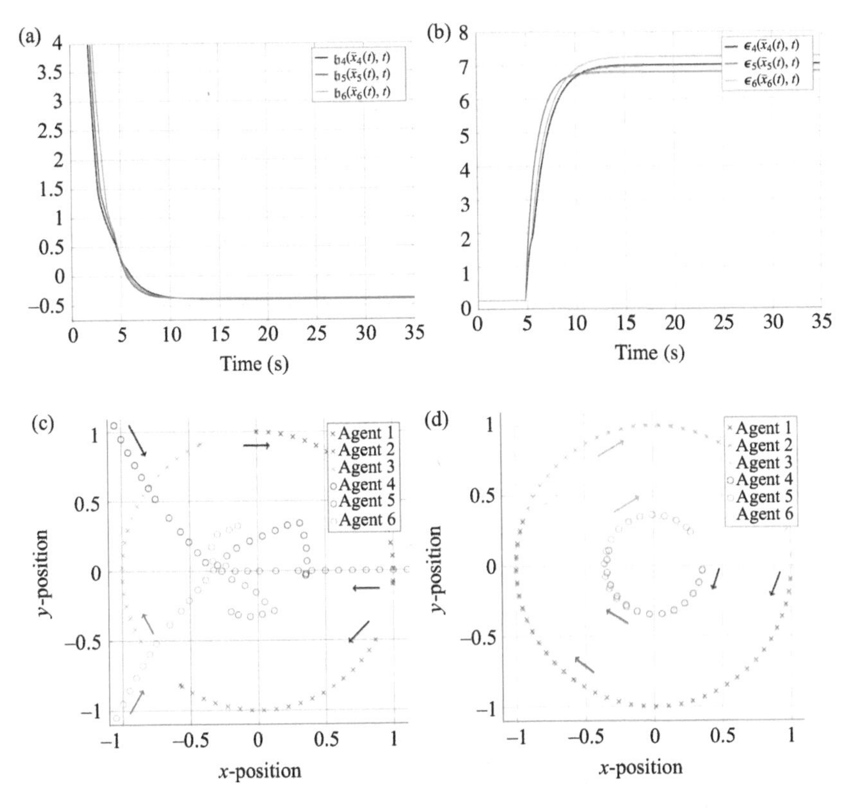


**Figure 5.4**
Barrier function evolution and agent trajectories for scenario 1. (a) Barrier function evolution; (b) slack variable evolution; (c) agent trajectories: 0–10 s; (d) agent trajectories: 10–35 s. Reprinted with permission from [171].

## 5.2.2 Locally Least-Violating Solutions with Collaboration

Online collaboration is initiated if a critical event (defined in this subsection) is detected by agent $i$. Recall the particular structure of $\mathfrak{b}_i$ as

$$\mathfrak{b}_i(\bar{x}_i, t) := -\frac{1}{\eta_i} \ln\Big(\sum_{l=1}^{p_i} \mathfrak{o}_i^l(t) \exp(-\eta_i \mathfrak{b}_i^l(\bar{x}_i^l, t))\Big),$$

where $\eta_i > 0$ is an accuracy parameter related to the approximation of the min operator, $\mathfrak{o}_i^l : \mathbb{R}_{\geq 0} \to \{0, 1\}$ is a switching function, and $p_i \in \mathbb{N}$ denotes the total number of CBFs $\mathfrak{b}_i^l$ encoded within $\mathfrak{b}_i$. Specifically, note that we used the state $\bar{x}_i^l$ to indicate the possibility that only knowledge of a subset of states from agents $\mathcal{V}_i^l \subseteq \mathcal{V}_i$ may be required to evaluate $\mathfrak{b}_i^l$. For instance, for the STL specification $\phi_1 := F_I(\|x_1 - x_2\| \leq \epsilon) \wedge G_I(\|x_1 - x_3\| \leq \epsilon)$, we obtain two functions $\mathfrak{b}_1^1$ and $\mathfrak{b}_1^2$, with $\bar{x}_1^1 := \begin{bmatrix} x_1^T & x_2^T \end{bmatrix}^T$ and $\bar{x}_1^2 := \begin{bmatrix} x_1^T & x_3^T \end{bmatrix}^T$, respectively. Effectively, this allows for collaboration only with the subset of agents in $\mathcal{V}_i^l$ if a critical event is detected for $\mathfrak{b}_i^l$, instead of all agents in $\mathcal{V}_i$. Let $\mathcal{A}_i(t)$ be such that $l \in \mathcal{A}_i(t)$ iff $\mathfrak{o}_i^l(t) = 1$. As previously stated in lemma 4.1, it holds that

$$\mathfrak{b}_i(\bar{x}_i, t) \leq \min_{l \in \mathcal{A}_i(t)} \mathfrak{b}_i^l(\bar{x}_i^l, t) \leq \mathfrak{b}_i(\bar{x}_i, t) + \frac{\ln(|\mathcal{A}_i(t)|)}{\eta_i}. \tag{5.18}$$

Based on the observation in equation (5.18), we now define a critical event.

**Definition 5.2** *A critical event happens at time $\tau > 0$ if*

$$\mathfrak{b}_i(\bar{x}_i(\tau), \tau) + \frac{\ln(|\mathcal{A}_i(\tau)|)}{\eta_i} < 0 \text{ and } \epsilon_i(\bar{x}_i(\tau), \tau) \geq \epsilon_{i,th},$$

*where $\epsilon_{i,th} > 0$ is a design parameter.*

If a critical event happens, there is at least one index $l \in \mathcal{A}_i(\tau)$ in $\mathfrak{b}_i$ such that $\mathfrak{b}_i^l(\bar{x}_i^l(\tau), \tau) < 0$ due to equation (5.18). In this case, we initiate collaboration with other agents $j \in \mathcal{V}_i^l \setminus \{i\}$ if these agents can communicate. Therefore, we model communication by the set $\mathcal{E} \subseteq \{1, \ldots, N\} \times \{1, \ldots, N\}$, where $(i, j) \in \mathcal{E}$ indicates that agent $j$ can receive information from agent $i$.

Collaboration requests are indicated by $\mathrm{cr}_{i,j}^l : \mathbb{R}_{\geq 0} \to \{\top, \bot\}$, where no collaboration is set by default via $\mathrm{cr}_{i,j}^l(t) := \bot$. Assume that a critical event is detected at time $t = \tau$ and $\mathfrak{b}_i^l(\bar{x}_i^l(\tau), \tau) < 0$ for $l \in \mathcal{A}_i(\tau)$. Then agent $i$ sends the function $\mathfrak{b}_i^l$ to agent $j \in \mathcal{V}_i^l \setminus \{i\}$ and sets $\mathrm{cr}_{i,j}^l(\tau) := \top$ if $(k, j) \in \mathcal{E}$ for each $k \in \mathcal{V}_i^l \setminus \{j\}$ (i.e., agent $j$ can receive information from all agents in $\mathcal{V}_i^l$). Define the sets $\mathcal{N}_i(t)$ and $\mathcal{L}_{i,j}(t)$ such that $j \in \mathcal{N}_i(t)$ and $l \in \mathcal{L}_{i,j}(t)$ iff $\mathrm{cr}_{j,i}^l(t') = \top$ for some $j \in \{1, \ldots, N\}$, $l \in \{1, \ldots, p_j\}$, and $t' \in [0, t]$. The set $\mathcal{N}_i(t) \subseteq \{1, \ldots, N\}$ is the set of agents from which a collaboration request has already been received, while $\mathcal{L}_{i,j}(t)$ is the set of corresponding indices $l$. Also define $\mathrm{CR}_i(t) := \sum_{j \in \mathcal{N}_i(t)} \|\mathcal{L}_{i,j}(t)\|$ as the number of received collaboration requests and let each pair $(j, l) \in \mathcal{N}_i(t) \times \mathcal{L}_{i,j}(t)$ be uniquely associated with a number $\nu_{j,l}(t) \in \{2, \ldots, \mathrm{CR}_i(t) + 1\}$. For $K_{i,k} \in [0, 1]$ with $\sum_{k=1}^{\mathrm{CR}_i(t)+2} K_{i,k} = 1$, agent $i$ solves the optimization problem:

$$\operatorname*{argmin}_{\hat{u}_i, \hat{\epsilon}_i} K_{i,1} \hat{u}_i^T \hat{u}_i + \sum_{k=2}^{\mathrm{CR}_i(t)+2} K_{i,k} \hat{\epsilon}_{i,k-1}^2 \tag{5.19a}$$

$$\text{such that } \frac{\partial \mathfrak{b}_i(\bar{x}_i,t)}{\partial x_i}(f_i(x_i)+g_i(x_i)\hat{u}_i)+\frac{\partial \mathfrak{b}_i(\bar{x}_i,t)}{\partial t} \geq -\alpha_i(\mathfrak{b}_i(\bar{x}_i,t)) + \left\|\frac{\partial \mathfrak{b}_i(\bar{x}_i,t)}{\partial \bar{x}_i}\right\|\tilde{C}_i - \hat{\epsilon}_{i,1} \tag{5.19b}$$

$$\frac{\partial \mathfrak{b}_j^l(\bar{x}_j^l,t)}{\partial x_i}(f_i(x_i)+g_i(x_i)\hat{u}_i)+\frac{\partial \mathfrak{b}_j^l(\bar{x}_j^l,t)}{\partial t} \geq -\alpha_j(\mathfrak{b}_j^l(\bar{x}_j^l,t)) + \left\|\frac{\partial \mathfrak{b}_j^l(\bar{x}_j^l,t)}{\partial \bar{x}_j^l}\right\|\tilde{C}_j - \hat{\epsilon}_{i,\nu_{j,l}(t)} \tag{5.19c}$$

for each $j \in \mathcal{N}_i(t)$, $l \in \mathcal{L}_{i,j}(t)$.

Collaboration in the optimization problem in equation (5.19) is indicated by equation (5.19c) to satisfy $\phi_j$ for each $j \in \mathcal{N}_i(t)$, while equation (5.19b) incentivizes agent $i$ to satisfy $\phi_i$. Collaboration may come at the cost of not satisfying $\phi_i$, depending on the ratio of the parameters $K_{i,k}$. Note that equation (5.19) is a convex quadratic program with $m_i+1+\mathrm{CR}_i(t)$ decision variables and $1+\mathrm{CR}_i(t)$ constraints. The following result is stated without proof and follows similarly to corollary 5.1.

**Theorem 5.6 (Locally Least-Violating Solutions for $\phi_i$ with Collaboration)** *Let the functions $f_i$, $g_i$, and $c_i$ that define the system in equation (5.1) be locally Lipschitz continuous, and let $\tilde{C}_i > 0$ be such that $\|\tilde{c}_i(x,t)\| \leq \tilde{C}_i$ for all $(x,t) \in \mathbb{R}^n \times \mathbb{R}_{\geq 0}$. Let the functions $\mathfrak{b}_i : \mathbb{R}^{\bar{n}_i} \times \mathbb{R}_{\geq 0} \to \mathbb{R}$ and $\alpha_i : \mathbb{R} \to \mathbb{R}$ satisfy assumption 5.2, where $\alpha_i$ is assumed to be a linear function. If each agent $i \in \{1,\ldots,N\}$ applies the control law $u_i(x_i,t) = \hat{u}_i$ where $\hat{u}_i$ is given by equation (5.19), then it holds that $\rho^{\phi_i}(\bar{x}_i,0) \geq \kappa_i \geq \alpha_i^{-1}(-\epsilon_{i,max}(\infty))$ with $\kappa_i := \inf_{t\geq 0} \mathfrak{b}_i(\bar{x}_i(t),t)$ and $\epsilon_{i,max}(\tau) := \sup_{t\in[0,\tau)} \epsilon_{i,1}(\bar{x}_i(t),t)$.*

Finally, we show a last simulation example. To illustrate that the method can be combined with CBFs for obstacle avoidance, such as in [18], consider $\mathfrak{h}(x_i) : \mathbb{R}^{n_i} \to \mathbb{R}$ with $\mathfrak{h}(x_i) < 0$ for $x_i \in \mathcal{O}_i$ and $\mathfrak{h}(x_i) \geq 0$ for $x_i \notin \mathcal{O}_i$, where $\mathcal{O}_i$ denotes a set of obstacles. For an extended class $\mathcal{K}$ function $\hat{\alpha}_i$, consider

$$\frac{\partial \mathfrak{h}_i(x_i)}{\partial x_i}(f_i(x_i,t)+g_i(x_i,t)\hat{u}_i) \geq -\hat{\alpha}_i(\mathfrak{h}_i(x_i)) + \left\|\frac{\partial \mathfrak{h}_i(x_i)}{\partial x_i}\right\| C_i. \tag{5.20}$$

The constraint in equation (5.20) is now simply added to the optimization problem in equation (5.19) within example 5.5.

---

**Example 5.5** To illustrate the use of collaboration requests, consider again example 5.3, but in a slightly altered setting with the STL formulas

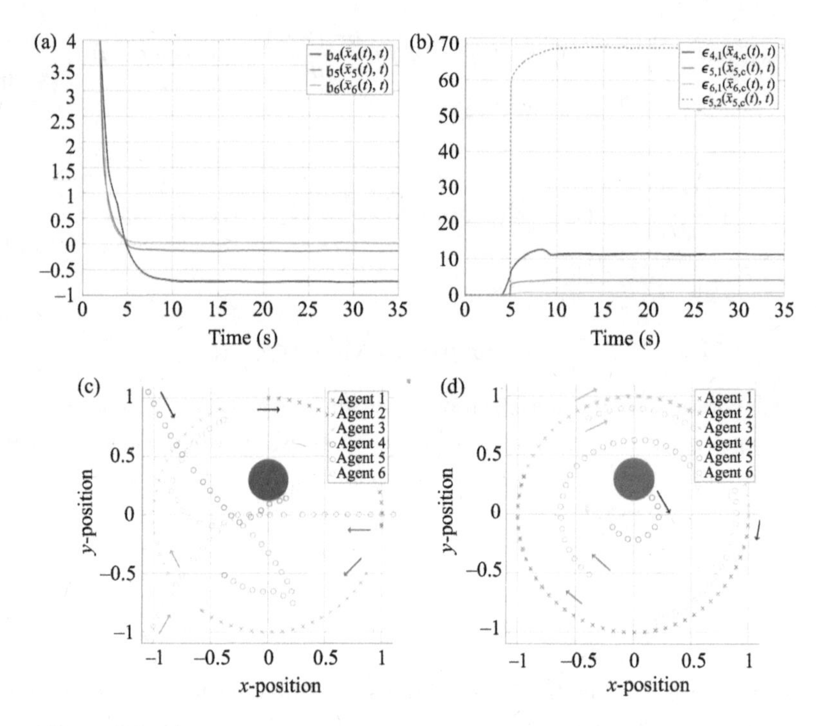


**Figure 5.5**
Barrier function evolution and agent trajectories for scenario 2 with an obstacle indicated in gray. (a) Barrier function evolution; (b) slack variable evolution; (c) agent trajectories: 0–10 s; (d) agent trajectories: 10–35 s. Reprinted with permission from [171].

$$\begin{aligned}
\phi_4' &:= \phi_4, \\
\phi_5' &:= G_{[5,\infty)}(\|x_5 - x_2\| \leq 0.3), \\
\phi_6' &:= G_{[5,\infty)}(\|x_6 - x_3\| \leq 0.3),
\end{aligned}$$

together with the edge set $\mathcal{E} := \{(1,4),(5,4),(6,4),(2,5),(4,5),(3,6)\}$ such that only agent 5 can colaborate with agent 4 in case of a critical event. Consider further one static obstacle $\mathcal{O}_i := \{o\}$ for each agent $i \in \{4,5,6\}$ with $o := \{x_i \in \mathbb{R}^{n_i} | \|x_i - \begin{bmatrix}0 & 0.3\end{bmatrix}^T \| \leq 0.2\}$; that is, placed such that it intersects the agents' trajectories. The simulation results are shown in figure 5.5. A critical event happens at $\tau = 4.927$ s, and then collaboration is established with agent 5 and the parameters $K_{5,1} := 0.1$, $K_{5,2} := 0.7$, and $K_{5,3} := 0.2$. Agent 5 deviates from its optimal trajectory, which would be similar to agent 6's trajectory (agent 6 cannot collaborate due to $\mathcal{E}$), in order to collaborate with agent 4. We have again used $\alpha_i(r) := 10r$, and the results from theorem 5.6 can be verified manually. The computation times are, on average for each agent and on an Intel Core i7-6600U with 16

GB of RAM, 2 ms without collaboration and 2.5 ms when collaboration is initiated.

---

## 5.3 Notes and References

The results from section 5.1 on decentralized collaborative control using time-varying CBFs are mainly taken from [175, 170]. Decentralized CBFs were initially presented in [316, 314] for safe multirobot navigation and obstacle avoidance. Other more recent works that consider static decentralized CBFs can be found in [72, 73, 289, 74], just to name a few. Lypunov-like barrier functions for multi-agent systems were presented in [224, 223]. Compositional approaches to CBFs for stochastic systems were presented in [137, 214, 19] and applied to temporal logic control design problems in [20]. In this section, we relied on nonsmooth analysis on which more detailed information can be found in [81, 104, 273].

The decentralized control strategies presented in section 5.2 that achieve locally least-violating solutions are mainly taken from [171]. The notion of locally least-violating solutions is inspired by the notion of input-to-state safety for CBFs that was introduced in [156, 253]. Noncooperative agent control using CBFs was presented in [229]. Partial satisfaction of STL specifications in multi-agent systems was considered in [64]. Finding least-violating solutions has also been studied for linear temporal logic (LTL) control synthesis in [124, 121]. Distributed planning of multi-agent systems under LTL specifications without task infeasibilities was considered in [299, 154, 121]. More generally, not with a specific focus on multi-agent systems, least-violating control synthesis for infeasible specifications were proposed in [301, 298, 308, 61, 65]. Finally, while local tasks may simply be assigned to the individual agents, various works have also considered the decomposition of a global task into local ones, as in [266, 150], which in some sense is a mix of a top-down and a bottom-up design approach.

## 5.4 Additional Proofs

*Proof of theorem 5.3:* Here, we will show, in three parts, that the optimization problem in equation (5.12) is always feasible (part 1), that the control law

$u_i$ is locally bounded and measurable so that there are Fillipov solutions to equation (5.1) (part 2), and theorem (5.2) can be applied (part 3).

For convenience, we first define the function

$$\omega_k(\bar{x}_k, t) := \frac{\partial \mathfrak{b}_k(\bar{x}_k, t)}{\partial t} + \alpha_k(\mathfrak{b}_k(\bar{x}_k, t)), \tag{5.21}$$

which is the term within the decentralized CBF condition in equation (5.12) that is multiplied by function $D_i$. As in the proof of theorem (5.2), it is sufficient to look at time intervals $(s_j^k, s_{j+1}^k)$ separately. Therefore, we define the set of points where the gradient $\frac{\partial \mathfrak{b}_k(\bar{x}_k,t)}{\partial \bar{x}_k}$ is zero within $(s_j^k, s_{j+1}^k)$ as

$$\mathfrak{B}_j^k := \Big\{(\bar{x}_k, t) \in \mathcal{D}_k \times (s_j^k, s_{j+1}^k) \Big| \frac{\partial \mathfrak{b}_k(\bar{x}_k, t)}{\partial \bar{x}_k} = 0\Big\}.$$

The domain $\mathcal{D}_k$ is assumed to be open and bounded and is, as in the centralized case, such that $\mathcal{D}_k \supseteq \mathcal{C}_k(t) := \{\bar{x}_k \in \mathbb{R}^{\bar{n}_k} | \mathfrak{b}_k(\bar{x}_k, t) \geq 0\}$. We further define the sets

$$\mathfrak{B}_{j,i}^k := \Big\{(\bar{x}_k, t) \in \mathcal{D}_k \times (s_j^k, s_{j+1}^k) \Big| \frac{\partial \mathfrak{b}_k(\bar{x}_k, t)}{\partial x_i} = 0\Big\} \setminus \mathfrak{B}_j^k$$

$$\bar{\mathfrak{B}}_{j,i}^k := \Big\{(\bar{x}_k, t) \in \mathcal{D}_k \times (s_j^k, s_{j+1}^k) \Big| \frac{\partial \mathfrak{b}_k(\bar{x}_k, t)}{\partial x_i} \neq 0\Big\},$$

where $\mathfrak{B}_{j,i}^k$ indicates points $(\bar{x}_k, t) \in \mathcal{D}_k \times (s_j^k, s_{j+1}^k)$, where the partial gradient $\frac{\partial \mathfrak{b}_k(\bar{x}_k,t)}{\partial x_i}$ is zero while the gradient $\frac{\partial \mathfrak{b}_k(\bar{x}_k,t)}{\partial \bar{x}_k}$ is not zero, and $\bar{\mathfrak{B}}_{j,i}^k$ indicates points where the partial gradient $\frac{\partial \mathfrak{b}_k(\bar{x}_k,t)}{\partial x_i}$ is not zero. We remark that $\mathfrak{B}_j^k \cup \mathfrak{B}_{j,i}^k \cup \bar{\mathfrak{B}}_{j,i}^k = \mathcal{D}_k \times (s_j^k, s_{j+1}^k)$ and $\mathfrak{B}_j^k$, $\mathfrak{B}_{j,i}^k$, and $\bar{\mathfrak{B}}_{j,i}^k$ are disjoint sets. To understand the details of the following proof, note that $\mathfrak{B}_j^k$ and $\mathfrak{B}_{j,i}^k$ cannot be closed sets (note that $(s_j^k, s_{j+1}^k)$ is open) and information regarding whether these sets are open is not available. We will, however, show and use the fact that $\bar{\mathfrak{B}}_{j,i}^k$ is open.

**Part 1—Feasibility of Equation (5.12):** We next show that equation (5.12) is always feasible and distinguish among three cases, indicated by $\mathfrak{B}_j^k$, $\mathfrak{B}_{j,i}^k$, and $\bar{\mathfrak{B}}_{j,i}^k$. It will turn out that $u_i$ may be discontinuous on the boundaries of $\mathfrak{B}_j^k$, $\mathfrak{B}_{j,i}^k$, and $\bar{\mathfrak{B}}_{j,i}^k$.

Case 1 applies when $(\bar{x}_k, t) \in \mathfrak{B}_j^k$. This is equivalent to $(\bar{x}_k, t) \in \mathcal{D}_k \times (s_j^k, s_{j+1}^k)$, such that $\frac{\partial \mathfrak{b}_k(\bar{x}_k,t)}{\partial \bar{x}_k} = 0$, which implies that $\frac{\partial \mathfrak{b}_k(\bar{x}_k,t)}{\partial x_i} = 0$. Therefore, the condition in equation (5.12b) reduces to $\omega_k(\bar{x}_k, t) \geq 0$ since $D_i(\bar{x}_k, t) = 1$ (note that $\sum_{v \in \mathcal{V}_k} \|\frac{\partial \mathfrak{b}_k(\bar{x}_k,t)}{\partial x_v}\|_1 = 0$). We now see that equation (5.12b) is satisfied due to assumption 5.2. Specifically, when assumption 5.2 holds, we know from the proof of theorem 4.5 that there is a constant $\epsilon_k > 0$ such that $\omega_k(\bar{x}_k, t) \geq \epsilon_k$ for each $(\bar{x}_k, t) \in \mathfrak{B}_j^k$. Hence, $u_i(\bar{x}_k, t) = 0$ is the optimal solution to equation (5.12).

Case 2 applies when $(\bar{x}_k, t) \in \mathfrak{B}^k_{j,i}$. This is equivalent to $(\bar{x}_k, t) \in \mathcal{D}_k \times (s^k_j, s^k_{j+1})$ such that $\frac{\partial \mathfrak{b}_k(\bar{x}_k,t)}{\partial \bar{x}_k} \neq 0$ and $\frac{\partial \mathfrak{b}_k(\bar{x}_k,t)}{\partial x_i} = 0$. The optimal solution to equation (5.12) is again $u_i(\bar{x}_k, t) = 0$ since equation (5.12b) is trivially satisfied as $D_i(\bar{x}_k, t) = 0$.

Case 3 applies when $(\bar{x}_k, t) \in \bar{\mathfrak{B}}^k_{j,i}$. This is equivalent to $(\bar{x}_k, t) \in \mathcal{D}_k \times (s^k_j, s^k_{j+1})$ such that $\frac{\partial \mathfrak{b}_k(\bar{x}_k,t)}{\partial x_i} \neq 0$, so that equation (5.12) is feasible, as $\frac{\partial \mathfrak{b}_k(\bar{x}_k,t)}{\partial x_i} \neq 0$ implies $\frac{\partial \mathfrak{b}_k(\bar{x}_k,t)}{\partial x_i} g_i(x_i) \neq 0$. In this case, $u_i$ is locally Lipschitz continuous on $\text{int}(\bar{\mathfrak{B}}^k_{j,i})$, where $\text{int}(\cdot)$ denotes the interior of $\bar{\mathfrak{B}}^k_{j,i}$. This follows by virtue of [18, theorem 3] and since all the functions in equation (5.12) are locally Lipschitz continuous on $\text{int}(\bar{\mathfrak{B}}^k_{j,i})$. In particular, $D_i$ is locally Lipschitz continuous on $\text{int}(\mathfrak{B}^k_{j,i}) \cup \text{int}(\bar{\mathfrak{B}}^k_{j,i})$ and $\frac{\partial \mathfrak{b}_k}{\partial x_i}$, $\frac{\partial \mathfrak{b}_k}{\partial t}$, $f_i$, $g_i$, and $\alpha_k \circ \mathfrak{b}_k$ are locally Lipschitz continuous on $\text{int}(\mathfrak{B}^k_j) \cup \text{int}(\mathfrak{B}^k_{j,i}) \cup \text{int}(\bar{\mathfrak{B}}^k_{j,i})$.

Therefore, the optimization problem in equation (5.12) is always feasible.

**Part 2—Existence of Filippov Solutions to Equation (5.1):** As indicated before, the control law $u_i$ may not be locally Lipschitz continuous on $\mathcal{D}_k \times (s^k_j, s^k_{j+1})$. However, we next show that $u_i$ is locally bounded and measurable on this domain. From the previous analysis, we know that $u_i$ is locally Lipschitz continuous, and hence locally bounded on $\text{int}(\mathfrak{B}^k_j)$, $\text{int}(\mathfrak{B}^k_{j,i})$, and $\text{int}(\bar{\mathfrak{B}}^k_{j,i})$. Therefore, we show that $u_i$ is locally bounded on the boundaries of $\mathfrak{B}^k_j$, $\mathfrak{B}^k_{j,i}$, and $\bar{\mathfrak{B}}^k_{j,i}$. The proof of part 2 is technical, and we invite the reader to directly go to part 3 if they are only interested in the main part of the proof.

We next investigate the cases where $(\bar{x}_k, t)$ is in $\{\text{bd}(\mathfrak{B}^k_j) \cap \text{bd}(\mathfrak{B}^k_{j,i})\} \setminus \text{bd}(\bar{\mathfrak{B}}^k_{j,i})$ (case 1 or 2), $\{\text{bd}(\mathfrak{B}^k_j) \cap \text{bd}(\bar{\mathfrak{B}}^k_{j,i})\} \setminus \text{bd}(\mathfrak{B}^k_{j,i})$ (case 1 or 3), $\{\text{bd}(\mathfrak{B}^k_{j,i}) \cap \text{bd}(\bar{\mathfrak{B}}^k_{j,i})\} \setminus \text{bd}(\mathfrak{B}^k_j)$ (case 2 or 3), and $\text{bd}(\mathfrak{B}^k_j) \cap \text{bd}(\mathfrak{B}^k_{j,i}) \cap \text{bd}(\bar{\mathfrak{B}}^k_{j,i})$ (case 1, 2, or 3), where $\text{bd}(\cdot)$ denotes the boundary of a set.

When $(\bar{x}_k, t) \in \{\text{bd}(\mathfrak{B}^k_j) \cap \text{bd}(\mathfrak{B}^k_{j,i})\} \setminus \text{bd}(\bar{\mathfrak{B}}^k_{j,i})$, either $(\bar{x}_k, t) \in \mathfrak{B}^k_j$ or $(\bar{x}_k, t) \in \mathfrak{B}^k_{j,i}$. Either way, due to continuity of $u_i$ on $\text{int}(\mathfrak{B}^k_j)$, $\text{int}(\mathfrak{B}^k_{j,i})$, and $\text{int}(\bar{\mathfrak{B}}^k_{j,i})$, there is a small neighborhood $\mathcal{U} \subseteq \{\mathfrak{B}^k_j \cup \mathfrak{B}^k_{j,i}\} \setminus \bar{\mathfrak{B}}^k_{j,i}$ around $(\bar{x}_k, t)$ such that $\|u_i(\bar{x}'_k, t')\| = 0$ for each $(\bar{x}'_k, t') \in \mathcal{U}$ (recall that $u_i(\bar{x}_i, t) = 0$ in cases 1 and 2). Consequently, $u_i$ is locally bounded on $\{\text{bd}(\mathfrak{B}^k_j) \cap \text{bd}(\mathfrak{B}^k_{j,i})\} \setminus \text{bd}(\bar{\mathfrak{B}}^k_{j,i})$.

When $(\bar{x}_k, t) \in \{\text{bd}(\mathfrak{B}^k_j) \cap \text{bd}(\bar{\mathfrak{B}}^k_{j,i})\} \setminus \text{bd}(\mathfrak{B}^k_{j,i})$, either $(\bar{x}_k, t) \in \mathfrak{B}^k_j$ or $(\bar{x}_k, t) \in \bar{\mathfrak{B}}^k_{j,i}$. Note that $\omega_k(\bar{x}_k, t) \geq \epsilon_k$ if $(\bar{x}_k, t) \in \mathfrak{B}^k_j$ due to assumption 5.2. Recall that $\omega_k$ is continuous on $\mathcal{D}_k \times (s^k_j, s^k_{j+1})$. Therefore, it follows that for a given $\epsilon_k > 0$, there is a $\delta_k > 0$ such that for each $(\bar{x}'_k, t')$ with $\| \begin{bmatrix} \bar{x}_k'^T & t' \end{bmatrix}^T - \begin{bmatrix} \bar{x}_k^T & t \end{bmatrix}^T \| < \delta_k$, it holds that $\omega_k(\bar{x}_k, t) - \epsilon_k < \omega_k(\bar{x}'_k, t') < \omega_k(\bar{x}_k, t) + \epsilon_k$. For the choice of $\epsilon_k$ following from assumption 5.2, it consequently holds that

$\omega_k(\bar{x}_k', t') \geq 0$. Hence, there is a small neighborhood $\mathcal{U} \subseteq \{\mathfrak{B}_j^k \cup \bar{\mathfrak{B}}_{j,i}^k\} \setminus \mathfrak{B}_{j,i}^k$ around $(\bar{x}_k, t)$ such that, for each $(\bar{x}_k', t') \in \mathcal{U}$, either $\|u_i(\bar{x}_k', t')\| = 0$ (if $(\bar{x}_k', t') \in \mathfrak{B}_j^k \cap \mathcal{U}$) or $\omega_k(\bar{x}_k', t') \geq 0$ (if $(\bar{x}_k', t') \in \bar{\mathfrak{B}}_{j,i}^k \cap \mathcal{U}$). For the latter case (i.e., $(\bar{x}_k', t') \in \bar{\mathfrak{B}}_{j,i}^k \cap \mathcal{U}$), note that a feasible (not necessarily optimal) and analytic control law for equation (5.12b) is

$$u_i^{\text{feas}}(\bar{x}_k', t') := g_i(x_i')^T G_i(x_i')^{-1}(-f_i(x_i') + v_i^{\text{feas}}(\bar{x}_k', t')),$$

where $x_i'$ is the component of agent $i$ within $\bar{x}_k'$, $v_i^{\text{feas}}$ is a free control input designed in the remainder, and where the inverse of $G_i(x_i') := g_i(x_i')g_i(x_i')^T$ exists due to assumption 5.1. We therefore know an upper bound of $u_i^{\text{feas}}$ as

$$\|u_i^{\text{feas}}(\bar{x}_k', t')\| \leq C_{g_i} C_{G_i}(C_{f_i} + \|v_i^{\text{feas}}(\bar{x}_k', t')\|),$$

where $C_{g_i}$, $C_{G_i}$, and $C_{f_i}$ are upper bounds on $\|g_i(x_i)\|$, $\|G_i(x_i)\|$, and $\|f_i(x_i)\|$ that follow due to the continuity of $g_i(x_i)$, $G_i(x_i)$, and $f_i(x_i)$ on the bounded domain $\mathcal{D}_k$. Specifically, $G_i(x_i)$ is upper-bounded by the inverse of the smallest singular value of $G_i(x_i)$ when the max matrix norm is used [130, chapters 5 and 6]. We next show how to select $v_i^{\text{feas}}$ so that equation (5.12b) is satisfied while finding an upper bound for $\|v_i^{\text{feas}}\|$. Using $u_i^{\text{feas}}(\bar{x}_k', t')$, equation (5.12b) reduces to

$$\frac{\partial \mathfrak{b}_k(\bar{x}_k', t')}{\partial x_i} v_i^{\text{feas}} \geq \left\| \frac{\partial \mathfrak{b}_k(\bar{x}_k, t)}{\partial x_i} \right\|_1 C - D_i(\bar{x}_k', t')\omega_k(\bar{x}_k', t'). \tag{5.22}$$

We select $v_i^{\text{feas}}(\bar{x}_k', t') := \text{sgn}\left(\frac{\partial \mathfrak{b}_k(\bar{x}_k', t')}{\partial x_i}\right)^T \kappa_i$, where $\text{sgn}(\cdot)$ is the element wise sign operator so equation (5.22) becomes

$$\left\| \frac{\partial \mathfrak{b}_k(\bar{x}_k', t')}{\partial x_i} \right\|_1 \left( \kappa_i - C + \frac{\omega_k(\bar{x}_k', t')}{\sum_{v \in \mathcal{V}_k} \left\| \frac{\partial \mathfrak{b}_k(\bar{x}_k', t')}{\partial x_v} \right\|_1} \right) \geq 0. \tag{5.23}$$

In particular, it holds that equation (5.23) is satisfied if $\kappa_i := C$ (recall that $\omega_k(\bar{x}_k', t') \geq 0$ if $(\bar{x}_k', t') \in \bar{\mathfrak{B}}_{j,i}^k \cap \mathcal{U}$), such that $\|v_i^{\text{feas}}(\bar{x}_k', t')\| \leq \sqrt{n_i} C$. Consequently, $\|u_i(\bar{x}_k', t')\| \leq \|u_i^{\text{feas}}(\bar{x}_k', t')\| \leq C_{g_i} C_{G_i}(C_{f_i} + \sqrt{n_i} C)$, and it follows that $u_i$ is locally bounded on $\{\text{bd}(\mathfrak{B}_j^k) \cap \text{bd}(\bar{\mathfrak{B}}_{j,i}^k)\} \setminus \text{bd}(\mathfrak{B}_{j,i}^k)$.

When $(\bar{x}_k, t) \in \{\text{bd}(\mathfrak{B}_{j,i}^k) \cap \text{bd}(\bar{\mathfrak{B}}_{j,i}^k)\} \setminus \text{bd}(\mathfrak{B}_j^k)$, either $(\bar{x}_k, t) \in \mathfrak{B}_{j,i}^k$ or $(\bar{x}_k, t) \in \bar{\mathfrak{B}}_{j,i}^k$, and a similar analysis can be made as before. In particular, then there is a neighborhood $\mathcal{U} \subseteq \{\mathfrak{B}_{j,i}^k \cup \bar{\mathfrak{B}}_{j,i}^k\} \setminus \mathfrak{B}_j^k$ around $(\bar{x}_k, t)$ such that, for each $(\bar{x}_k', t') \in \mathcal{U}$, either $\|u_i(\bar{x}_k', t')\| = 0$ (if $(\bar{x}_k', t') \in \mathfrak{B}_{j,i}^k \cap \mathcal{U}$) or $\sum_{v \in \mathcal{V}_k} \|\frac{\partial \mathfrak{b}_k(\bar{x}_k', t')}{\partial x_v}\|_1 \geq \nu$ for some $\nu > 0$ (if $(\bar{x}_k', t') \in \bar{\mathfrak{B}}_{j,i}^k \cap \mathcal{U}$) since $(\bar{x}_k', t') \notin \mathfrak{B}_j^k$, and again this is due to continuity. In the latter case, selecting $\kappa_i := C - \frac{\omega_k(\bar{x}_k', t')}{\nu}$ satisfies equation (5.23). The same arguments as before then show that $u_i$ is locally bounded on $\{\text{bd}(\mathfrak{B}_{j,i}^k) \cap \text{bd}(\bar{\mathfrak{B}}_{j,i}^k)\} \setminus \text{bd}(\mathfrak{B}_j^k)$.

When $(\bar{x}_k,t)\in \mathrm{bd}(\mathfrak{B}_j^k)\cap \mathrm{bd}(\mathfrak{B}_j^k)\cap \mathrm{bd}(\bar{\mathfrak{B}}_{j,i}^k)$, it can again be shown that $u_i$ is locally bounded on $\mathrm{bd}(\mathfrak{B}_j^k)\cap \mathrm{bd}(\mathfrak{B}_{j,i}^k)\cap \mathrm{bd}(\bar{\mathfrak{B}}_{j,i}^k)$. The proof is straighforward using the same arguments as in the previous discussion, so that it has been omitted here.

It follows that $u_i$ is locally bounded on $\mathrm{bd}(\mathfrak{B}_j^k)$, $\mathrm{bd}(\mathfrak{B}_{j,i}^k)$, and $\mathrm{bd}(\bar{\mathfrak{B}}_{j,i}^k)$ and is consequently also locally bounded on $\mathfrak{B}_j^k\cup\mathfrak{B}_{j,i}^k\cup\bar{\mathfrak{B}}_{j,i}^k=\mathcal{D}_k\times(s_j^k,s_{j+1}^k)$. To see that $u_i$ is measureable, note that $\mathfrak{B}_j^k$, $\mathfrak{B}_{j,i}^k$, and $\bar{\mathfrak{B}}_{j,i}^k$ are measurable sets. The product of measurable functions is measurable and the indicator function (here used to indicate cases 1, 2, and 3) defined on measurable sets is measurable such that $u_i$ is measurable. Consequently, the system in equation (5.1) admits Filippov solutions $x:[0,\tau_{\max})\to\mathbb{R}^n$ from each initial condition in $\mathcal{D}\times\mathbb{R}_{\geq 0}$ where $\mathcal{D}:=\mathcal{D}_1\times\ldots\times\mathcal{D}_K$.

**Part 3—Application of Theorem (5.2):** We have shown in part 1 that the optimization problem in equation (5.12) is always feasible. For each $\mathfrak{b}_k$, the individual solutions $u_i$ to equation (5.12) for agents $i\in\mathcal{V}_k$ now result in

$$\sum_{i\in\mathcal{V}_k}\frac{\partial\mathfrak{b}_k(\bar{x}_k,t)}{\partial x_i}(f_i(x_i)+g_i(x_i)u_i(\bar{x}_k,t))\geq$$
$$\sum_{i\in\mathcal{V}_k}\Big(-D_i(\bar{x}_k,t)\omega_k(\bar{x}_k,t)+\Big\|\frac{\partial\mathfrak{b}_k(\bar{x}_k,t)}{\partial x_i}\Big\|_1 C\Big)$$
$$\Leftrightarrow\frac{\partial\mathfrak{b}_k(\bar{x}_k,t)}{\partial\bar{x}_k}(\bar{f}_k(\bar{x}_k)+\bar{g}_k(\bar{x}_k)\bar{u}_k(\bar{x}_k,t))\geq-\omega_k(\bar{x}_k,t)+\Big\|\frac{\partial\mathfrak{b}_k(\bar{x}_k,t)}{\partial\bar{x}_k}\Big\|_1 C, \tag{5.24}$$

following the same reasoning as previously applied in section 5.1.1.

In our analysis that follows, it is crucial to note that $\bar{\mathfrak{B}}_{j,i}^k$ is open, as shown next. Denote by $\mathrm{inv}\Big(\frac{\partial\mathfrak{b}_k(\bar{x}_k,t)}{\partial x_i}(\mathcal{O}^{n_i})\Big)$ the inverse image of $\frac{\partial\mathfrak{b}_k(\bar{x}_k,t)}{\partial x_i}$ under $\mathcal{O}^{n_i}$, where $\mathcal{O}:=\mathbb{R}\setminus\{0\}$. Now, $\mathrm{inv}\Big(\frac{\partial\mathfrak{b}_k(\bar{x}_k,t)}{\partial x_i}(\mathcal{O}^{n_i})\Big)$ is open since $\mathcal{O}^{n_i}$ is open, and since the inverse image of a continuous function under an open set is open [26, proposition 1.4.4]. It then holds that

$$\bar{\mathfrak{B}}_{j,i}^k=\{\mathcal{D}_k\times(s_j^k,s_{j+1}^k)\}\cap\mathrm{inv}\Big(\frac{\partial\mathfrak{b}_k(\bar{x}_k,t)}{\partial x_i}(\mathcal{O}^{n_i})\Big)$$

is open since the intersection of open sets is open.

We next show that equation (5.24) implies equation (5.8) such that theorem (5.2) can be applied. Since $\bar{f}_k(\bar{x}_k)$ is locally Lipschitz continuous, it follows that

$$\hat{\mathcal{L}}_{F[\bar{f}_k]}\mathfrak{b}_k(\bar{x}_k,t):=\{\frac{\partial\mathfrak{b}_k(\bar{x}_k,t)}{\partial\bar{x}_k}\bar{f}_k(\bar{x}_k)\}.$$

For $\hat{\mathcal{L}}_{F[\bar{g}_k\bar{u}_k]}\mathfrak{b}_k(\bar{x}_k,t)$, we have to distinguish among these three cases. First, note that, if for each $i\in\mathcal{V}_k$ we have $(\bar{x}_k,t)\in\bar{\mathfrak{B}}_{j,i}^k$ (case 3), then

$$\hat{\mathcal{L}}_{F[\bar{g}_k\bar{u}_k]}\mathfrak{b}_k(\bar{x}_k,t)=\{\frac{\partial\mathfrak{b}_k(\bar{x}_k,t)}{\partial\bar{x}_k}\bar{g}_k(\bar{x}_k)\bar{u}_k(\bar{x}_k,t)\}.$$

This in particular follows since $\bar{\mathfrak{B}}^k_{j,i}$ is open, so $u_i$ as well as $\bar{g}_k$ are locally Lipschitz continuous on $\text{int}(\bar{\mathfrak{B}}^k_{j,i})=\bar{\mathfrak{B}}^k_j$. If for each $i\in\mathcal{V}_k$ we have $(\bar{x}_k,t)\in\mathfrak{B}^k_{j,i}$ (case 2), then

$$\hat{\mathcal{L}}_{F[\bar{g}_k\bar{u}_k]}\mathfrak{b}_k(\bar{x}_k,t)=\{0\}$$

since $\frac{\partial\mathfrak{b}_k(\bar{x}_k,t)}{\partial x_i}=0$ for each $i\in\mathcal{V}_k$. If for some agents $(\bar{x}_k,t)\in\bar{\mathfrak{B}}^k_{j,i}$ while for others $(\bar{x}_k,t)\in\mathfrak{B}^k_{j,i}$ (i.e., a mix of cases 2 and 3), the resulting $\hat{\mathcal{L}}_{F[\bar{g}_k\bar{u}_k]}\mathfrak{b}_k(\bar{x}_k,t)$ will still be a singleton. If we have $(\bar{x}_k,t)\in\mathfrak{B}^k_j$ (case 1), then

$$\hat{\mathcal{L}}_{F[\bar{g}_k\bar{u}_k]}\mathfrak{b}_k(\bar{x}_k,t)=\{0\}$$

since $\frac{\partial\mathfrak{b}_k(\bar{x}_k,t)}{\partial\bar{x}_k}=0$. Since $\hat{\mathcal{L}}_{F[\bar{f}_k]}\mathfrak{b}k(\bar{x}_k,t)$, $\hat{\mathcal{L}}_{F[\bar{g}_k\bar{u}_k]}\mathfrak{b}_k(\bar{x}_k,t)$, and $\frac{\partial\mathfrak{b}_k(\bar{x}_k,t)}{\partial t}$ are singletons, equation (5.24) is equivalent to

$$\begin{aligned}\min\Big\{\hat{\mathcal{L}}_{F[\bar{f}_k]}\mathfrak{b}_k(\bar{x}_k,t)\oplus\hat{\mathcal{L}}_{F[\bar{g}_k\bar{u}_k]}\mathfrak{b}_k(\bar{x}_k,t)\oplus\{\frac{\partial\mathfrak{b}_k(\bar{x}_k,t)}{\partial t}\}\Big\}\\ \geq-\alpha_k(\mathfrak{b}_k(\bar{x}_k,t))+\Big\|\frac{\partial\mathfrak{b}_k(\bar{x}_k,t)}{\partial\bar{x}_k}\Big\|_1C.\end{aligned}\tag{5.25}$$

Due to lemma 5.2, it holds that

$$\mathcal{L}_{F[\bar{f}_k+\bar{g}_k\bar{u}_k]}\mathfrak{b}_k(\bar{x}_k,t)=\hat{\mathcal{L}}_{F[\bar{f}_k]}\mathfrak{b}_k(\bar{x}_k,t)\oplus\hat{\mathcal{L}}_{F[\bar{g}_k\bar{u}_k]}\mathfrak{b}_k(\bar{x}_k,t)\oplus\{\frac{\partial\mathfrak{b}_k(\bar{x}_k,t)}{\partial t}\}$$

such that equation (5.25) implies equation (5.8). By theorem (5.2), it follows that $(x,0)\models\phi_1\wedge\ldots\wedge\phi_K$.

# III

# Multi-Agent Funnel Control for Spatiotemporal Constraints

# Chapter 6

# Centralized Funnel Control

In this chapter, we examine ideas from funnel control, as introduced in section 2.2.4, to design feedback control laws that satisfy signal temporal logic (STL) specifications. Specifically, we will constrain the quantitative semantics associated with the STL specification by a funnel. In this way, we enforce a system behavior that is guided by the specification. By the design of the funnel, which will encode the specification, this will lead to a satisfaction of the specification at hand. As in the previous two chapters, we design feedback controllers that are computationally tractable and provide satisfaction guarantees for a fragment of STL specifications. In contrast to CBFs, funnel-based control laws are model-free, in the sense that no model of the system is needed under the same controllability assumption made in chapters 4 and 5.

At time $t$, let $x(t)\in\mathbb{R}^n$ and $w(t)\in\mathbb{R}^n$ denote the state and additive noise of our dynamical system. The system is modeled as

$$\dot{x}(t)=f(x(t))+g(x(t))u(x(t),t)+w(t),\ x(0):=x_0, \tag{6.1}$$

where $x_0$ is the initial state and $u:\mathbb{R}^n\times\mathbb{R}_{\geq 0}\to\mathbb{R}^m$ is the control law that we wish to design. The functions $f:\mathbb{R}^n\to\mathbb{R}^n$ and $g:\mathbb{R}^n\to\mathbb{R}^{m\times n}$ are locally Lipschitz continuous, and the function $w:\mathbb{R}_{\geq 0}\to\mathbb{R}^n$ is piecewise continuous. The strength of funnel control, compared to CBFs, is that the functions $f$ and $w$ need not be known. In the case that state and input dimensions are equal (i.e., when $m=n$), the function $g$ need not be known either, so that the control design becomes fully model-free. However, we make the following controllability and boundedness assumption about $g$ and $w$, respectively.

**Assumption 6.1** *Let the function $g$ that describes the input dynamics of the system in equation (6.1) be such that $g(x)g(x)^T$ is positive definite for each*

*$x \in \mathbb{R}^n$. Further, let the function $w$ that describes the additive noise be contained within a bounded set $\mathcal{W} \subseteq \mathbb{R}^n$ (i.e., $w(t) \in \mathcal{W}$ for all $t \in \mathbb{R}_{\geq 0}$).*

The first part of assumption 6.1 (i.e., the controllability assumption), is the same assumption that was made in chapters 4 and 5. Similar to the case of CBFs, we can consider relative degree systems (e.g., double integrator or mechanical systems), using the same ideas that will be presented in this section by using a backstepping-like approach. For the sake of simplicity, however, we omit this extension and refer the reader to [37]. We could also again consider unicycle dynamics using a near-identity diffeomorphism. For the second part of assumption 6.1, we do not need to know an explicit upper bound for the control design (i.e., we do not need to know the set $\mathcal{W}$), as opposed to the case of CBFs.

To construct valid funnel-based control laws for the dynamical system in equation (6.1), we consider again a fragment of STL specifications. First, we consider a set of atomic temporal logic formulas that are defined as

$$\psi ::= \top \mid \mu \mid \psi_1 \wedge \psi_2 \tag{6.2a}$$

$$\phi ::= F_{[a,b]}\psi \mid G_{[a,b]}\psi \mid F_{[\underline{a},\underline{b}]}G_{[\bar{a},\bar{b}]}\psi, \tag{6.2b}$$

where $\psi_1$ and $\psi_2$ are Boolean formulas constructed according to equation (6.2a), and where the constants $a, b, \underline{a}, \underline{b}, \bar{a}, \bar{b} \in \mathbb{R}_{\geq 0}$ are such that $a \leq b$, $\underline{a} \leq \underline{b}$ and $\bar{a} \leq \bar{b}$. We refer to $\psi$ in equation (6.2a) as *nontemporal formulas* (i.e., Boolean formulas), and to $\phi$ in equation (6.2b) as *atomic temporal formulas* due to the use of eventually and always operators. Note the special structure for the formulas in the fragment in equation (6.2) that again does not permit negations and disjunctions.

---

**Remark 6.1** As previously argued, disjunctions cannot be satisfied (at least globally from every initial condition) by continuous feedback control laws unless the formula is trivially simplified as illustrated next. Consider the system $\dot{x}(t) = u(x(t))$ and the formula $\phi := F_{[a,b]}(\mu_1 \vee \mu_2)$, where $\mu_1$ and $\mu_2$ are associated with $h_1(x) := 0.5 - |x+1|$ and $h_2(x) := 0.5 - |x-1|$, respectively. For an initial condition of $x(0) := -\epsilon$ (or $x(0) := \epsilon$) for a small $\epsilon > 0$, the control law $u_1(x) := -(x+1)$ (or $u_2(x) := -(x-1)$) drives $x(t)$ toward $\mu_1$ (or $\mu_2$). Note, however, that there is no feedback control law, or a combination of the control laws $u_1$ and $u_2$, that is continuous at $x = 0$ and satisfies $\phi := F_{[a,b]}(\mu_1 \vee \mu_2)$. Instead, one could manually decide to only satisfy either $u_1$ or $u_2$, which technically simplifies the formula $\phi$ to $F_{[a,b]}\mu_1$ or $F_{[a,b]}\mu_2$, respectively. To avoid such simplifications, higher-level decision making, such as using switched systems resulting in discontinuous control laws, can be used (see [166, chapter 4]). Similarly, and more generally, the automata-based approach that will be presented in part IV can be used for disjunctions as well as for negations.

---

We will show that the atomic temporal formula $\phi$ from the fragment in equation (6.2b) can be encoded by a single funnel. By using several funnels, we will be able to satisfy formulas from the more general fragment of STL formulas as follows:

$$\theta ::= F_{[a,b]}\psi \mid G_{[a,b]}\psi \mid F_{[\underline{a},\underline{b}]}G_{[\bar{a},\bar{b}]}\psi \mid \psi' \, U_{[a,b]} \, \psi'' \mid \theta_1 \wedge \theta_2, \tag{6.3}$$

where $\theta_1$ and $\theta_2$ are recursively constructed from equation (6.3). STL formulas $\theta$ from equation (6.3) are referred to as *temporal formulas.* Note that temporal formulas permit the until operator and a nesting of atomic temporal formulas via conjunction.

We proceed as follows. In section 6.1, we will use the transient properties of a performance function as introduced in definition 2.10 to encode atomic temporal logic formulas $\phi$ within a single funnel. In section 6.2, we will derive funnel-based feedback control laws that lead to a satisfaction of $\phi$. We then propose hybrid control strategies in section 6.3 for temporal formulas $\theta$ by using multiple funnels.

# 6.1 Encoding Signal Temporal Logic

Recall that we can associate quantitative semantics $\rho^\psi$ and $\rho^\phi$ with nontemporal and atomic temporal formulas $\psi$ and $\phi$, respectively. Note that $\rho^\psi(x,t)$ and $\rho^\phi(x,t)$ map from $\mathcal{T} \times \mathbb{R}_{\geq 0}$ to $\mathbb{R}$, where $\mathcal{T}$ is the set of all $n$-dimensional signals with time domain $\mathbb{R}_{\geq 0}$. For nontemporal formulas $\psi$, however, we will drop the explicit dependence on time and equivalently write $\rho^\psi(x(t))$ by a slight change of notation. This notation is introduced to highlight that $t$ is contained in $\rho^\psi$ only through the composition of $\rho^\psi$ with $x$. We also replace the nonsmooth quantitative semantics of $\psi_1 \wedge \psi_2$ by a smooth approximation, and we denote the smooth version of the quantitative semantics as $\bar{\rho}^\phi(x,t)$. Formally, we define

$$\begin{aligned}
\bar{\rho}^{\mu}(x(t)) &:= h(x(t))\\
\bar{\rho}^{\neg\mu}(x(t)) &:= -\bar{\rho}^{\mu}(x(t))\\
\bar{\rho}^{\psi_1\wedge\psi_2}(x(t)) &:= -\frac{1}{\eta}\ln\big(\exp\big(-\eta\bar{\rho}^{\psi_1}(x(t))\big)+\exp\big(-\eta\bar{\rho}^{\psi_2}(x(t))\big)\big)
\end{aligned}$$

$$\bar{\rho}^{F_{[a,b]}\psi}(x,t) := \max_{t_1 \in [t+a,t+b]} \bar{\rho}^{\psi}(x(t_1))$$
$$\bar{\rho}^{G_{[a,b]}\psi}(x,t) := \min_{t_1 \in [t+a,t+b]} \bar{\rho}^{\psi}(x(t_1))$$
$$\bar{\rho}^{F_{[\underline{a},\underline{b}]}G_{[\bar{a},\bar{b}]}\psi}(x,t) := \max_{t_1 \in [t+\underline{a},t+\underline{b}]} \min_{t_2 \in [t_1+\bar{a},t_1+\bar{b}]} \bar{\rho}^{\psi}(x(t_2)),$$

where $\eta > 0$ again determines the accuracy by which $\bar{\rho}^{\psi_1 \wedge \psi_2}(x(t))$ approximates $\rho^{\psi_1 \wedge \psi_2}(x(t))$. Similar to lemma 4.1, it holds that $\bar{\rho}^{\psi_1 \wedge \psi_2}(x(t)) = \rho^{\psi_1 \wedge \psi_2}(x(t))$ as $\eta \to \infty$, and again we obtain an underapproximation such that $\bar{\rho}^{\psi_1 \wedge \psi_2}(x(t)) \leq \rho^{\psi_1 \wedge \psi_2}(x(t))$, which we will use for control design. Note that each formula $\psi$ in equation (6.2a) can be written as $\psi := \wedge_{i=1}^{q} \mu_i$ so that, in direct analogy to lemma 4.1, we can quantify how good the approximation is.

**Lemma 6.1** *For $\eta > 0$ and a set of $q$ predicates $\mu_i$ that result in the nontemporal formula $\psi := \wedge_{i=1}^{q} \mu_i$, it holds that*

$$\bar{\rho}^{\psi}(x(t)) \leq \rho^{\psi}(x(t)) \leq \bar{\rho}^{\psi}(x(t)) + \frac{\ln(q)}{\eta}.$$

**Satisfiability of $\psi$.** Finally, recall that there is a direct connection between semantics and quantitative semantics such that $(x,t) \models \psi$ if $\rho^{\psi}(x(t)) > 0$ and $(x,t) \not\models \psi$ if $\rho^{\psi}(x(t)) < 0$, as previously alluded to in chapter 3. For the smooth approximation $\bar{\rho}^{\psi}(x(t))$, it holds that $(x,t) \models \psi$ if $\bar{\rho}^{\psi}(x(t)) > 0$ due to lemma 6.1. However, $\bar{\rho}^{\psi}(x(t)) < 0$ can imply both $(x,t) \models \psi$ and $(x,t) \not\models \psi$, and we have that $(x,t) \not\models \psi$ only if $\bar{\rho}^{\psi}(x(t)) + \frac{\ln(q)}{\eta} < 0$. A large-enough $\eta$ may provide a sufficiently accurate approximation, but there may still be some conservatism that we need to address. Therefore, let the optimum of $\rho^{\psi}$ and $\bar{\rho}^{\psi}$ (for a given $\eta$) be

$$\rho^{\psi}_{\text{opt}} := \sup_{x \in \mathbb{R}^n} \rho^{\psi}(x),$$
$$\bar{\rho}^{\psi}_{\text{opt}} := \sup_{x \in \mathbb{R}^n} \bar{\rho}^{\psi}(x).$$

Note that there is always an $\eta$ such that $\bar{\rho}^{\psi}_{\text{opt}} > 0$ if $\rho^{\psi}_{\text{opt}} > 0$, since $\bar{\rho}^{\psi}(x) = \rho^{\psi}(x)$ as $\eta \to \infty$. We remark that $\bar{\rho}^{\psi}_{\text{opt}}$ will be easy to compute in subsequent sections of this chapter, as we will assume that $\bar{\rho}^{\psi}$ is concave and that $\bar{\rho}^{\psi}_{\text{opt}}$ is bounded (as formally stated in assumption 6.3). We then find $\eta$ by solving a convex feasibility problem; that is, we find $\eta$ such that $\bar{\rho}^{\psi}(x) > 0$ for some $x \in \mathbb{R}^n$, which consequently implies that $\bar{\rho}^{\psi}_{\text{opt}} > 0$. The case that $\bar{\rho}^{\psi}_{\text{opt}} > 0$ means that $\psi$ (and hence $\phi$ as well) are satisfiable. Most of our results will operate under this assumption.

**Assumption 6.2** *The maximum of the smooth quantitative semantics* $\bar{\rho}^{\psi}$ *is such that* $\bar{\rho}^{\psi}_{opt} > 0$.

Otherwise (i.e., if $\rho^{\psi}_{\text{opt}} \leq 0$), then there is no $\eta$ such that $\bar{\rho}^{\psi}_{\text{opt}} > 0$; that is, assumption 6.2 cannot be satisfied. Instead, we select $\eta := 1$ (in fact, any $\eta > 0$ can be selected). We will deal with this case in the next section.

**Main Idea.** We aim to achieve $r \leq \bar{\rho}^{\phi}(x, 0)$ where $r \in \mathbb{R}$ has a minimum robustness and $x : \mathbb{R}_{\geq 0} \to \mathbb{R}^n$ is the solution of the system in equation (6.1). In particular, note that $(x, 0) \models \phi$ if $r > 0$. For technical reasons, we introduce $\rho_{\max} \in \mathbb{R}$ with $r < \rho_{\max}$ and also aim for an upper bound $\bar{\rho}^{\phi}(x, 0) \leq \rho_{\max}$. As will be argued later, this will not introduce any conservatism. For a given atomic temporal formula $\phi$, let $\psi$ be the nontemporal formula appearing in $\phi$. Then, we aim to achieve

$$r \leq \bar{\rho}^{\phi}(x, 0) \leq \rho_{\max}$$

by assigning a temporal behavior to the quantitative semantics of the nontemporal formula $\bar{\rho}^{\psi}\big(x(t)\big)$ by means of a performance function $\gamma : \mathbb{R}_{\geq 0} \to \mathbb{R}_{>0}$, which we previously introduced in definition 2.10. Specifically, we will achieve this objective by a specific design of $\gamma$ and the funnel

$$-\gamma(t) + \rho_{\max} < \bar{\rho}^{\psi}\big(x(t)\big) < \rho_{\max}. \tag{6.4}$$

In other words, by assigning a behavior to $\bar{\rho}^{\psi}\big(x(t)\big)$, we will achieve $\bar{\rho}^{\phi}(x, 0) \geq r$. The connection between the nontemporal $\bar{\rho}^{\psi}\big(x(t)\big)$ and the temporal $\bar{\rho}^{\phi}(x, 0)$ is made by the choice of the performance function $\gamma$.

**Funnel Design.** Before we explain in detail how to select $\gamma$ and $\rho_{\max}$ so that satisfaction of the funnel in equation (6.4) for all times $t \geq 0$ implies $r \leq \bar{\rho}^{\phi}(x, 0) \leq \rho_{\max}$, we provide some intuition in example 6.1.

---

**Example 6.1** Consider the two STL formulas

$$\phi_1 := G_{[0,7]}\mu_1,$$
$$\phi_2 := F_{[0,7]}\mu_2 \wedge \mu_3,$$

where, for $x := \begin{bmatrix} x_1 & x_2 \end{bmatrix}^T \in \mathbb{R}^2$, the predicate functions $h_1(x) := 1 - \|x\|^2$, $h_2(x) := x_1$, and $h_3(x) := x_2 + 0.5$ define the predicates $\mu_1$, $\mu_2$, and $\mu_3$, respectively. Figures 6.1a and 6.1b show the funnels in equation (6.4) (dashed lines) prescribing the required temporal behavior to satisfy $\phi_1$ and $\phi_2$, respectively. If now a signal $x : \mathbb{R}_{\geq 0} \to \mathbb{R}^2$, which could be a solution to equation (6.1), is such that $\bar{\rho}^{\psi_1}(x(t))$ is contained within its funnel,

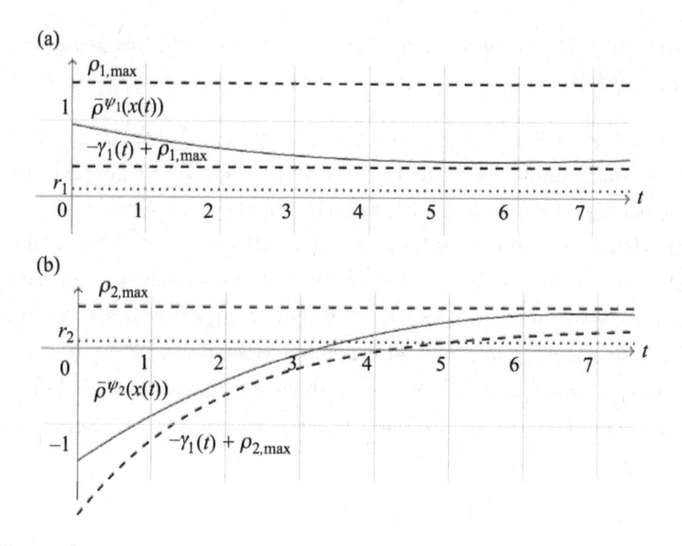


**Figure 6.1**
Illustration of the connection between the quantitative semantics of nontemporal and atomic temporal formula $\bar{\rho}^{\psi}(x(t))$ and $\bar{\rho}^{\phi}(x,0)$ in example 6.1. (a) Funnel $(-\gamma_1(t)+\rho_{1,\max}, \rho_{1,\max})$ (dashed lines) for $\phi_1 = G_{[0,7]}\psi_1$ such that $\bar{\rho}^{\phi_1}(x,0) \geq 0.1 =: r_1$ (dotted line) and $\bar{\rho}^{\psi_1}(x(t))$ (solid line). Reprinted with permission from [177, 185]. (b) Funnel $(-\gamma_2(t)+\rho_{2,\max}, \rho_{2,\max})$ (dashed lines) for $\phi_2 = F_{[0,7]}\psi_2$ such that $\bar{\rho}^{\phi_2}(x,0) \geq 0.1 =: r_2$ (dotted line) and $\bar{\rho}^{\psi_2}(x(t))$ (solid line). Reprinted with permission from [177, 185].

it holds that $(x,0) \models \phi_1$. Similarly, if $\bar{\rho}^{\psi_2}(x(t))$ is contained within its funnel, we have $(x,0) \models \phi_2$. In particular, this follows from the choice of $\gamma_1$, $\gamma_2$, $\rho_{1,\max}$, and $\rho_{2,\max}$, and since $\bar{\rho}^{\psi_1}(x(t)) \in (-\gamma_1(t)+\rho_{1,\max}, \rho_{1,\max})$ and $\bar{\rho}^{\psi_2}(x(t)) \in (-\gamma_2(t)+\rho_{2,\max}, \rho_{2,\max})$ for all $t \in \mathbb{R}_{\geq 0}$; that is, equation (6.4) is satisfied for all $t \in \mathbb{R}_{\geq 0}$. Note also that, for instance in figure 6.1b, the lower funnel $-\gamma_2(t)+\rho_{2,\max}$ enforces $\bar{\rho}^{\psi_2}(x(t)) \geq r_2 := 0.1$ for all $t \geq 6$. Thus, $\phi_2$ is robustly satisfied with $\bar{\rho}^{\phi_2}(x,0) \geq r_2$.

Based on this example, let us now formalize the design of the performance function $\gamma$. First, we select the upper bound $\rho_{\max}$ as

$$\rho_{\max} \in (\bar{\rho}^{\psi}_{\text{opt}}, \infty). \tag{6.5}$$

The intuition here is that we would like to select a value for $\rho_{\max}$ that is larger than $\bar{\rho}^{\psi}_{\text{opt}}$ so that we do not introduce any conservatism. Next, we define the variable

$$t_* \in \begin{cases} \{a\} & \text{if } \phi = G_{[a,b]}\psi \\ [a,b] & \text{if } \phi = F_{[a,b]}\psi \\ [\underline{a}+\bar{a}, \underline{b}+\bar{a}] & \text{if } \phi = F_{[\underline{a},\underline{b}]}G_{[\bar{a},\bar{b}]}\psi \end{cases} \tag{6.6}$$

that will determine when the nontemporal formula $\psi$ is satisfied. Our goal is to enforce $r \leq \bar{\rho}^{\psi}(x(t)) \leq \rho_{\max}$ for all $t \geq t_*$ by the choice of $\gamma$ and the use of a funnel-based feedback control law. By this specific choice of $t_*$, it will then hold that $r \leq \bar{\rho}^{\phi}(x,0) \leq \rho_{\max}$. Furthermore, we select the robustness $r$ according to

$$r \in \begin{cases} (0, \bar{\rho}^{\psi}_{\text{opt}}) & \text{if } t_* > 0 \\ (0, \bar{\rho}^{\psi}(x_0)) & \text{if } t_* = 0. \end{cases} \tag{6.7}$$

Note that this choice is made under assumption 6.2, where $\bar{\rho}^{\psi}_{\text{opt}} > 0$, which implies that $r > 0$ is a feasible choice. Consequently, it holds that $0 < r < \bar{\rho}^{\psi}_{\text{opt}} < \rho_{\max}$. Note the special case $t_* = 0$, where we have to select $\bar{\rho}^{\psi}(x_0) > r$. Due to this dependency on the initial condition $x_0$, it is advisable to select $t_* > 0$ whenever possible. In this regard, we say that the formula $\psi$ is feasible with respect to $r$, $x_0$, and $t_*$ if $\bar{\rho}^{\psi}(x_0) > r$ in case that $t_* = 0$. This will make it possible to achieve $r \leq \bar{\rho}^{\psi}(x(t)) \leq \rho_{\max}$ for all $t \geq t_*$ by the choice of $\gamma$ and the later proposed funnel-based feedback control law in equation (6.14).

For the design of $\gamma$, assume now that $\psi$ is feasible with regard to $r$, $x_0$, and $t_*$, and recall that the performance function in definition 2.10 was defined as

$$\gamma(t) := (\gamma_0 - \gamma_\infty)\exp(-lt) + \gamma_\infty.$$

First, it is crucial to design $\gamma_0$ such that the funnel in equation (6.4) is initially satisfied; that is, that $-\gamma_0 + \rho_{\max} < \bar{\rho}^{\psi}(x_0)$. It should also hold that $r \leq -\gamma_0 + \rho_{\max}$ if $t_* = 0$ since we want that $r \leq \bar{\rho}^{\psi}(x(t))$ for all $t \geq t_*$. This is illustrated in figure 6.1a, where $t_* = 0$ and $r := 0.2$, and it should therefore hold that the lower funnel satisfies $r := 0.2 \leq -\gamma_0 + \rho_{\max}$. Therefore, $\gamma_0$ is selected as

$$\gamma_0 \in \begin{cases} (\rho_{\max} - \bar{\rho}^{\psi}(x_0), \infty) & \text{if } t_* > 0 \\ (\rho_{\max} - \bar{\rho}^{\psi}(x_0), \rho_{\max} - r] & \text{if } t_* = 0. \end{cases} \tag{6.8}$$

At $t = \infty$, it is required that $\max(-\gamma_0 + \rho_{\max}, r) \leq -\gamma_\infty + \rho_{\max} < \bar{\rho}^{\psi}_{\text{opt}}$ where the first inequality enforces that $-\gamma + \rho_{\max}$ is a nondecreasing function, which in turn leads to $\gamma$ being nonincreasing, and the second inequality ensures that the funnel is always feasible. The smaller $\gamma_\infty$ is selected, the tighter the funnel in equation (6.4) will become as $t \to \infty$. The parameter $\gamma_\infty$ is hence selected as

$$\gamma_\infty \in \left(\rho_{\max} - \bar{\rho}^{\psi}_{\text{opt}}, \min\left(\gamma_0, \rho_{\max} - r\right)\right). \tag{6.9}$$

Finally, we select the remaining parameter $l$ as

$$l \in \begin{cases} 0 & \text{if } -\gamma_0 + \rho_{\max} \geq r \\ -\frac{\ln\left(\frac{r+\gamma_\infty-\rho_{\max}}{-(\gamma_0-\gamma_\infty)}\right)}{t_*} & \text{if } -\gamma_0 + \rho_{\max} < r, \end{cases} \tag{6.10}$$

which ensures that $-\gamma(t_*) + \rho_{\max} = r$ if $-\gamma_0 + \rho_{\max} < r$, and $-\gamma(t_*) + \rho_{\max} \geq r$ otherwise. We summarize the correctness of this encoding in the following result.

**Theorem 6.1 (Soundness of $\gamma$ in Encoding $\phi$)** *Let $\phi$ be an atomic temporal formula $\phi$ from equation (6.2b), and let the nontemporal formula $\psi$ contained within $\phi$ have smooth quantitative semantics $\bar{\rho}^{\psi}$ that satisfy assumptions 6.2. Further, let $\rho_{max}$, $t_*$, $r$, $\gamma_0$, $\gamma_\infty$, and $l$ be selected as in equations (6.5)–(6.10). If a signal $x : \mathbb{R}_{\geq 0} \to \mathbb{R}^n$ is such that the funnel in equation (6.4) is satisfied for all $t \geq 0$, then it follows that $0 < r \leq \bar{\rho}^{\phi}(x, 0) \leq \rho_{max}$ (i.e., that $(x, 0) \models \phi$).*

*Proof:* *The result follows from the construction of the upper bound $\rho_{max}$ and the performance function $\gamma$, which adheres to the quantitative semantics of $\phi$. Specifically, if the funnel in equation (6.4) is satisfied for all $t \geq 0$, then we have that $r \leq \bar{\rho}^{\psi}(x(t))$ for all $t \geq t_*$, which leads to $r \leq \bar{\rho}^{\phi}(x, 0)$ by the choice of $t_*$. Since $r > 0$, it follows that $(x, 0) \models \phi$.*

---

**Example 6.2** Consider the STL formula $\phi := G_{[0,7]}\psi$ with $\psi := \mu$, which is associated with the predicate function $h(x) := 1 - \|x\|^2$ (this is the same formula as $\phi_1$ in example 6.1), for which $\bar{\rho}^{\psi}_{\text{opt}} = 1$ (note that $\bar{\rho}^{\psi}_{\text{opt}}$ is independent of $\eta$ in this case) and let $x_0 := \begin{bmatrix} -0.05 & 0 \end{bmatrix}^T$. We hence select, as instructed, $\rho_{\max} := 1.5 \in (\bar{\rho}^{\psi}_{\text{opt}}, \infty)$, $r := 0.1 \in (0, \bar{\rho}^{\psi}_{\text{opt}})$, and $t_* := 0$, as well as $\gamma_0 := 1.1 \in (\rho_{\max} - \bar{\rho}^{\psi}(x_0), \rho_{\max} - r] = (0.55, 1.4]$, $\gamma_\infty := 1 \in (\rho_{\max} - \bar{\rho}^{\psi}_{\text{opt}}, \min(\gamma_0, \rho_{\max} - r)) = (0.5, 1.1)$, and $l := 0$. Note that these exact parameters are plotted in figure 6.1a.

---

## 6.2 Control Laws Based on a Single Funnel

To design a control law $u$ that ensures that the funnel in equation (6.4) is satisfied for all times $t \geq 0$, we make two assumptions on nontemporal formulas $\psi$ that are contained within atomic temporal formulas $\phi$, described next.

**Assumption 6.3** *Each nontemporal formula of class $\psi$ from equation (6.2a) that is contained in the atomic temporal formula $\phi$ from equation (6.2b) is such that $\bar{\rho}^{\psi}: \mathbb{R}^n \to \mathbb{R}$ is concave and the superlevel sets of $\bar{\rho}^{\psi}$ are bounded.*

A sufficient condition for concavity of $\bar{\rho}^{\psi}$ as required in assumption 6.3 is that all predicate functions $h_i: \mathbb{R}^n \to \mathbb{R}$ associated with predicates $\mu_i$ in $\psi := \wedge_{i=1}^{q} \mu_i$ are concave. This fact can be shown following the same idea as in lemma 4.2, where we used the log-sum approximation for CBFs. Note hence that assumption 6.3 directly resembles assumption 4.2, which we imposed in the CBF case. Concavity of $\bar{\rho}^{\psi}$ is needed since the proposed funnel controller will use the gradient $\frac{\partial \bar{\rho}^{\psi}(x)}{\partial x}$ so that local minima or saddle points may again lead to a loss of controllability.

The assumption of bounded level sets is of technical nature. For a constant $c \in \mathbb{R}$, the superlevel set of $\bar{\rho}^{\psi}$ at level $c$ is given as $L_c := \{x \in \mathbb{R}^n | \bar{\rho}^{\psi}(x) \geq c\}$. For each constant $c \in \mathbb{R}$, this assumption guarantees that there is another constant $C \geq 0$ such that $\|x\| \leq C$ for all $x \in L_c$. This property will ensure in theorem 6.2 that solutions $x: \mathcal{I} \to \mathbb{R}^n$ to equation (6.1) are bounded and hence complete (i.e., that $\mathcal{I} = \mathbb{R}_{\geq 0}$). This assumption is not restrictive in practice since $\mu_{\text{bounded}} := (\|x\|^2 \leq C)$, for a sufficiently large $C$, can be combined with the nontemporal formula $\psi$ as $\psi \wedge \mu_{\text{bounded}}$, which is used instead.

Now that we have fixed these assumptions, let us follow section 2.2.4 and define the one-dimensional error $e: \mathbb{R}^n \to \mathbb{R}$, the normalized error $\xi: \mathbb{R}^n \times \mathbb{R}_{\geq 0} \to \mathbb{R}$, and the transformed error $\epsilon: \mathbb{R}^n \times \mathbb{R}_{\geq 0} \to \mathbb{R}$ of the funnel in equation (6.4) as follows, respectively:

$$e(x) := \bar{\rho}^{\psi}(x) - \rho_{\max} \tag{6.11}$$

$$\xi(x,t) := \frac{e(x)}{\gamma(t)} \tag{6.12}$$

$$\epsilon(x,t) := S(\xi(x,t)) = \ln\Big(-\frac{\xi(x,t)+1}{\xi(x,t)}\Big), \tag{6.13}$$

where $S$ is the transformation function from definition 2.11, with $\underline{M} := 1$ and $\overline{M} := 0$. When referring to the solution $x: \mathbb{R}_{\geq 0} \to \mathbb{R}^n$ to equation (6.1) we use the notation $e(t) := e(x(t))$, $\xi(t) := \xi(x(t), t)$, and $\epsilon(t) := \epsilon(x(t), t)$, while we use $e(x)$, $\xi(x,t)$, and $\epsilon(x,t)$ when we want to emphasize dependence on the state $x \in \mathbb{R}^n$. The funnel equation (6.4) can now be written as the constrained problem

$$-\gamma(t) < e(t) < 0,$$

which can further be written as

$$-1 < \xi(t) < 0.$$

Applying the transformation function $S$ then results in the unconstrained problem

$$-\infty < \epsilon(t) < \infty.$$

As argued before, if now a control law $u$ exists such that $\epsilon(t)$ is bounded for all $t \geq 0$, then equation (6.4) holds for all $t \geq 0$. By the design of the funnel via the parameters $\gamma$ and $\rho_{\max}$ that encode the specification $\phi$, as discussed in section 6.1, it would then follow that $r \leq \bar{\rho}^{\phi}(x,0) \leq \rho_{\max}$. Hence, our objective is to design a continuous feedback control law $u$ such that $\epsilon(t)$ is bounded for all $t \in \mathbb{R}_{\geq 0}$.

**Theorem 6.2 (Funnel Controller Satisfying $\phi$)** *Let the functions $f$, $g$, and $w$ that describe the system in equation (6.1) be locally Lipschitz continuous in $x$ and piecewise continuous in $t$, and let $g$ and $w$ satisfy assumption 6.1. Further, let $\phi$ be an atomic temporal formula from equation (6.2b), and let the nontemporal formula $\psi$ contained within $\phi$ have smooth quantitative semantics $\bar{\rho}^{\psi}$ that satisfy assumptions 6.2 and 6.3. If $\rho_{max}$, $t_*$, $r$, $\gamma_0$, $\gamma_\infty$, and $l$ are selected as in equations (6.5)–(6.10), then the control law*

$$u(x,t) := -\epsilon(x,t) g(x)^T \frac{\partial \bar{\rho}^{\psi}(x)}{\partial x}^T \tag{6.14}$$

*guarantees that the solution $x : \mathcal{I} \to \mathbb{R}^n$ to equation (6.1) satisfies the funnel in equation (6.4) for all $t \geq 0$ and is hence such that $0 < r \leq \bar{\rho}^{\phi}(x,0) \leq \rho_{max}$ (i.e., $(x,0) \models \phi$).*

*Proof:* *We proceed in two steps (A and B). In step A, we use lemma 2.1 and show that there is a unique and maximal solution $x : \mathcal{I} \to \mathbb{R}^n$ to the system in equation (6.1) such that $\xi(t) \in \Omega_\xi := (-1,0)$ for all $t \in \mathcal{I}$ where $\mathcal{I} := [0, \tau_{max}) \subseteq \mathbb{R}_{\geq 0}$ with $\tau_{max} > 0$. Step B then consists of using lemma 2.2 to show that the solution $x$ is complete (i.e., that $\tau_{max} = \infty$). Using equations (6.11), (6.12), and (6.13), we can derive the transformed error dynamics of $\epsilon$ as*

$$\dot{\epsilon} = \frac{\partial \epsilon}{\partial \xi} \dot{\xi} = -\frac{1}{\gamma \xi (1+\xi)} \Big( \frac{\partial \bar{\rho}^{\psi}(x)}{\partial x} \dot{x} - \xi \dot{\gamma} \Big),$$

*where we used $\frac{\partial \epsilon}{\partial \xi} = -\frac{1}{\xi(1+\xi)}$ and $\dot{\xi} = \frac{1}{\gamma}(\dot{e} - \xi \dot{\gamma})$, and further, $\dot{e} = \frac{\partial e(x)}{\partial x} \dot{x}$ and $\frac{\partial e(x)}{\partial x} = \frac{\partial \bar{\rho}^{\psi}(x)}{\partial x}$.*

*Step A: Define the stacked vector $y := \begin{bmatrix} x^T & \xi \end{bmatrix}^T$. Consider the closed-loop system that is obtained by inserting the control law in equation (6.14) into* *the system dynamics in equation (6.1), resulting in the closed-loop dynamics* $\dot{x} = H_1(x,\xi,t)$*, where*

$$H_1(x,\xi,t) := f(x) - \ln\Big(-\frac{\xi+1}{\xi}\Big) g(x) g(x)^T \frac{\partial \bar{\rho}^{\psi}(x)}{\partial x}^T + w(t).$$

*We also obtain the normalized error dynamics* $\dot{\xi} = H_2(x,\xi,t)$*, where*

$$H_2(x,\xi,t) := \frac{1}{\gamma(t)}\Big(\frac{\partial \bar{\rho}^{\psi}(x)}{\partial x} H_1(x,\xi,t) - \xi\dot{\gamma}(t)\Big),$$

*which finally results in the stacked dynamics* $\dot{y} = H(y,t)$*, where*

$$H(y,t) := \Big[H_1(x,\xi,t)^T \quad H_2(x,\xi,t)\Big]^T.$$

*By construction of* $\rho_{max}$ *and* $\gamma_0$*, we get that* $\bar{\rho}^{\psi}_{opt} < \rho_{max}$ *and* $-\gamma_0 + \rho_{max} < \bar{\rho}^{\psi}(x_0)$*. Consequently, the initial condition* $x_0$ *is such that* $\xi(x_0,0) \in \Omega_\xi := (-1,0)$ *so that* $x_0 \in \Omega_x(0)$*, where the set of feasible states within the funnel* $\Omega_x(t)$ *is defined as*

$$\Omega_x(t) := \Big\{x \in \mathbb{R}^n \,|\, -1 < \xi(x,t) = \frac{\bar{\rho}^{\psi}(x) - \rho_{max}}{\gamma(t)} < 0\Big\}.$$

*The time-varying set* $\Omega_x$ *is shrinking over time since* $\gamma$ *is nonincreasing; that is, it holds that* $\Omega_x(t_1) \supseteq \Omega_x(t_2)$ *for* $t_1 < t_2$*. By construction of* $\gamma_\infty$*, note that* $\Omega_x(t)$ *is always nonempty since* $-\gamma_\infty + \rho_{max} < \bar{\rho}^{\psi}_{opt} < \rho_{max}$*, so that the funnel in equation (6.4) is always well defined. Further, note that* $\Omega_x(t)$ *is bounded due to bounded superlevel sets of* $\bar{\rho}^{\psi}$ *according to assumption 6.3.*

*We note that the inverse image of an open (closed) set under a continuous function is open (closed); for instance, see [26, proposition 1.4.4]. With* $\xi_0(x) := \xi(x,0)$*, it hence holds that the inverse image* ${\xi_0}^{-1}(\Omega_\xi) = \Omega_{x_0} := \Omega_x(0)$ *is open. Define the open, bounded, and nonempty set* $\Omega_y := \Omega_{x_0} \times \Omega_\xi$ *for which it holds that* $y_0 := \Big[{x_0}^T \quad \xi(x_0,0)\Big]^T \in \Omega_y$*. Next, we need to check the conditions of lemma 2.1 for the initial value problem (IVP)* $\dot{y} = H(y,t)$ *with* $y_0 \in \Omega_y$ *and* $H : \Omega_y \times \mathbb{R}_{\geq 0} \to \mathbb{R}^{n+1}$*. We can see that* $H(y,t)$ *is (1) locally Lipschitz continuous in* $y$ *since* $f(x)$*,* $g(x)$*,* $\frac{\partial \bar{\rho}^{\psi}(x)}{\partial x}$*, and* $\epsilon = \ln\big(-\frac{\xi+1}{\xi}\big)$ *are locally Lipschitz continuous in* $y$*; and (2) continuous in* $t$ *(for each fixed* $y \in \Omega_y$*) due to continuity of* $\gamma(t)$ *and* $\dot{\gamma}(t)$*. Since* $\Omega_y$ *is nonempty and open, there is a unique and maximal solution* $y : \mathcal{I} \to \Omega_y$ *with* $\mathcal{I} := [0, \tau_{max})$ *and* $\tau_{max} > 0$ *so that* $\xi(t) \in \Omega_\xi$ *and* $x(t) \in \Omega_{x_0}$ *for all* $t \in \mathcal{I}$*.*

*Step B: From step A, we know that* $y(t) \in \Omega_y$ *for all* $t \in \mathcal{I}$*. We now show that* $y$ *is complete (i.e., that* $\tau_{max} = \infty$*), by contradiction of lemma 2.2. Therefore, assume that* $\tau_{max} < \infty$*.*

*First, consider all states $x \in \Omega_{x_0}$ such that $\frac{\partial \bar{\rho}^{\psi}(x)}{\partial x} = 0$. It then holds that $\bar{\rho}^{\psi}(x) = \bar{\rho}^{\psi}_{opt}$ since $\bar{\rho}^{\psi}$ is concave by assumption. Recall that $-\gamma_{\infty} + \rho_{max} < \bar{\rho}^{\psi}_{opt} < \rho_{max}$ by design, so that $x$ is in the interior of the set $\Omega_x(t)$ for all $t \in \mathcal{I}$. Hence, there is a neighborhood $\mathcal{U}(x)$ around $x$ and some positive constant $\zeta > 0$ with $\frac{\zeta}{\gamma_0} \in (0,1)$, such that $-\gamma(t) + \rho_{max} \leq \bar{\rho}^{\psi}(x') - \zeta$ and $\bar{\rho}^{\psi}(x') + \zeta \leq \rho_{max}$ for each $x' \in \mathcal{U}(x)$. Consequently, it can be concluded that $\epsilon$ is bounded on $\mathcal{U}(x)$; that is, that $|\epsilon(x',t)| \leq \max\left(|S(-1+\frac{\zeta}{\gamma_0})|, |S(-\frac{\zeta}{\gamma_0})|\right) =: \bar{\epsilon}_1$ for all $t \in \mathcal{I}$.*

*Next, consider all states $x \in \Omega_{x_0} \setminus \overline{\mathcal{U}}$ where $\overline{\mathcal{U}}$ is the union of the regions $\mathcal{U}(x')$ from the former case for points $x' \in \Omega_{x_0}$, such that $\frac{\partial \bar{\rho}^{\psi}(x')}{\partial x} = 0$. Let us now define the Lyapunov function candidate $V(\epsilon) := \frac{1}{2}\epsilon^2$ with*

$$
\begin{aligned}
\dot{V} = \epsilon\dot{\epsilon} &= \epsilon\Big(-\frac{1}{\gamma\xi(1+\xi)}\big(\frac{\partial \bar{\rho}^{\psi}(x)}{\partial x}\dot{x} - \xi\dot{\gamma}\big)\Big) \\
&= \epsilon\alpha\Big(\frac{\partial \bar{\rho}^{\psi}(x)}{\partial x}(f(x) + g(x)u + w) - \xi\dot{\gamma}\Big),
\end{aligned}
$$

*where $\alpha := -\frac{1}{\gamma\xi(1+\xi)}$ satisfies $\alpha(t) \in [\frac{4}{\gamma_0}, \infty) \in \mathbb{R}_{>0}$ for all $t \in \mathcal{I}$. This follows since $\frac{4}{\gamma_0} \leq -\frac{1}{\gamma_0\xi(1+\xi)} \leq -\frac{1}{\gamma\xi(1+\xi)} \leq -\frac{1}{\gamma_{\infty}\xi(1+\xi)} < \infty$ for $\xi \in \Omega_{\xi}$. Consequently,*

$$
\begin{aligned}
\dot{V} &\leq |\epsilon|\alpha\big(\|\frac{\partial \bar{\rho}^{\psi}(x)}{\partial x}\|\|f(x) + w\| + |\xi\dot{\gamma}|\big) + \epsilon\alpha\frac{\partial \bar{\rho}^{\psi}(x)}{\partial x}g(x)u \\
&\leq |\epsilon|\alpha k_1 + \epsilon\alpha\frac{\partial \bar{\rho}^{\psi}(x)}{\partial x}g(x)u, \qquad (6.15)
\end{aligned}
$$

*where the positive constant $k_1$ is derived as follows: $\|f(x)\|$ and $\|\frac{\partial \bar{\rho}^{\psi}(x)}{\partial x}\|$ are upper bounded due to the continuity of $f(x)$ and $\frac{\partial \bar{\rho}^{\psi}(x)}{\partial x}$ on $cl(\Omega_{x_0})$, where $cl$ denotes the closure operation, and since $cl(\Omega_{x_0})$ is a compact set. Note that $x(t) \in \Omega_x(t) \subseteq \Omega_{x_0}$ for all $t \in \mathcal{I}$. Furthermore, $w \in \mathcal{W}$, $\xi \in \Omega_{\xi}$, and $\dot{\gamma}$ are bounded and thus the upper bound $k_1$ follows. Next, insert the control law in equation (6.14) into equation (6.15), which results in*

$$
\dot{V} \leq |\epsilon|\alpha k_1 - \epsilon^2\alpha\frac{\partial \bar{\rho}^{\psi}(x)}{\partial x}g(x)g(x)^T\frac{\partial \bar{\rho}^{\psi}(x)}{\partial x}^T \leq |\epsilon|\alpha\big(k_1 - |\epsilon|\lambda_{min}\|\frac{\partial \bar{\rho}^{\psi}(x)}{\partial x}\|^2\big),
$$

*where $\lambda_{min} > 0$ is the minimum eigenvalue of $g(x)g^T(x)$, which is positive according to assumption 6.1. It further holds that $0 < k_2 \leq \|\frac{\partial \bar{\rho}^{\psi}(x)}{\partial x}\|^2$ for a positive constant $k_2$ since we consider, at this point, only $x \in \Omega_{x_0} \setminus \overline{\mathcal{U}}$. Hence,*

$$
\dot{V} \leq |\epsilon|\alpha(k_1 - |\epsilon|\lambda_{min}k_2),
$$

*such that $\dot{V} \leq 0$ if $\bar{\epsilon}_2 := \frac{k_1}{\lambda_{min}k_2} \leq |\epsilon|$. Consequently, the solution $|\epsilon(t)|$ cannot grow beyond $\bar{\epsilon}_2$ in the region $\Omega_{x_0} \setminus \overline{\mathcal{U}}$. It can be concluded, therefore, that $|\epsilon(t)|$ is bounded as $|\epsilon(t)| \leq \max(|\epsilon(0)|, \bar{\epsilon}_1, \bar{\epsilon}_2)$, which leads to the conclusion that $\epsilon(t)$ is* *upper- and lower-bounded by some constants $\epsilon_u$ and $\epsilon_l$ (i.e., that $\epsilon_l \leq \epsilon(t) \leq \epsilon_u$ for all $t \in \mathcal{I}$). By using the inverse of $S$, the normalized error $\xi(t)$ can be bounded as $-1 < \xi_l := -\frac{1}{\exp(\epsilon_l+1)} \leq \xi(t) \leq \xi_u := -\frac{1}{\exp(\epsilon_u+1)} < 0$ for all $t \in \mathcal{I}$; that is, that*

$$\xi(t) \in [\xi_l, \xi_u] =: \Omega'_\xi \subset \Omega_\xi \text{ for all } t \in \mathcal{I},$$

*and consequently implies that $x(t) \in \Omega'_x(t)$ for all $t \in \mathcal{I}$, where*

$$\Omega'_x(t) := \Big\{x \in \Omega_x | \xi_l \leq \xi(x,t) = \frac{\bar{\rho}^\psi(x) - \rho_{max}}{\gamma(t)} \leq \xi_u\Big\}.$$

*Note in particular that $\Omega'_x(t) \subset \Omega_x(t) \subseteq \Omega_{x0}$ and $\Omega'_x(t)$ is nonempty for each $t \geq 0$ since $\Omega_x(t)$ is nonempty for each $t \geq 0$ and $\xi(x,t)$ is continuous. Also, note that it can be shown that $\Omega'_x(t)$ is compact by again using [26, proposition 1.4.4]. By defining $\Omega'_{x0} := \Omega'_x(0)$, it can be concluded that*

$$x(t) \in \Omega'_{x0} \subset \Omega_{x0} \text{ for all } t \in \mathcal{I}.$$

*Define the compact set $\Omega'_y := \Omega'_{x0} \times \Omega'_\xi$ and notice that $\Omega'_y \subset \Omega_y$, by which it follows that there is no $t \in \mathcal{I} := [0, \tau_{max})$ such that $y \notin \Omega'_y$. By contradiction of lemma 2.2, it follows that $\tau_{max} = \infty$ (i.e., that $\mathcal{I} = \mathbb{R}_{\geq 0}$). The control law $u(x,t)$ is locally Lipschitz continuous and bounded because $\bar{\rho}^\psi(x)$, $\epsilon(x,t)$, $g(x)$, and $\gamma(t)$ are locally Lipschitz continuous and bounded on $\Omega'_x$ and $\mathcal{I}$.*

The funnel control law in equation (6.14) resembles a high gain controller in which the gain $\epsilon(x,t)$ increases as the boundary of the funnel is approached. The gradient $\frac{\partial \bar{\rho}^\psi(x)}{\partial x}$, on the other hand, determines the directionality in which the system should be moving.

Some remarks are in order here. If the input dimension is equal to the state dimension (i.e., if $m = n$), then it can be shown that the control law in equation (6.14) in theorem 6.2 can be replaced by $u(x,t) := -\epsilon(x,t) \frac{\partial \bar{\rho}^\psi(x)}{\partial x}^T$, which does not use any information of the input dynamics $g$ and is thus model-free. We also recall that the disturbance $w$ has to be bounded, and larger disturbances $w$ may result in larger control inputs as indicated by the constant $k_1$ in the proof of theorem 6.2. In the same spirit, note that a steep performance function $\gamma$ may result in a large, although bounded control input. Therefore, it may be advisable to select $t_*$ to be as big as possible and $\gamma_0$ and $l$ to be as small as possible. We next present a simple simulation example, including code, that can be used to reproduce the results.

---

**Example 6.3** Consider $N := 3$ agents with single integrator dynamics that deploy a consensus protocol [209] with additional free inputs. In other words, each agent $i \in \{1, \dots, N\}$ is subject to the dynamics $\dot{x}_i = -\sum_{j\in\mathcal{N}_i}(x_i - x_j) + u_i$ with position $x_i := \begin{bmatrix} p_{x,i} & p_{y,i}\end{bmatrix}^T \in \mathbb{R}^2$, where $\mathcal{N}_i$ denotes the neighborhood of agent $i$. Using the graph Laplacian $L$ [209], we can express the stacked agent dynamics compactly as

$$\dot{x}(t) = -(L \otimes I_2)x(t) + u(t).$$

In this case, let the Laplacian and the initial positions be $L = \begin{bmatrix} 1 & -1 & 0 \\ -1 & 2 & -1 \\ 0 & -1 & 1 \end{bmatrix}$ and $x_1(0) := \begin{bmatrix}1.1 & 3.1\end{bmatrix}^T$, $x_2(0) := \begin{bmatrix}2 & 0.5\end{bmatrix}^T$, and $x_3(0) := \begin{bmatrix}7 & 1.5\end{bmatrix}^T$. We consider one goal position $A$, located at $p_A := \begin{bmatrix}6 & 4\end{bmatrix}^T$, and use $(\|x_i - p_A\|_\infty < c) = \big(|p_{x,i} - p_{x,A}| < c\big) \wedge \big(|p_{y,i} - p_{y,A}| < c\big) = \big(p_{x,i} - p_{x,A} < c\big) \wedge \big(-p_{x,i} + p_{x,A} < c\big) \wedge \big(p_{y,i} - p_{y,A} < c\big) \wedge \big(-p_{y,i} + p_{y,A} < c\big)$ to ensure that $\|x_i - p_A\|_\infty = \max(|p_{x,i} - p_{x,A}|, |p_{y,i} - p_{y,A}|) < c$.

Robot 1 is supposed to move to goal position A within $7-10$ seconds. This is encoded via the specification $\phi_1 := F_{[7,10]}\psi_1$, with $\psi_1 := \big(\|x_1 - p_A\|_\infty < 0.1\big) \wedge \psi$ where $\psi := (\|x\|_\infty < 100)$ enforces assumption 6.3. The simulation result is shown in figure 6.2. Specifically, figure 6.2(a) shows the robot trajectories, where robot 1 reaches region $A$ while the other agents follow due to the consensus protocol. The funnel itself is illustrated in figure 6.2(b).

We present the following code to reproduce these simulations where we, for simplicity, have not included the formula $\psi$. The code can be used as a starting point for more sophisticated simulations, such as presented in section 6.3. We note that we have included a gain $k$ in line 24 of the code.

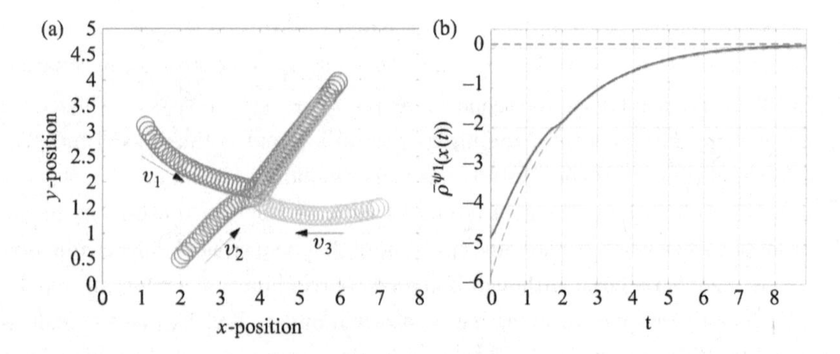


**Figure 6.2**
Simulation results for example 6.3 for a single funnel. (a) Robot trajectories for the specification $\phi_1$. (b) Smooth quantitative semantics $\rho^{\psi_1}(x(t))$. Reprinted with permission from [185].

The reason for this is that we would like to account for errors resulting from the digital implementation of the continuous-time feedback control law in equation (6.14). For instance, a zero-order hold implementation may result in $\bar{\rho}^{\psi_1}$ jumping outside the funnel where the control input $u$ is not defined. It turns out that inserting a sufficiently large gain $k$ into the control law in equation (6.14) as $u(x,t) := -\kappa\epsilon(x,t)g(x)^T \frac{\partial\bar{\rho}^{\psi}(x)}{\partial x}^T$ prevents such issues without affecting the theoretical guarantees in theorem 6.2.

```
1  clc; clear all; close all;
2
3  t          = linspace(0, 9, 901); % time
4  step       = t(2)-t(1); % time increment
5  [A,B,]     = load_system(); % system
6  x(:,1)     = [110 ; 310 ; 200 ; 50 ; 700 ; 150 ]; % initial
        condition
7
8  % Compute funnel parameters
9  rho_max        = 11;
10 t_star         = 8.5;
11 r              = 5;
12 [rho,drhodx]=  get_rho(x);
13 gamma_0        = rho_max-rho+100;
14 gamma_inf      = 0.05;
15 l              = -log(-(r+gamma_inf-rho_max)/(gamma_0-gamma_inf)
        )/t_star;
16
17 for n=2:length(t)
18     % Calculate control input
19     gamma    = (gamma_0-gamma_inf)*exp(-l*t(n))+gamma_inf;
20     [ rho,drhodx ] = get_rho(x(:,n-1));
21     e        = rho-rho_max; % error
22     xi       = e/gamma; % normalized error
23     epsilon  = log(-(xi+1)/xi); % transformed error
24     k        = 50; % gain
25     u(:,n-1)= -epsilon*k*B.'*drhodx;
26
27     % Euler step
28     x(:,n)   = x(:,n-1)+step*(A*x(:,n-1)+B*u(:,n-1));
29 end
30
31 function [ rho,drhodx ] = get_rho( x )
32     c    = 10; ax = 600; ay = 400;
33     z1   = c+ax-x(1);
34     z2   = c-ax+x(1);
35     z3   = c+ay-x(2);
36     z4   = c-ay+x(2);
37     rho  = -log(exp(-z1)+exp(-z2)+exp(-z3)+exp(-z4));
38     den  = exp(-z1)+exp(-z2)+exp(-z3)+exp(-z4);
39
40     drhodx = [-(exp(-z1)-exp(-z2))/den; -(exp(-z3)-exp(-z4))/
           den ; zeros(4,1)];
41 end
```

Finally, let us discuss the case where assumption 6.2 does not hold. In this case, recall that the formula $\phi$ is not satisfiable. Therefore, let $\rho_{\text{gap}} > 0$ be a parameter indicating how much $\phi$ may be violated. If the solution $x : \mathbb{R}_{\geq 0} \to \mathbb{R}^n$

to the system in equation (6.1) is such that $\bar{\rho}^{\psi}_{\text{opt}} - \rho_{\text{gap}} \leq \bar{\rho}^{\phi}(x,0) < \bar{\rho}^{\psi}_{\text{opt}}$, we say that $x$ is a *least-violating solution* with a gap of $\rho_{\text{gap}}$.

**Corollary 6.1 (Least-Violating Solutions via Funnel Control)** *Let all conditions from theorem 6.2 hold except for assumption 6.2 and the choice of $r$. Instead, let $r \in [\bar{\rho}^{\psi}_{opt} - \rho_{gap}, \bar{\rho}^{\psi}_{opt})$, where $\rho_{gap} > 0$. Then the control law in equation (6.14) guarantees that the solution $x : \mathcal{I} \to \mathbb{R}^n$ to (6.1) is such that $r \leq \bar{\rho}^{\phi}(x,0) \leq \rho_{max}$ (i.e., a least-violating solution with a gap of $\rho_{gap}$ is found).*

## 6.3 Control Laws Based on Multiple Funnels

We will now design control strategies that use multiple funnels to satisfy more complex STL specifications. In the first subsection, we consider specifications that can be satisfied by sequentially applying funnel control laws according to equation (6.14), such as when we consider conjunctions of atomic temporal formulas without overlapping time intervals. In the second subsection, we consider the case where multiple funnels have to be enforced at the same time; that is, when we consider until operators or conjunctions of atomic temporal formulas with overlapping time intervals.

### 6.3.1 Sequential Application of Funnel Control Laws

A straightforward idea is to sequentially apply the funnel control law in equation (6.14). For instance, we can do so for sequential formulas as follows:

$$\theta := F_{[a_1,b_1]}(\psi_1 \wedge F_{[a_2,b_2]}(\psi_2 \wedge F_{[a_3,b_3]}\psi_3)).$$

Note here that the main challenge is to design the funnel for the formula $F_{[a_2,b_2]}\psi_2$ online once the formula $F_{[a_1,b_1]}\psi_1$ is satisfied, and then do the same for the funnel for the formula $F_{[a_3,b_3]}\psi_3$ once the formula $F_{[a_2,b_2]}\psi_2$ is satisfied.

---

**Example 6.4** Consider the setting from example 6.3 and let us also consider the goal positions $B$, $C$, $D$, and $E$ that are located at $p_B := \begin{bmatrix}1.2 & 9\end{bmatrix}^T$, $p_C := \begin{bmatrix}1.2 & 7\end{bmatrix}^T$, $p_D := \begin{bmatrix}1.2 & 5\end{bmatrix}^T$, and $p_E := \begin{bmatrix}8 & 7\end{bmatrix}^T$. The robots are subject to the following sequential tasks: (1) robot 1 moves to $A$ within $7-10$

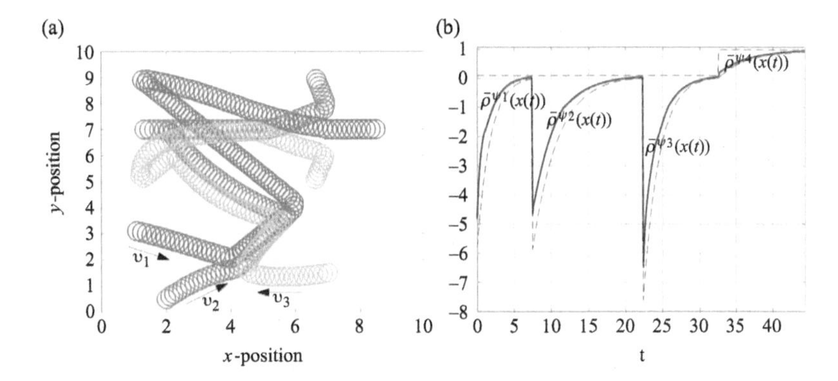


**Figure 6.3**
Simulation results for example 6.4 for multiple sequential funnels. (a) Robot trajectories for task $\theta$. (b) Smooth quantitative semantics. Reprinted with permission from [185].

seconds; (2) within the next $10-20$ seconds, robots 1, 2, and 3 move to $B$, $C$, and $D$, respectively; (3) then robot 1 moves to $E$ within the next $5-15$ seconds, while robots 1, 2, and 3 form a triangular formation; and (4) for the next 12 seconds, robots 2 and 3 always keep a distance of at least 1 unit from robot 1 and disperse. More specifically, we have

$$\theta := F_{[7,10]}(\psi_1 \wedge F_{[10,20]}(\psi_2 \wedge F_{[5,15]}(\psi_3 \wedge \phi_4))),$$

with $\psi_1 := (\|x_1 - p_A\|_\infty < 0.1)$, $\psi_2 := (\|x_1 - p_B\|_\infty < 0.1) \wedge (\|x_2 - p_C\|_\infty < 0.1) \wedge (\|x_3 - p_D\|_\infty < 0.1)$, $\psi_3 := (\|x_1 - p_E\|_\infty < 0.1) \wedge (1 < p_{x,1} - p_{x,2} < 1.2) \wedge (1 < p_{x,1} - p_{x,3} < 1.2) \wedge (1 < p_{y,2} - p_{y,1} < 1.2) \wedge (1 < p_{y,1} - p_{y,3} < 1.2)$, and $\phi_4 := G_{[0,12]}((1 < p_{x,1} - p_{x,2}) \wedge (1 < p_{y,2} - p_{y,1}) \wedge (1 < p_{x,1} - p_{x,3}) \wedge (1 < p_{y,1} - p_{y,3}))$.

The simulation result for all four tasks is displayed in figure 6.3, where the trajectories are shown in figure 6.3(a), while the funnels are shown in figure 6.3(b). We see that $\theta$ is satisfied with a robustness of $r > 0.05$. Note that due to the precision that we chose, such as 0.1 in $\phi_1 = F_{[7,10]}(\|\boldsymbol{x}_1 - p_A\|_\infty < 0.1)$, $r$ cannot exceed 0.1.

---

In a similar spirit, we can satisfy temporal formulas $\theta$ from the fragment in equation (6.3) if the time intervals are not overlapping. Therefore, let us consider the formulas

$$\phi_k ::= F_{[a_k,b_k]}\psi_k \mid G_{[a_k,b_k]}\psi_k \mid F_{[\underline{a}_k,\underline{b}_k]}G_{[\bar{a}_k,\bar{b}_k]}\psi_k$$
$$\theta ::= \bigwedge_{k=1}^{K} \phi_k,$$

where we assume that $b_k < a_{k+1}$ such that time intervals are not overlapping. We introduce the quantitative semantics of $\theta$ as

$$\bar{\rho}^{\theta}(x,t) := \min_{k\in\{1,\ldots,K\}} \bar{\rho}^{\phi_k}(x,t).$$

The idea is again to develop a switching strategy to apply and combine the previously introduced, funnel-based control laws, while computing the new funnel online.

---

**Example 6.5** We consider a modified version of the Lotka-Volterra equations for predator-prey models. Let $x_1$, $x_2$, and $x_3$ be three species, where $x_3$ is a predator hunting $x_1$ and $x_2$ and $x_2$ is a predator hunting $x_1$. To have a more specific example, let $x_1$, $x_2$, and $x_3$ represent fish population densities within an aquarium ($x_1$: nonpredatory fish, $x_2$: trout, and $x_3$: pike). The dynamics are given by

$$\begin{aligned}\dot{x}_1 &= \beta_1 x_1 - \hat{\beta}_{1,2} x_1 x_2 - \hat{\beta}_{1,3} x_1 x_3 + u_1\\ \dot{x}_2 &= \beta_{2,1} x_1 x_2 - \hat{\beta}_{2,3} x_2 x_3 + u_2\\ \dot{x}_3 &= \beta_{3,1} x_1 x_3 + \beta_{3,2} x_2 x_3 - \hat{\beta}_3 x_3 + u_3,\end{aligned}$$

where $\beta_1 x_1$ is the growth rate of $x_1$, while $\beta_{2,1} x_1$ and $\beta_{3,1} x_1 + \beta_{3,2} x_2$ are the growth rates of $x_2$ and $x_3$, respectively. The terms $\hat{\beta}_{1,2} x_1 x_2$ and $\hat{\beta}_{1,3} x_1 x_3$ measure the impact of predation by $x_2$ and $x_3$ on the population of $x_1$, while $\hat{\beta}_{2,3} x_2 x_3$ measures the impact of predation by $x_3$ on the population of $x_2$. The parameter $\hat{\beta}_3$ indicates the death rate of $x_3$. The system, with stacked states $x := \begin{bmatrix} x_1 & x_2 & x_3 \end{bmatrix}^T$ and inputs $u := \begin{bmatrix} u_1 & u_2 & u_3 \end{bmatrix}^T$, can be written as $\dot{x} = f(x) + g(x)u$, where

$$f(x) := \begin{bmatrix} \beta_1 x_1 - \hat{\beta}_{1,2} x_1 x_2 - \hat{\beta}_{1,3} x_1 x_3 \\ \beta_{2,1} x_1 x_2 - \hat{\beta}_{2,3} x_2 x_3 \\ \beta_{3,1} x_1 x_3 + \beta_{3,2} x_2 x_3 - \hat{\beta}_3 x_3 \end{bmatrix} \quad \text{and} \quad g(x) := I_3,$$

with $I_3$ being the $3 \times 3$-identity matrix. The inputs $u_1$, $u_2$, and $u_3$ represent the actions of adding fish. Let the initial state be $x_0 := \begin{bmatrix} 100 & 50 & 10 \end{bmatrix}^T$, while the parameters $\beta_1 := 21.8$, $\hat{\beta}_{1,2} := \beta_{2,1} := 0.03$, $\hat{\beta}_{1,3} := \beta_{3,1} := 0.02$, $\hat{\beta}_{2,3} := \beta_{3,2} := 0.0055$, and $\hat{\beta}_3 := 4$ are selected such that the resulting trajectory is a periodic orbit when no control is used ($u := 0$); see figure 6.4. Note that this system exhibits different equilibria at $\{x \in \mathbb{R}^3 | x_1 = 0, x_3 = 0\}$ and $\{x \in \mathbb{R}^3 | x_1 = 200, x_2 = 0, x_3 = 1090\}$, and different initial conditions may not lead to a periodic orbit.

We consider an unsatisfiable STL specification $\theta$ with no overlapping time intervals, and we show how least-violating solutions with a gap of $\rho_{\text{gap}} := 30$ can be found. Assume that we want to establish certain fish populations during the first six months of the year. By accident, however,

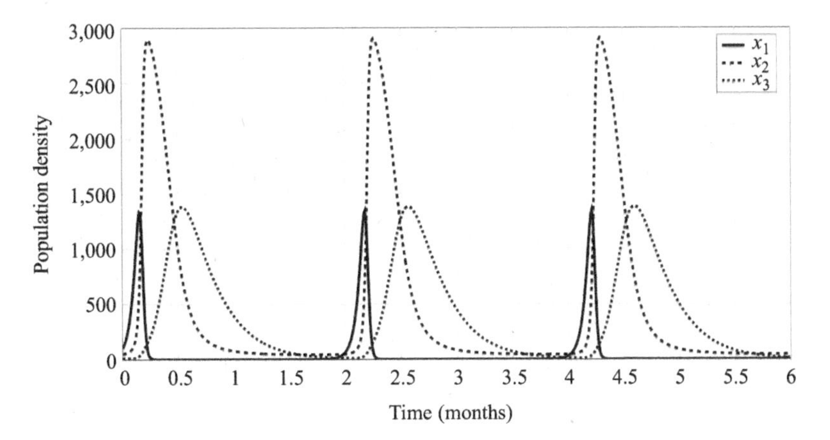


**Figure 6.4**
Population densities when no control is applied. Reprinted with permission from [177].

we impose an unsatisfiable STL specification. During all six months, there should not be too many or too few fish in the lake. During the first two months, we want to eventually establish a roughly equal density among all the fish populations. Between months two and three, it should eventually hold that $x_2 \geq 1{,}100$, while between months three and four, it should hold that $x_2 \geq 1{,}100$ and $x_3 \geq 1{,}000$. For the last two months, it should eventually hold that $x_1 \geq 300$, $x_2 \leq 200$, and $x_3 \leq 900$. Formally, the STL specification is

$$\theta := \phi_1 \wedge \phi_2 \wedge \phi_3 \wedge \phi_4$$

with the atomic temporal formulas $\phi_1 := F_{[0,2]}\psi_1$, $\phi_2 := F_{[2,3]}\psi_2$, $\phi_3 := G_{[3,4]}\psi_3$, and $\phi_4 := F_{[4,6]}\psi_4$, where the nontemporal formulas are as follows:

$$\psi_1 := (\|x - \begin{bmatrix}500 & 500 & 500\end{bmatrix}^T\| \leq 500),$$

$$\psi_2 := (\|x - \begin{bmatrix}500 & 500 & 500\end{bmatrix}^T\| \leq 500) \wedge (x_2 \geq 1{,}100),$$

$$\psi_3 := (\|x - \begin{bmatrix}500 & 500 & 500\end{bmatrix}^T\| \leq 500) \wedge (x_2 \geq 1{,}100) \wedge (x_3 \geq 1{,}000),$$

$$\psi_4 := (\|x - \begin{bmatrix}500 & 500 & 500\end{bmatrix}^T\| \leq 500) \wedge (x_1 \geq 300) \wedge (x_2 \leq 200) \wedge (x_3 \leq 900).$$

Note that the formula $\|x - \begin{bmatrix}500 & 500 & 500\end{bmatrix}^T\| \leq 500$ that is contained in $\psi_1$, $\psi_2$, $\psi_3$, and $\psi_4$ ensures that assumption 6.3 is satisfied. For $\eta = 1$, it holds that $\bar{\rho}_{\text{opt}}^{\psi_1} = 500$, $\bar{\rho}_{\text{opt}}^{\psi_2} = -50.69$, $\bar{\rho}_{\text{opt}}^{\psi_3} = -117.84$, and $\bar{\rho}_{\text{opt}}^{\psi_4} = -1.08$ (note that $\rho_{\text{opt}}^{\psi_4} = 0$). We select $t_{*,1} := 1.7$, $t_{*,2} := 2.7$, $t_{*,3} := 3$, and $t_{*,4} := 6$, respectively. Since $\rho_{\text{gap}} := 30$, we select $r_1 := 450$, $\rho_{\max,1} := 600$, $r_2 := -80$, $\rho_{\max,2} := -20$, $r_3 := -145$, $\rho_{\max,3} := -50$, $r_4 := -30$, and $\rho_{\max,4} := 5$. Based on these values, we design funnels $\gamma_1$, $\gamma_2$, $\gamma_3$, and $\gamma_4$ for the formulas $\phi_1$, $\phi_2$, $\phi_3$, and $\phi_4$, respectively.

The simulation results are shown in figure 6.5. From figure 6.5a, it follows that $\theta$ is not satisfied, which is no surprise since the formula is not satisfiable. However, we find a least-violating solution with a gap of $\rho_{\text{gap}}$ since the third subformula is the most-violating subformula with $\bar{\rho}_{\text{opt}}^{\psi_3} = -117.84$, and also since we selected $r_3 := -145$. It can be concluded that $r := -145 \leq \bar{\rho}^{\theta}(x, 0)$, which is also illustrated in figure 6.5b. Figure 6.5c also shows the bounded and piecewise continuous control inputs. The couplings in $f$ are nonlinear and quite strong, especially when the STL formula requires $x_1$ to increase. Increasing $x_1$ automatically leads to an increase of $x_2$ and $x_3$ (predators), which in turn decrease $x_1$ (prey). All simulations have been performed in real time on a four-core, 1.4-GHz central processing unit with 8 gigabytes (GB) of random access memory (RAM). The computation of $u$ took, on average, 0.1 ms as $u$ is given in closed form.

## 6.3.2 Simultaneous Satisfaction of Multiple Funnels

We now deal with until operators and conjunctions of atomic temporal formulas with overlapping time intervals by simultaneously enforcing multiple funnels.

**Until Operator.** So far, we have not dealt with the until operator

$$\theta := \psi' \, U_{[a,b]} \, \psi'',$$

where $\psi'$ and $\psi''$ are nontemporal formulas as in equation (6.2a). To satisfy $\theta$, we will consider two funnels simultaneously, one for each nontemporal formula. We will then combine these two funnels into one funnel based on which we design a funnel control law. To do so, let us decompose $\psi' \, U_{[a,b]} \, \psi''$ into $G_{[0,t_*]}\psi'$ and $F_{[t_*,t_*]}\psi''$, where $t_* \in [a, b]$. In direct analogy to subsection 6.3.1, we select

$$\rho_{\max} \in (\max(\bar{\rho}_{\text{opt}}^{\psi'}, \bar{\rho}_{\text{opt}}^{\psi''}), \infty) \tag{6.16}$$

$$r \in (0, \bar{\rho}_{\text{opt}}^{\theta}), \tag{6.17}$$

where $\bar{\rho}_{\text{opt}}^{\theta} := \sup_{x \in \mathbb{R}^n} \min \left(\bar{\rho}^{\psi'}(x), \bar{\rho}^{\psi''}(x)\right)$ is the maximum robustness of both nontemporal formulas that is naturally smaller than their individual maxima; that is, it holds that $\bar{\rho}_{\text{opt}}^{\theta} \leq \max(\bar{\rho}_{\text{opt}}^{\psi'}, \bar{\rho}_{\text{opt}}^{\psi''})$. Consequently, we will have that $r < \rho_{\max}$. We implicitly assume that $\bar{\rho}_{\text{opt}}^{\theta} > 0$ throughout this section–if not, a least-violating solution with a gap can be found similar to corollary 6.1. We also define the performance functions

$$\gamma'(t) := (\gamma_0' - \gamma_\infty') \exp(-l't) + \gamma_\infty' \text{ and}$$
$$\gamma''(t) := (\gamma_0'' - \gamma_\infty'') \exp(-l''t) + \gamma_\infty''$$

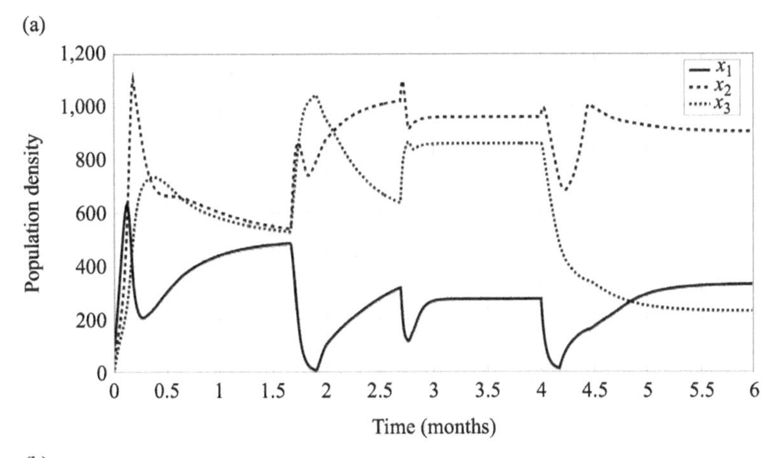


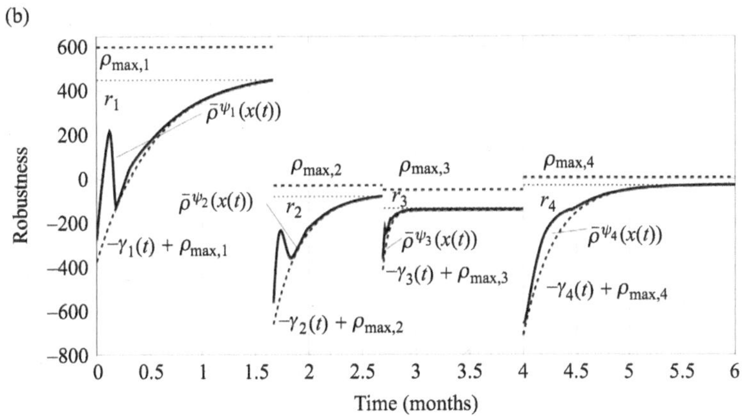


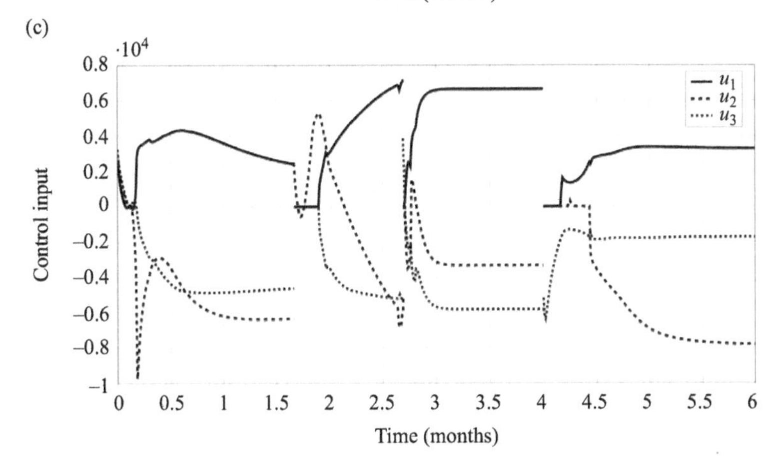


**Figure 6.5**
Simulation results for example 6.5. (a) Population densities; (b) STL robustness; (c) control inputs. Reprinted with permission from [177].

for the nontemporal formulas $\psi'$ and $\psi''$, respectively. These funnels now constrain the smooth quantitative semantics of $\bar{\rho}^{\psi'}$ and $\bar{\rho}^{\psi''}$ as

$$-\gamma'(t)+\rho_{\max}<\bar{\rho}^{\psi'}(x(t))<\rho_{\max} \tag{6.18a}$$

$$-\gamma''(t)+\rho_{\max}<\bar{\rho}^{\psi''}(x(t))<\rho_{\max} \tag{6.18b}$$

for all $t\in[0,b]$. Now, let us fold the first funnel into the second funnel by defining

$$\hat{\rho}^{\psi''}(x,t):=\bar{\rho}^{\psi''}(x)-\gamma'(t)+\gamma''(t).$$

Instead of considering the smooth quantitative semantics $\bar{\rho}^{\psi'}$ and $\bar{\rho}^{\psi''}$, we will use $\bar{\rho}^{\psi'}$ and $\hat{\rho}^{\psi''}$. In this regard, we combine $\bar{\rho}^{\psi'}$ and $\hat{\rho}^{\psi''}$ by again using a smooth approximation of the min operator as

$$\hat{\rho}(x,t):=-\frac{1}{\hat{\eta}}\ln\big(\exp\big(-\hat{\eta}\bar{\rho}^{\psi'}(x)\big)+\exp\big(-\hat{\eta}\hat{\rho}^{\psi''}(x,t)\big)\big),$$

where $\hat{\eta}>0$ determines the accuracy of this approximation. Note that $\hat{\rho}(x,t)\leq\min(\bar{\rho}^{\psi'}(x),\hat{\rho}^{\psi''}(x,t))$ with $\hat{\rho}(x,t)=\min(\bar{\rho}^{\psi'}(x),\hat{\rho}^{\psi''}(x,t))$ as $\hat{\eta}\to\infty$. We also define the optimal value $\hat{\rho}_{\text{opt}}(t):=\sup_{x\in\mathbb{R}^n}\hat{\rho}(x,t)$ for each time $t\in[0,b]$.

If we now design a control law so the single funnel

$$-\gamma'(t)+\rho_{\max}<\hat{\rho}(x(t),t)<\rho_{\max} \tag{6.19}$$

is satisfied for all $t\in[0,b]$, then we see that both funnels in equation (6.18) are satisfied for all $t\in[0,b]$ since we have $\hat{\rho}(x,t)\leq\min(\bar{\rho}^{\psi'}(x),\hat{\rho}^{\psi''}(x,t))$.

To design a funnel control law for the single funnel in equation (6.19), let us again define the error, the normalized error, and the transformed error as

$$\hat{e}(x,t):=\hat{\rho}(x,t)-\rho_{\max},$$

$$\hat{\xi}(x,t):=\frac{\hat{e}(x,t)}{\gamma'(t)},$$

$$\hat{\epsilon}(x,t):=S(\hat{\xi}(x,t))=\ln\Big(-\frac{\hat{\xi}(x,t)+1}{\hat{\xi}(x,t)}\Big).$$

We can now design a funnel control law in the same spirit as presented before, as formally stated in the next theorem of which the proof is given in section 6.5.

**Theorem 6.3 (Funnel Controller for Until Operators)** *Let the functions $f$, $g$, and $w$ that describe the system in equation (6.1) be locally Lipschitz* *continuous in $x$ and piecewise continuous in $t$, and let $g$ be such that assumption 6.1 holds. Further, let $\theta := \psi' U_{[a,b]} \psi''$, where the nontemporal formulas $\psi'$ and $\psi''$ have smooth quantitative semantics $\bar{\rho}^{\psi'}$ and $\bar{\rho}^{\psi''}$ that satisfy assumption 6.3. Let $\rho_{max}$ and $r$ be selected as in equations (6.16) and (6.17), and $\gamma'$ and $\gamma''$ be constructed as in equations (6.8)–(6.10) for $G_{[0,t_*]}\psi'$ and $F_{[t_*,t_*]}\psi''$, respectively, with $t_* \in [a,b]$. If $-\gamma'_0 + \rho_{max} < \hat{\rho}(x_0, 0)$ and $-\gamma'(t) + \rho_{max} < \hat{\rho}_{opt}(t)$ for all $t \in [0,b]$, then the control law*

$$u(x,t) := -\hat{\epsilon}(x,t) g(x)^T \frac{\partial \hat{\rho}(x,t)}{\partial x}^T \tag{6.20}$$

*guarantees that the solution $x : \mathcal{I} \to \mathbb{R}^n$ to equation (6.1) satisfies the funnel in equation (6.19) for all $t \in [0,b]$ and is hence such that $0 < r \leq \bar{\rho}^{\theta}(x,0) \leq \rho_{max}$ (i.e., $(x,0) \models \theta$).*

The important condition in theorem 6.3 is that the funnel in equation (6.19) is initially satisfied (via $-\gamma'_0 + \rho_{\max} < \hat{\rho}(x_0, 0)$) and always feasible (via $-\gamma'(t) + \rho_{\max} < \hat{\rho}_{\text{opt}}(t)$) by appropriate choices of the parameters $\rho_{\max}$, $r$, $\gamma'$, and $\gamma''$. We illustrate this in example 6.6.

---

**Example 6.6** Consider the temporal operator $\theta := \psi' U_{[0,7]} \psi''$ with $\psi' := \mu_1$ and $\psi'' := \mu_2 \wedge \mu_3 \wedge \mu_{\text{bounded}}$, where $\mu_1$, $\mu_2$, and $\mu_3$ are as in example 6.1, $\mu_{\text{bounded}} := \mu_1$ is to satisfy assumption 6.3, and where again $x_0 := \begin{bmatrix} -0.05 & 0 \end{bmatrix}^T$ as in example 6.2 so that $(x_0, 0) \models \psi'$ but $(x_0, 0) \not\models \psi''$. For $\eta := 5$, we have $\bar{\rho}^{\psi''}_{\text{opt}} = 0.31$ and $\bar{\rho}^{\psi''}(x_0) = -0.06$ so that we select $\rho_{\max} := 1.5 \in (\max(\bar{\rho}^{\psi'}_{\text{opt}}, \bar{\rho}^{\psi'}_{\text{opt}}), \infty) = (1, \infty)$. Furthermore, $\bar{\rho}^{\theta}_{\text{opt}} := \sup_{x \in \mathbb{R}^n} \min\left(\bar{\rho}^{\psi'}(x), \bar{\rho}^{\psi''}(x)\right) = \bar{\rho}^{\psi''}_{\text{opt}} = 0.31$ so that, as instructed, we select $r := 0.1 \in (0, \bar{\rho}^{\theta}_{\text{opt}})$. Let $\gamma'(t) := \gamma'_0 := 1.1$ be as derived in example 6.2 so that it remains to define $\gamma''(t)$. We select $t_* := 7 \in [0,7]$ as well as $\gamma''_0 := 3 \in (\rho_{\max} - \bar{\rho}^{\psi''}(x_0), \infty) = (1.56, \infty)$, $\gamma''_\infty := 1.25 \in (\rho_{\max} - \bar{\rho}^{\psi''}_{\text{opt}}, \min(\gamma''_0, \rho_{\max} - r)) = (1.19, 1.4)$, and $l'' := -\ln\left(\frac{r + \gamma''_\infty - \rho_{\max}}{-(\gamma''_0 - \gamma''_\infty)}\right)/t_* = 0.51$. Let $\hat{\eta} := 5$ so that $-\gamma'_0 + \rho_{\max} < \hat{\rho}(x_0, 0)$ is satisfied, and further, for all $t \in [0,7]$, it holds that $-\gamma'(t) + \rho_{\max} < \hat{\rho}_{\text{opt}}(t)$.

---

**Atomic Temporal Formulas with Overlapping Time Intervals.** Let us now look at temporal formulas $\theta$ in equation (6.3) with overlapping time intervals. The basic idea is to combine the previous results and to sequentially apply control laws that are designed to satisfy multiple simultaneous funnels. For simplicity, we treat each atomic temporal formula $\phi_k$ in $\theta := \wedge_{k=1}^{K} \phi_k$ either as an eventually or as an always formula; that is, we let $\phi_k := F_{[a_k,b_k]}\psi_k$ or $\phi_k := G_{[a_k,b_k]}\psi_k$.[1]

First, we have to determine a sequence in which we would like to process atomic temporal formulas $\phi_k$. Note that we may have to deal with multiple formulas $\phi_k$ simultaneously, e.g., in case of multiple always operators with overlapping time intervals. We will introduce algorithm 6.1 shortly, which proposes a computationally efficient heuristic for this task. Specifically, the algorithm constructs a switching sequence, a sequence of times at which the funnel control law changes, along with the times $t_{k,*}$ that are associated with $\phi_k$ and that resemble $t_*$ for $\phi$ as in section 6.2. We denote this sequence as $\mathcal{S} := \{s_0 := 0, s_1, \ldots, s_q\}$ and remark that $q$ will be finite and $q \leq K$. Next, define

$$\mathfrak{i}_k := \begin{cases} 0 & \text{if } \phi_k := F_{[a_k,b_k]}\psi_k \\ 1 & \text{if } \phi_k := G_{[a_k,b_k]}\psi_k \end{cases}$$

and denote $\mathfrak{i}_k[a_k, b_k] := \emptyset$ if $\mathfrak{i}_k = 0$ and $\mathfrak{i}_k[a_k, b_k] := [a_k, b_k]$ if $\mathfrak{i}_k = 1$.

Since atomic temporal formulas $\phi_k$ may be independent of each other, we separate them into independent subsets. The set $\Xi \subseteq \{1, \ldots, K\}$ is a dependency cluster if there is no other set $\bar{\Xi} \subseteq \{1, \ldots, K\}$ such that the satisfaction of $\phi_k$ for some $k \in \Xi$ depends on the satisfaction of $\phi_{\bar{k}}$ for some $\bar{k} \in \bar{\Xi}$. Without loss of generality, let $\Xi_1$, $\Xi_2$, ... correspond to $\{\phi_1, \ldots, \phi_{|\Xi_1|}\}$, $\{\phi_{|\Xi_1|+1}, \ldots, \phi_{|\Xi_1|+|\Xi_2|}\}$, ..., respectively, and assume that $L$ dependency clusters exist in total. Algorithm 6.1 iterates through each dependency cluster $\Xi_l$ with $l \in \{1, \ldots, L\}$ (lines 2–16) to determine, for each $\phi_k$ for which $k \in \Xi_l$, candidate $t_{k,*}$ for $\phi_k = F_{[a_k,b_k]}\psi_k'$ (lines 5–12), while $t_{k,*} := a_k$ if $\phi_k = G_{[a_k,b_k]}\psi_k'$ (line 15). The idea is, for the eventually operators, to find a $t_{k,*}$ that does not intersect with other time intervals of temporal operators in this cluster (lines 6–8). In particular, $\mathcal{D}_1$ denotes the union of all time intervals of always operators in $\Xi_l$, while $\mathcal{D}_2$ denotes the union of all $t_{k,*}$ already determined for eventually operators in $\Xi_l$. The switching sequence $\mathcal{S}$ is constructed in lines 1, 13, and 16.

**Algorithm 6.1** *1:* $\mathcal{S} = \{0\}$

*2:* ***for*** $l := 1$, $l := l+1$, *while* $l \leq L$ ***do***

*3:* $N := \sum_{j=1}^{l-1} |\Xi_j|$

[1] We avoid considering $F_{[\underline{a}_k,\underline{b}_k]}G_{[\bar{a}_k,\bar{b}_k]}\psi_k$ or $\psi_k' \, U_{[a_k,b_k]} \, \psi_k''$ because they can be treated as a combination of eventually and always operators, as argued before.

4: *for* $k := N+1,\ k := k+1,$ *while* $k \leq N + |\Xi_l|$ *do*

5: *if* $\mathrm{i}_k = 0$ *then*

6: $\mathcal{D}_1 := \bigcup_{j \in \Xi_l \setminus \{k\}} \mathrm{i}_j [a_j, b_j]$

7: $\mathcal{D}_2 := \bigcup_{j=N+1}^{k-1} (1 - \mathrm{i}_j) t_{j,*}$

8: $\mathcal{K} := [a_k, b_k] \setminus \{\mathcal{D}_1 \cup \mathcal{D}_2\}$

9: *if* $\mathcal{K} \neq \emptyset$ *then*

10: $t_{k,*} \in \mathcal{K}$

11: *else*

12: $t_{k,*} \in [a_k, b_k]$

13: $\mathcal{S} := \mathcal{S} \cup \{t_{k,*}\}$

14: *else*

15: $t_{k,*} := a_k$

16: $\mathcal{S} := \mathcal{S} \cup \{b_k\}$

Based on the switching sequence $\mathcal{S}$, we now propose a switched control law $u(x,t)$; that is, $u(x,t)$ is discontinuous at $t = s_j$ for each $j \in \{0, 1, \ldots, q\}$. At $s_j$, we consider, for the calculation of $u(x,t)$, all atomic temporal formulas that are active during $(s_j, s_{j+1}]$. Let $\mathcal{A}_j$ denote the set of indices corresponding to active atomic temporal formulas during $[s_j, s_{j+1}]$. In particular, $k \in \mathcal{A}_j$ if $[a_k, b_k] \cap (s_j, s_{j+1}] \neq \emptyset$ if $\phi_k := G_{[a_k, b_k]} \psi_k$, or if $t_{k,*} \cap (s_j, s_{j+1}] \neq \emptyset$ if $\phi_k := F_{[a_k, b_k]} \psi_k$.

**Example 6.7** Consider $x := \begin{bmatrix} x_1 & x_2 & x_3 & x_4 \end{bmatrix}^T \in \mathbb{R}^4$ and the atomic temporal formulas $\phi_1 := G_{[0,5]} \mu_1$, $\phi_2 := F_{[2,6]} \mu_2$, $\phi_3 := F_{[1,7]} \mu_3$, and $\phi_4 := G_{[6,7]} \mu_4$ with associated predicate functions $h_1(x) := x_1 - 2$, $h_2(x) := x_1 - x_2$, $h_3(x) := x_3 - x_4$, and $h_4(x) := 2 - x_4$. The dependency clusters are hence $\Xi_1 := \{v_1, v_2\}$ and $\Xi_2 := \{v_3, v_4\}$. By using algorithm 6.1, we obtain, for instance, $t_{1,*} := 0$, $t_{2,*} := 5.5$, $t_{3,*} := 2$, and $t_{4,*} := 6$ and correspondingly $\mathcal{S} := \{0, 2, 5, 5.5, 7\}$. Furthermore, $\mathcal{A}_0 = \{1, 3\}$, $\mathcal{A}_1 = \{1\}$, $\mathcal{A}_2 = \{2\}$, and $\mathcal{A}_3 = \{4\}$.

What remains to be done is to design the funnel control law that is applied during each time interval $[s_j, s_{j+1})$. Similar to how we dealt with the until operator, define

$$\rho^j_{\max} \in (\max_{k \in \mathcal{A}_j} \bar{\rho}^{\psi_k}_{\text{opt}}, \infty),$$

$$r^j \in (0, \bar{\rho}^j_{\text{opt}}),$$

where $\bar{\rho}^j_{\text{opt}} := \sup_{x \in \mathbb{R}^n} \min_{k \in \mathcal{A}_j} \bar{\rho}^{\psi_k}(x)$. We again implicitly assume that $\bar{\rho}^j_{\text{opt}} > 0$ throughout this section—if not, a least-violating solution with a gap can be found in a similar way as corollary 6.1. We consider the performance function $\gamma_k^j(t) := (\gamma_{k,0}^j - \gamma_{k,\infty}^j)\exp(-l_k^j t) + \gamma_{k,\infty}^j$ for the atomic temporal formula $\phi_k$ for each $k \in \mathcal{A}_j$ with

$$\gamma_{k,0}^j \in \begin{cases} (\rho_{\max}^j - \bar{\rho}^{\psi_k}(x(s_j)), \infty) & \text{if } t_{k,*} > s_j \\ (\rho_{\max}^j - \bar{\rho}^{\psi_k}(x(s_j)), \rho_{\max}^j - r^j] & \text{if } t_{k,*} = s_j \end{cases}$$

$$\gamma_{k,\infty}^j \in \left(\rho_{\max}^j - \bar{\rho}_{\text{opt}}^j, \min\left(\gamma_{k,0}^j, \rho_{\max}^j - r^j\right)\right)$$

$$l^j \in \begin{cases} 0 & \text{if } -\gamma_{k,0}^j + \rho_{\max}^j \geq r^j \\ -\dfrac{\ln\left(\frac{r^j + \gamma_{k,\infty}^j - \rho_{\max}^j}{-(\gamma_{k,0}^j - \gamma_{k,\infty}^j)}\right)}{t_{k,*} - s_j} & \text{if } -\gamma_{k,0}^j + \rho_{\max}^j < r^j. \end{cases}$$

Similar to the funnel in equation (6.18) that we considered for the until operator, we would now like that the funnel

$$-\gamma_k^j(t) + \rho_{\max}^j < \bar{\rho}^{\psi_k}(x(t)) < \rho_{\max}^j$$

is satisfied for each atomic temporal formula $k \in \mathcal{A}_j$ and for all times $t \in [s_j, s_{j+1})$. We select a single index $k' \in \mathcal{A}_j$ and define, for each $k \in \mathcal{A}_j$, the function

$$\hat{\rho}^{\psi_k}(x,t) := \bar{\rho}^{\psi_k}(x) - \gamma_{k'}^j(t) + \gamma_k^j(t)$$

for $t \in [s_j, s_{j+1})$. Similar to the funnel in equation (6.19) for the until operator, we define

$$\hat{\rho}(x,t) := -\frac{1}{\hat{\eta}} \ln\left(\exp \sum_{k \in \mathcal{A}_j} \left(-\hat{\eta}\hat{\rho}^{\psi_k}(x,t)\right)\right),$$

where $\hat{\eta} > 0$, and we aim at satisfying the single funnel

$$-\gamma_{k'}^j(t) + \rho_{\max}^j < \hat{\rho}(x(t),t) < \rho_{\max}^j \tag{6.21}$$

for all times $t \in [s_j, s_{j+1})$. Inspired by theorem 6.3, we select the control law

$$u(x,t) := -\hat{\epsilon}(x, t - s_j) g(x)^T \frac{\partial \hat{\rho}(x,t)}{\partial x}^T \tag{6.22}$$

for $s_j \leq t < s_{j+1}$, where $\hat{\epsilon}(x,t)$ is the transformed error for the funnel in equation (6.21). Note that we combined two funnels into one funnel for the until operator, while there are now $|\mathcal{A}_j|$ funnels combined into one funnel. Nonetheless, under the same assumptions as in theorem 6.3, the control law in equation (6.22) is sequentially valid.

---

**Example 6.8** Consider again the modified Lotka-Volterra equations of the predator-prey model in example 6.5. We now impose the satisfiable STL specification

$$\theta := \phi_1 \wedge \phi_2 \wedge \phi_3 \wedge \phi_4 \wedge \phi_5,$$

where $\phi_1$, $\phi_2$, $\phi_3$, $\phi_4$, and $\phi_5$ have overlapping time intervals. Specifically, we have $\phi_1 := G_{[2,5]}\psi_1$, $\phi_2 := G_{[2.5,3.5]}\psi_2$, $\phi_3 := F_{[3.5,3.5]}\psi_3$, $\phi_4 := F_{[0,6]}\psi_4$, and $\phi_5 := F_{[3,6]}\psi_5$, where the corresponding nontemporal formulas are

$$\begin{aligned}
\psi_1 &:= (x_1 \leq 500) \wedge (x_2 \leq 300) \wedge (x_3 \leq 200), \\
\psi_2 &:= (0 \leq x_2) \wedge (x_2 \leq 200) \wedge (0 \leq x_3) \wedge (x_3 \leq 200), \\
\psi_3 &:= (x_1 \geq 300), \\
\psi_4 &:= (\|x - \begin{bmatrix}100 & 100 & 100\end{bmatrix}^T\| \leq 100), \\
\psi_5 &:= (\|x - \begin{bmatrix}400 & 400 & 400\end{bmatrix}^T\| \leq 100).
\end{aligned}$$

Algorithm 6.1 returns $\mathcal{S} := \{s_0 := 0, s_1 := 1.3, s_2 := 3.5, s_3 := 5, s_4 := 6\}$ with $t_{1,*} := 2$, $t_{2,*} := 2.5$, $t_{3,*} := 3.5$, $t_{4,*} := 1.3$, and $t_{5,*} := 6$. Note also that $\mathcal{A}_0 = \{4\}$, $\mathcal{A}_1 = \{1,2,3\}$, $\mathcal{A}_2 = \{1\}$, and $\mathcal{A}_3 = \{5\}$ so three funnel-based constraints are active between $[s_1, s_2]$ (illustrated in particular in figure 6.6c). We set $\rho^j_{\max} := 600$ for each $j \in \{0,1,2,3\}$, while $\bar{\rho}^j_{\text{opt}} = 100$ and $r^j := 50$. The results are shown in figure 6.6, and it can be verified, in both figures 6.6a and 6.6b, that $50 \leq \bar{\rho}^\theta(x,0) \leq 600$.

---

## 6.4 Notes and References

This chapter was mainly based on our works [185, 177]. Funnel control laws, as presented in this chapter, have been used as a guide in reinforcement learning to satisfy complex STL specifications [303, 304, 265] and have also been applied to manipulation tasks [269]. Recent work also considered the control design under LTL specifications by decomposing the LTL specification, using the automata representation of the specification, into simpler reachability problems that are sequentially implemented by funnel controllers [135]. As opposed to such hard funnel constraints, control laws for soft funnel constraints that may be violated were presented in [204].

Prescribed performance control was originally developed for output tracking control problems [33, 34]. These works use a similar controllability assumption that we made throughout this chapter (see assumption 6.1). It is straightforward to extend the prescribed performance control methodology to relative

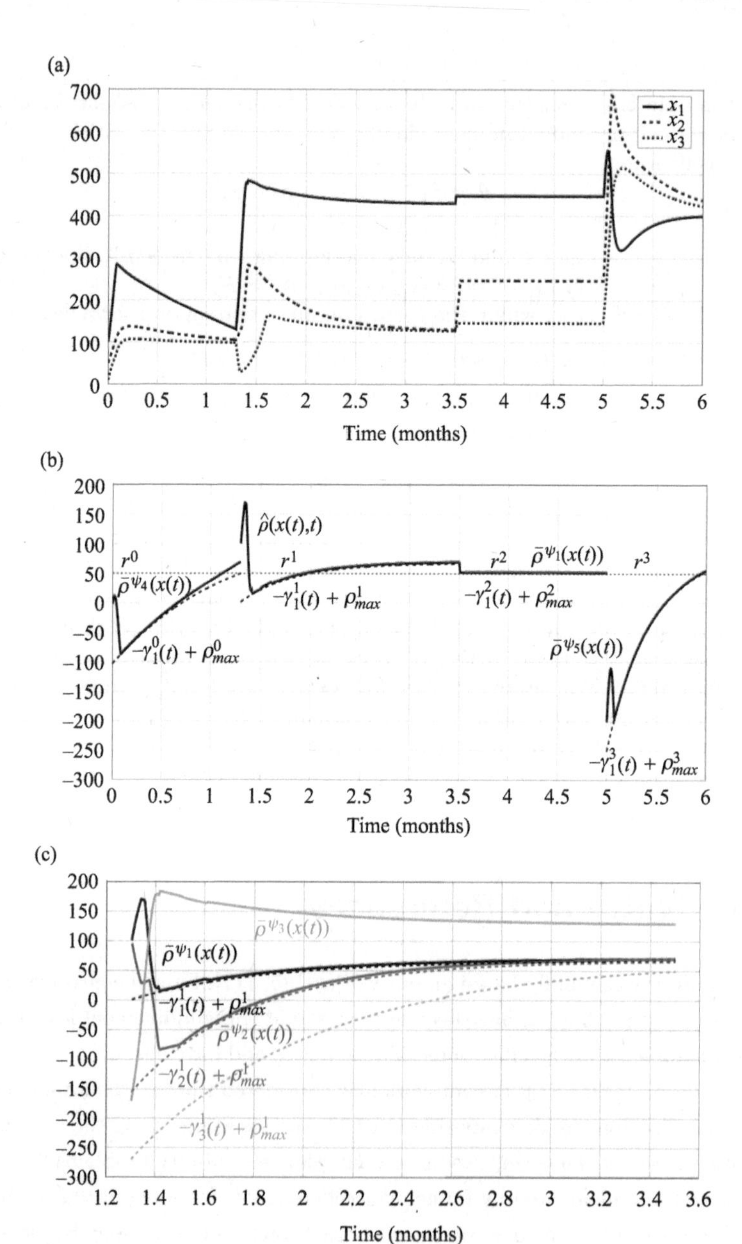


**Figure 6.6**
Simulation results for example 6.8. (a) Population densities, (b) STL robustness, (c) components of $\hat{\rho}(x(t), t)$ between $[s_1, s_2]$, (d) control inputs. Reprinted with permission from [177].

(d)

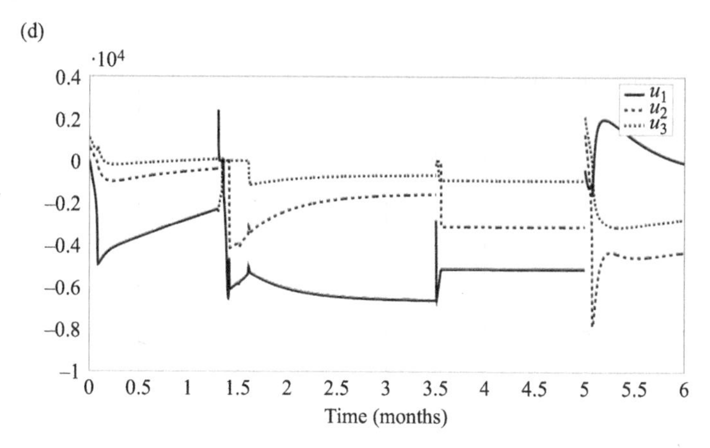


**Figure 6.6**
(continued)

degree systems (see e.g., [36, 37, 337]). Various authors have used the prescribed conformance methodology in adaptive control [35, 213] and in applications for robotics [148, 149, 237, 310] including space robotic [318, 91]. Finally, prescribed performance control has lately incorporated input constraints in addition to output constraints (see e.g., [309, 210, 297, 296]). Much in the same spirit of this chapter, but without explicitly considering STL specifications, we recently presented a prescribed performance control methodology for multiple time-varying output constraints [206]. Finally, besides prescribed performance control, other funnel control strategies have been proposed; for instance, the interested reader may refer to [75, 276, 46, 167].

## 6.5 Additional Proofs

*Proof of theorem 6.3:* The proof follows similar steps as the proof of theorem 6.2 and is given in short. The dynamics of the transformed error $\hat{\epsilon}$ is given by

$$\dot{\hat{\epsilon}} = \frac{\partial \hat{\epsilon}}{\partial \hat{\xi}}\dot{\hat{\xi}} = -\frac{1}{\gamma'\hat{\xi}(1+\hat{\xi})}\Big(\frac{\partial \hat{\rho}(x,t)}{\partial x}\dot{x} + \frac{\partial \hat{\rho}(x,t)}{\partial t} - \hat{\xi}\dot{\gamma}'\Big),$$

which can be derived since, again, $\frac{\partial \hat{\epsilon}}{\partial \hat{\xi}} = -\frac{1}{\hat{\xi}(1+\hat{\xi})}$ and $\dot{\hat{\xi}} = \frac{1}{\gamma'}(\dot{\hat{e}} - \hat{\xi}\dot{\gamma}')$, but now with $\dot{\hat{e}} = \frac{\partial \hat{e}(x,t)}{\partial x}\dot{x} + \frac{\partial \hat{e}(x,t)}{\partial t}$ where $\frac{\partial \hat{e}(x,t)}{\partial x} = \frac{\partial \hat{\rho}(x,t)}{\partial x}$ and $\frac{\partial \hat{e}(x,t)}{\partial t} = \frac{\partial \hat{\rho}(x,t)}{\partial t}$.

Note first that $\hat{\rho}_{\text{opt}}(t) < \rho_{\max}$ for all $t \in [0,b]$ due to the choice of $\rho_{\max} \in (\max(\bar{\rho}_{\text{opt}}^{\psi'}, \bar{\rho}_{\text{opt}}^{\psi''}), \infty)$ and since $\hat{\rho}(x,t) \leq \min(\bar{\rho}^{\psi'}(x), \hat{\rho}^{\psi''}(x,t))$. By assumption, $-\gamma'(t) + \rho_{\max} < \hat{\rho}_{\text{opt}}(t)$ for all $t \in [0,b]$ which ensures that equation (6.19) is feasible for all $t \in [0,b]$; that is, $\hat{\Omega}_x(t) := \{x \in \mathbb{R}^n | -1 < \hat{\xi}(x,t) = \frac{\hat{\rho}(x,t) - \rho_{\max}}{\gamma'(t)} < 0\}$ is nonempty for each $t \in [0,b]$. In addition, the assumption $-\gamma_0' + \rho_{\max} < \hat{\rho}(x_0, 0)$ ensures that $x_0 \in \hat{\Omega}_x(0) =: \hat{\Omega}_{x_0}$. As in step A in the proof of theorem 6.2, we define $y := \begin{bmatrix} x^T & \hat{\xi} \end{bmatrix}^T$ and see that there is a unique and maximal solution $y : \mathcal{I} \to \Omega_y$, with $\mathcal{I} := [0, \tau_{\max}) \subseteq [0,b]$ and $\tau_{\max} > 0$ such that $\hat{\xi}(t) \in \Omega_\xi$ and $x(t) \in \hat{\Omega}_{x_0}$ for all $t \in \mathcal{I}$. As in step B in the proof of theorem 6.2, define $V(\hat{\epsilon}) := \frac{1}{2}\hat{\epsilon}^2$ and note that

$$
\begin{aligned}
\dot{V} = \hat{\epsilon}\dot{\hat{\epsilon}} &= \hat{\epsilon}\Big(\frac{-1}{\gamma'\hat{\xi}(1+\hat{\xi})}\Big(\frac{\partial \hat{\rho}(x,t)}{\partial x}\dot{x} + \frac{\partial \hat{\rho}(x,t)}{\partial t} - \hat{\xi}\dot{\gamma}'\Big)\Big) \\
&= \hat{\epsilon}\hat{\alpha}\Big(\frac{\partial \hat{\rho}(x,t)}{\partial x}(f(x) + g(x)u + w) + \frac{\partial \hat{\rho}(x,t)}{\partial t} - \xi\dot{\gamma}\Big),
\end{aligned}
$$

where $\hat{\alpha} := -\frac{1}{\gamma'\hat{\xi}(1+\hat{\xi})}$ satisfies $\hat{\alpha}(t) \in [\frac{4}{\gamma_0'}, \infty) \in \mathbb{R}_{>0}$ for all $t \in \mathcal{I}$. It can again be shown that $\dot{V} \leq |\hat{\epsilon}|\hat{\alpha}k_1 + \hat{\epsilon}\hat{\alpha}\frac{\partial \hat{\rho}(x,t)}{\partial x}g(x)u$, where $k_1$ now, compared to equation (6.15), also depends on an upper bound of $\left|\frac{\partial \hat{\rho}(x,t)}{\partial t}\right|$, which exists since $\frac{\partial \hat{\rho}(x,t)}{\partial t}$ is continuous and $[0,b]$ is compact. Following the same steps as in the proof of theorem 6.2, we can show that $\mathcal{I} = [0,b]$ so that the funnel in equation (6.19) is satisfied for all $t \in [0,b]$. We conclude the proof by recalling that equation (6.19) implies equation (6.18).

# Chapter 7

# Decentralized Funnel Control

In this chapter, we start with the same motivation as in chapter 5, which was to design decentralized control laws for multi-agent systems under global and local signal temporal logic (STL) specifications. However, here our goal is to obtain decentralized funnel control laws, as opposed to decentralized CBFs.

Consider again $N$ agents in which agent $i$ is modeled as

$$\dot{x}_i = f_i(x_i) + g_i(x_i)u_i + w_i(x,t), \tag{7.1}$$

where $x_i \in \mathbb{R}^{n_i}$, $u_i \in \mathbb{R}^{m_i}$, $f_i : \mathbb{R}^{n_i} \to \mathbb{R}^{n_i}$, $g_i : \mathbb{R}^{n_i} \to \mathbb{R}^{n_i \times m_i}$, and $w_i : \mathbb{R}^n \times \mathbb{R}_{\geq 0} \to \mathbb{R}^{n_i}$ denote the state, control input, internal dynamics, input dynamics, and dynamical couplings of agent $i$, respectively, where $x := \begin{bmatrix} x_1^T \dots x_N^T \end{bmatrix}^T$ denotes the multi-agent state with dimension $n := n_1 + \dots + n_N$. The functions $f_i$, $g_i$, and $w_i$ are locally Lipschitz continuous in $x$ and piecewise continuous in $t$.[1] Recall that in the case of decentralized CBFs, the functions $f_i$ and $g_i$ had to be known by agent $i$. Using decentralized funnel control, the function $f_i$ need not be known and, if $m_i = n_i$, the function $g_i$ need not be known either, so that the control design becomes fully model-free. As for centralized funnel control, we make the following controllability and boundedness assumption on $g_i$ and $w_i$.

**Assumption 7.1** *Let the function $g_i$ that describes the input dynamics of the system in equation (7.1) be such that $g_i(x)g_i(x)^T$ is positive definite for each*

[1] As before, the functions $f_i$ and $g_i$ could depend on time as well.

*$x_i \in \mathbb{R}^{n_i}$. Further, let the function $w_i$ that describes additive noise and agent couplings be contained within a bounded set $\mathcal{W} \subseteq \mathbb{R}^n$ (i.e., $w_i(x,t) \in \mathcal{W}$ for all $(x,t) \in \mathbb{R}^n \times \mathbb{R}_{\geq 0}$).*

With regard to the second part of assumption 7.1, we note that we do not need to know an explicit upper bound of $w$ for the control design (i.e., we do not need to know the set $\mathcal{W}$), unlike in the case of decentralized CBFs.

For a global STL specification $\phi$, each agent $i$ could now compute the stacked control input $u := \left[u_1^T, \ldots, u_N^T\right]^T$ by using the centralized funnel control law in equation (6.14) for the stacked multi-agent dynamics, and then apply the control portion $u_i$. As the funnel control law is given in closed form in equation (6.14) and due to its special high-gain structure, we will show that it can trivially be decentralized such that agent $i$ can directly compute $u_i$ for the global specification $\phi$. We will then consider local STL specifications $\phi_i$ that are again individually assigned to each agent. We recall that it may not be possible to satisfy all the local specifications. Hence, we devise decentralized funnel control laws that aim to find locally least-violating solutions by maximizing the quantitative semantics associated with the local specification. Our approach consists of an online detection and repair scheme in which the level of satisfaction (i.e., the robustness) of the specification is gradually decreased when collaboration between agents is initiated.

## 7.1 Decentralized Collaborative Funnel Control

As before, we consider the STL fragment from chapter 6 as per equation (6.2). For simplicity, we only consider atomic temporal formulas $\phi$ as in equation (6.2) and do not consider temporal formulas $\theta$ of the form equation (6.3). We can deal with these formulas in the same way as illustrated in section 6.3 by using multiple funnels. Formally, each agent $i$ is assigned a local task $\phi_i$ from the STL fragment:

$$\phi_i ::= F_{[a_i,b_i]}\psi_i \mid G_{[a_i,b_i]}\psi_i \mid F_{[\underline{a}_i,\underline{b}_i]}G_{[\bar{a}_i,\bar{b}_i]}\psi_i \tag{7.2}$$

where $\psi_i$ is a nontemporal formula. For these formulas, we again consider the quantitative semantics $\rho^{\phi_i}$, as well as the smooth quantitative semantics $\bar{\rho}^{\phi_i}$ as previously defined in section 6.1 by using the smooth approximation of the min operator. The satisfaction of $\phi_i$ depends again on the behavior of agent $i$ and possibly other agents $j \neq i$. This set of agents is denoted by $\mathcal{V}_i \subseteq \{1, \ldots, N\}$, for which it has to hold that $|\mathcal{V}_i| \geq 1$ since $i \in \mathcal{V}_i$. If the satisfaction of $\phi_i$ depends on the behavior of agent $j$, we also say that agent $j$ is participating in $\phi_i$. Specifically, we characterize $\phi_i$ as a noncollaborative formula if $|\mathcal{V}_i| = 1$ and a collaborative formula otherwise.

Similarly to section 5.1, in chapter 5 we now define the stacked state vector of all agents within this set $j_1, \ldots, j_{|\mathcal{V}_i|} \in \mathcal{V}_i$ as

$$\bar{x}_i(t) := \begin{bmatrix} x_{j_1}^T(t) & \ldots & x_{j_{|\mathcal{V}_i|}}^T(t) \end{bmatrix}^T \in \mathbb{R}^{\bar{n}_i},$$

with state dimension $\bar{n}_i := n_{j_1} + \ldots + n_{j_{|\mathcal{V}_i|}}$. We also compute the maximum of the quantitative semantics $\rho^{\psi_i}$ and the smooth quantitative semantics $\bar{\rho}^{\psi_i}$ as

$$\rho_i^{\text{opt}} := \sup_{\bar{x}_i \in \mathbb{R}^{\bar{n}_i}} \rho^{\psi_i}(\bar{x}_i),$$

$$\bar{\rho}_i^{\text{opt}} := \sup_{\bar{x}_i \in \mathbb{R}^{\bar{n}_i}} \bar{\rho}^{\psi_i}(\bar{x}_i).$$

We further distinguish between $\bar{\rho}_i^{\text{opt}} > 0$ and $\bar{\rho}_i^{\text{opt}} \leq 0$ and will mostly assume that $\bar{\rho}_i^{\text{opt}} > 0$ for convenience throughout this chapter.

**Assumption 7.2** *The maximum of the smooth quantitative semantics $\bar{\rho}^{\phi_i}$ is such that $\bar{\rho}_i^{\text{opt}} > 0$.*

Assumption 7.2 implies that $\psi_i$, and hence $\phi_i$, are locally satisfiable by the agents in $\mathcal{V}_i$. However, in our setting, this may not be possible if $|\mathcal{V}_i| > 1$, as other agents $j \in \mathcal{V}_i \setminus \{i\}$ have to satisfy their own local task $\phi_j$. Therefore, let $\rho_i^{\text{gap}} > 0$ be a parameter that indicates how much $\phi_i$ is violated. If the signal $\bar{x}_i : \mathbb{R}_{\geq 0} \to \mathbb{R}^{\bar{n}_i}$ is such that $r_i \leq \bar{\rho}^{\phi_i}(\bar{x}_i, 0)$ for $r_i$ with $\bar{\rho}_i^{\text{opt}} - \rho_i^{\text{gap}} \leq r_i$, we say that $\bar{x}_i$ is a locally least-violating solution with a gap of $\rho_i^{\text{gap}}$.

Our goal is to derive a funnel control law $u_i$ such that $r_i \leq \bar{\rho}^{\phi_i}(\bar{x}_i, 0) \leq \rho_i^{\max}$, where $r_i \in \mathbb{R}$ is a robustness measure, while $\rho_i^{\max} \in \mathbb{R}$ is an upper bound with $\rho_i^{\max} > \bar{\rho}_i^{\text{opt}}$ as before. To derive such a control law, we pose an assumption that resembles assumption 6.3 in chapter 6.

**Assumption 7.3** *Each nontemporal formula of class $\psi_i$ that is contained in the atomic temporal formula $\phi_i$ from equation (7.2) is such that $\bar{\rho}^{\psi_i} : \mathbb{R}^{\bar{n}_i} \to \mathbb{R}$ is concave and the superlevel sets of $\bar{\rho}^{\psi}$ are bounded.*

**Designing Local Funnels.** Following the funnel control strategy from chapter 6, we first define the local performance function

$$\gamma_i(t) := (\gamma_i^0 - \gamma_i^\infty)\exp(-l_i t) + \gamma_i^\infty,$$

where we let $\gamma_i^0, \gamma_i^\infty \in \mathbb{R}_{>0}$ and $l_i \in \mathbb{R}_{\geq 0}$ encode $\phi_i$ as done in equations (6.8)–(6.10) in section 6.2. Under assumption 7.2, we thus select

$$\rho_i^{\max} \in \left(\bar{\rho}_i^{\text{opt}}, \infty\right) \tag{7.3}$$

$$t_i^* \in \begin{cases} [a_i, b_i] & \text{if } \phi_i = F_{[a_i,b_i]}\psi_i \\ \{a_i\} & \text{if } \phi_i = G_{[a_i,b_i]}\psi_i \\ [\underline{a}_i + \bar{a}_i, \underline{b}_i + \bar{a}_i] & \text{if } \phi_i = F_{[\underline{a}_i,\underline{b}_i]}G_{[\bar{a}_i,\bar{b}_i]}\psi_i, \end{cases} \tag{7.4}$$

$$r_i \in \begin{cases} (0, \rho_i^{\max}) & \text{if } t_i^* > 0 \\ (0, \bar{\rho}^{\psi_i}(\bar{x}_i(0)) & \text{if } t_i^* = 0 \end{cases} \tag{7.5}$$

$$\gamma_i^0 \in \begin{cases} \left(\rho_i^{\max} - \bar{\rho}^{\psi_i}(\bar{x}_i(0)), \infty\right) & \text{if } t_i^* > 0 \\ \left(\rho_i^{\max} - \bar{\rho}^{\psi_i}(\bar{x}_i(0)), \rho_i^{\max} - r_i\right] & \text{else} \end{cases} \tag{7.6}$$

$$\gamma_i^\infty \in \left(\rho_i^{\max} - \bar{\rho}_i^{\text{opt}}, \min\left(\gamma_i^0, \rho_i^{\max} - r_i\right)\right] \tag{7.7}$$

$$l_i = \begin{cases} 0 & \text{if } -\gamma_i^0 + \rho_i^{\max} \geq r_i \\ -\ln\left(\frac{r_i + \gamma_i^\infty - \rho_i^{\max}}{-(\gamma_i^0 - \gamma_i^\infty)}\right)/t_i^* & \text{else.} \end{cases} \tag{7.8}$$

**Decentralized Funnel Control.** Recall that we aim to achieve

$$r_i \leq \bar{\rho}^{\phi_i}(\bar{x}_i, 0) \leq \rho_i^{\max}$$

by prescribing a temporal behavior to $\bar{\rho}^{\psi_i}(\bar{x}_i(t))$ through $\gamma_i$ and $\rho_i^{\max}$ as

$$-\gamma_i(t) + \rho_i^{\max} < \bar{\rho}^{\psi_i}(\bar{x}_i(t)) < \rho_i^{\max}. \tag{7.9}$$

Let us now define the error, the normalized error, and the transformed error as

$$e_i(\bar{x}_i) := \bar{\rho}^{\psi_i}(\bar{x}_i) - \rho_i^{\max},$$

$$\xi_i(\bar{x}_i, t) := \frac{e_i(\bar{x}_i)}{\gamma_i(t)},$$

$$\epsilon_i(\bar{x}_i, t) := S\big(\xi_i(\bar{x}_i, t)\big),$$

where $S(\xi_i) := \ln\left(-\frac{\xi_i+1}{\xi_i}\right)$ is the transformation function. Now, it again follows that equation (7.9) is equivalent to $-\infty < \epsilon_i(t) < \infty$, such that our goal is to keep $\epsilon_i(t)$ bounded for all times $t \geq 0$. By the choice of $\gamma_i$, we then have $0 < r_i \leq \bar{\rho}^{\phi_i}(\bar{x}_i, 0) \leq \rho_i^{\max}$.

We will first present two trivial cases where the local specifications $\phi_i$ are not conflicting: global specifications or local noncollaborative specifications.

**Global Specification.** The first case is a global specification $\phi$ in the sense that $\phi_i := \phi$; that is, all agents share the same specification.

**Theorem 7.1 (Global Specification)** *Let the functions $f_i$, $g_i$, and $w_i$ that define the system in equation (7.1) be locally Lipschitz continuous in $x$ and piecewise continuous in $t$, and let $g_i$ and $w_i$ satisfy assumption 7.1. Let $\phi_i$ be an atomic temporal formula from equation (6.2b), and let the nontemporal formula $\psi_i$ contained within $\phi_i$ have smooth quantitative semantics $\bar{\rho}^{\psi_i}$ that satisfy assumptions 7.2 and 7.3. For all agents $i, j \in \{1, \ldots, N\}$, assume that $\phi_i = \phi_j$, and that $\rho_i^{max} = \rho_j^{max}$, $t_i^* = t_j^*$, $r_i = r_j$, $\gamma_i^0 = \gamma_j^0$, $\gamma_i^\infty = \gamma_j^\infty$, and $l_i = l_j$ are selected as in equations (7.3)–(7.8). If each agent $i \in \{1, \ldots, N\}$ applies the control law*

$$u_i(x,t) := -\epsilon_i(x,t) g_i(x_i)^T \frac{\partial \bar{\rho}^{\psi_i}(x)}{\partial x_i}^T, \tag{7.10}$$

*then each solution $x : \mathcal{I} \to \mathbb{R}^n$ to equation (7.1) is such that $0 < r_i \leq \bar{\rho}^{\phi_i}(x, 0) \leq \rho_i^{max}$; that is, $(x, 0) \models \phi_i$.*

*Proof:* *By the fact that all agents are assigned the same formula, we can define the global specification $\phi := \phi_i$ and apply theorem 6.2 with the performance function $\gamma := \gamma_i$ and funnel parameter $\rho_{max} := \rho_i^{max}$ to the stacked multi-agent system dynamics $\dot{x} = f(x) + g(x)u + w(x,t)$, where*

$$f(x) := \begin{bmatrix} f_1(x_1)^T & \ldots & f_N(x_N)^T \end{bmatrix}^T,$$
$$g(x) := diag\big(g_1(x_1), \ldots, g_N(x_N)\big),$$
$$w(x,t) := \begin{bmatrix} w_1(x,t)^T & \ldots & w_N(x,t)^T \end{bmatrix}^T.$$

*We thus design a decentralized funnel control law according to equation (6.2) as $u(x,t) := -\epsilon(x,t) g(x)^T \frac{\partial \bar{\rho}^{\psi}(x)}{\partial x}^T$. Since $\epsilon(x,t) = \epsilon_i(x,t)$, it is now easy to see that $u(x,t) = \begin{bmatrix} u_1(x,t)^T & \ldots & u_N(x,t)^T \end{bmatrix}^T$, i.e., stacking the individual control laws in equation (7.10) gives us the central funnel control law in equation (6.2).*

Note that all $N$ agents are assumed to be subject to the same specification in theorem 7.1. The same result holds if there are independent sets of agents that share the same specification (e.g., the specification $\phi_1 = \phi_2$ only depends on agents 1 and 2 and the specification $\phi_3 = \phi_4$ only depends on agents 3 and 4, while $\phi_1 = \phi_2 \neq \phi_3 = \phi_4$).[2] This result follows, as we can simply apply theorem 7.1 to each set of independent agents separately. The technical argument for why this is possible is that the coupling term $w_i$ between agents is bounded and absorbed into the constant $k_1$ in the proof of theorem 6.2.

---

**Example 7.1** We consider $N := 8$ omni-directional robots with two-dimensional position $p_i$ and orientation $\theta_i$. The state of agent $i$ is thus denoted by $x_i := \begin{bmatrix} p_i & \theta_i \end{bmatrix} = \begin{bmatrix} p_{x,i} & p_{y,i} & \theta_i \end{bmatrix}$, and the agent dynamics are

$$\dot{x}_i = \begin{bmatrix} \cos(\theta_i) & -\sin(\theta_i) & 0 \\ \sin(\theta_i) & \cos(\theta_i) & 0 \\ 0 & 0 & 1 \end{bmatrix} \left( B_i^T \right)^{-1} R_i u_i,$$

where $R_i := 0.02$ is the wheel radius and

$$B_i := \begin{bmatrix} 0 & \cos(\pi/6) & -\cos(\pi/6) \\ -1 & \sin(\pi/6) & \sin(\pi/6) \\ L_i & L_i & L_i \end{bmatrix}$$

describes geometrical constraints with $L_i := 0.2$ as the radius of the robot body.

We consider the goal regions $p_A := \begin{bmatrix} 50 & 50 \end{bmatrix}$, $p_B := \begin{bmatrix} 110 & 40 \end{bmatrix}$, $p_C := \begin{bmatrix} 40 & 70 \end{bmatrix}$, and $p_D := \begin{bmatrix} 55 & 70 \end{bmatrix}$. Here, the sets of agents $\{1, 2, 3\}$, $\{4, 5, 6\}$, and $\{7, 8\}$ are subject to the same formula so we get the global formula $\phi$. Specifically, we have $\phi_1 := \phi_2 := \phi_3 := F_{[10,15]}\psi'$ with $\psi' := (\|p_1 - p_2\| < 2) \wedge (\|p_1 - p_3\| < 2) \wedge (\|p_2 - p_3\| < 2) \wedge (\|p_1 - p_A\| < 2)$. For the second set of agents, we have $\phi_4 := \phi_5 := \phi_6 := F_{[10,15]}\psi''$ with $\psi'' := (\|p_5 - p_B\| < 5) \wedge (27 < p_{x,5} - p_{x,4} < 33) \wedge (27 < p_{x,5} - p_{x,6} < 33) \wedge (27 < p_{y,4} - p_{y,5} < 33) \wedge (27 < p_{y,5} - p_{y,6} < 33) \wedge (|\deg(\theta_4) + 45| < 5) \wedge (|\deg(\theta_5) - 180| < 5) \wedge (|\deg(\theta_6) - 45| < 5)$, where $\deg(\cdot)$ converts radians into degrees. The third set of agents uses $\phi_7 := \phi_8 := F_{[10,15]}\psi'''$, with $\psi''' := (\|p_7 - p_8\| < 10) \wedge (\|p_7 - p_C\| < 10) \wedge (\|p_8 - p_D\| < 10) \wedge (|\deg(\theta_7) + 90| < 5) \wedge (|\deg(\theta_5) + 90| < 5)$. The simulation result is shown in figure 7.1. Note that all tasks are satisfied within the time interval $[10, 15]$.

---

In the case that assumption 7.2 does not hold (i.e., when $\bar{\rho}_i^{\text{opt}} \leq 0$), we can find a locally least-violating solution with a gap of $\rho_i^{\text{gap}}$ if we select $r_i$ instead

[2] This resembles section 5.1 in chapter 5, where we considered $K \leq N$ collaborative specifications that in conjunction form the global specification $\phi$.

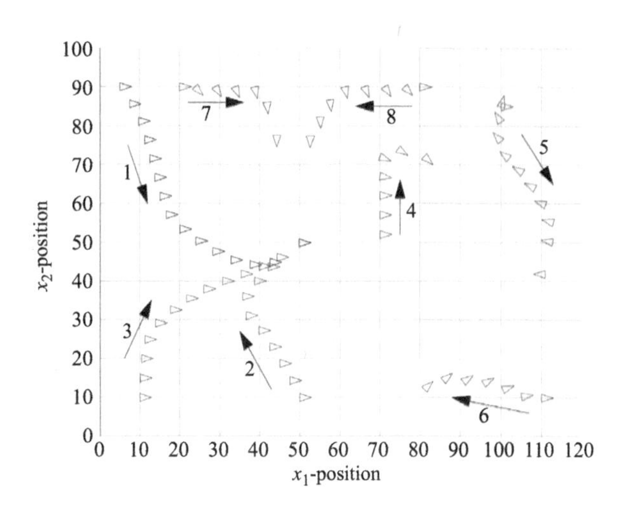


**Figure 7.1**
Agent trajectories for example 7.1. Reprinted with permission from [170].

of equation (7.5) as

$$r_i \in [\bar{\rho}_i^{\text{opt}} - \rho_i^{\text{gap}}, \bar{\rho}_i^{\text{opt}}).$$

**Local Specifications.** In the case of local and noncollaborative specifications $\phi_i$ (i.e., when $|\mathcal{V}_i| = 1$), we can now immediately see that we can use the control law in equation (7.10) by defining the global specification $\phi := \phi_1 \wedge \ldots \wedge \phi_N$.

**Corollary 7.1 (Local Noncollaborative Specifications)** *Let the functions $f_i$, $g_i$, and $w_i$ that define the system in equation (7.1) be locally Lipschitz continuous in $x$ and piecewise continuous in $t$, and let $g_i$ and $w_i$ satisfy assumption 7.1. Let $\phi_i$ be a noncollaborative atomic temporal formula from equation (6.2b), and let the nontemporal formula $\psi_i$ contained within $\phi_i$ have smooth quantitative semantics $\bar{\rho}^{\psi_i}$ that satisfy assumptions 7.2 and 7.3. If $\rho_i^{max}$, $t_i^*$, $r_i$, $\gamma_i^0$, $\gamma_i^\infty$, and $l_i$ are selected as in equations (7.3)–(7.8) and each agent $i \in \{1, \ldots, N\}$ applies the control law*

$$u_i(x_i, t) := -\epsilon_i(x_i, t) g_i(x_i)^T \frac{\partial \bar{\rho}^{\psi_i}(x_i)}{\partial x_i}^T, \tag{7.11}$$

*then each solution $x : \mathcal{I} \to \mathbb{R}^n$ to (7.1) is such that $0 < r_i \le \bar{\rho}^{\phi_i}(x_i, 0) \le \rho_i^{max}$, i.e., $(x_i, 0) \models \phi_i$.*

## 7.2 Decentralized Funnel Control under Conflicting Local Specifications

In this section, we will deal with local and potentially conflicting specifications $\phi_i$. Motivated by results from section 7.1, we propose that each agent $i \in \{1, \ldots, N\}$ initially applies the control law

$$u_i(\bar{x}_i, t) := -\epsilon_i(\bar{x}_i, t) g_i(x_i)^T \frac{\partial \bar{\rho}^{\psi_i}(\bar{x}_i)}{\partial x_i}^T \tag{7.12}$$

with funnel parameters $\rho_i^{\max}$ and $\gamma_i$ selected as previously instructed. As already observed, funnel control laws such as equation (7.12) are high-gain controllers. Specifically, note that equation (7.12) consists of two components, one determining the control gain and one the control direction. The higher the control gain $\epsilon_i(\bar{x}_i, t)$ gets when the funnel boundary in equation (7.9) is approached, the larger the control input $u_i(\bar{x}_i, t)$ will become; that is, $\|u_i(\bar{x}_i, t)\| \to \infty$ as $\epsilon_i(\bar{x}_i, t) \to \{-\infty, \infty\}$. The control direction is given by $-\frac{\partial \bar{\rho}^{\psi_i}(\bar{x}_i)}{\partial x_i}$, and we thus reason that applying equation (7.12) is a good initial choice despite potential conflicts. However, the resulting trajectory $\bar{x}_i$ may still lead to $\epsilon_i(\bar{x}_i(t), t)$ approaching $\{-\infty, \infty\}$ for some time $t \geq 0$ leading to a violation of equation (7.9). The next two examples exhibit such behavior.

---

**Example 7.2** Consider the agents $\{1, 2, 3, 4\}$ with the collaborative formula

$$\phi_1 := G_{[0,15]}((\|x_1 - x_2\|^2 \leq 25^2) \wedge (\|x_1 - x_3\|^2 \leq 25^2) \wedge (\|x_1 - x_4\|^2 \leq 25^2))$$

for agent 1 and the noncollaborative formulas

$$\phi_2 := G_{[10,15]}(\|x_2 - [10 \quad 90]^T\|^2 \leq 5^2),$$
$$\phi_3 := G_{[10,15]}(\|x_3 - [10 \quad 10]^T\|^2 \leq 5^2),$$
$$\phi_4 := G_{[10,15]}(\|x_4 - [45 \quad 20]^T\|^2 \leq 5^2).$$

for the other agents. The set of formulas $\{\phi_1, \phi_2, \phi_3, \phi_4\}$ is not satisfiable,[3] but each formula $\phi_i$ is satisfiable individually. Under equation (7.12), agents 2, 3, and 4 move to $[10 \quad 90]^T$, $[10 \quad 10]^T$, and $[45 \quad 20]^T$, respectively. Agent 1 can hence not satisfy $\phi_1$ and will violate the funnel constraint in equation (7.9) for some time $t \geq 0$. A solution is to decrease the robustness online such that $r_1 < 0$ to achieve a locally least-violating solution $r_1 \leq \rho^{\phi_1}(\bar{x}_1, 0) \leq 0$.

---

[3] By that, we meant that the formula $\phi_1 \wedge \phi_2 \wedge \phi_3 \wedge \phi_4$ is not satisfiable.

Even if the set $\{\phi_1, \ldots, \phi_N\}$ is satisfiable, the resulting trajectory may violate the funnel constraint in equation (7.9) for some time $t \geq 0$, as illustrated in example 7.3.

---

**Example 7.3** Consider the agents $\{5, 6, 7\}$ with the collaborative formula

$$\phi_5 := F_{[5,10]}\big((\|x_5 - x_6\|^2 \leq 10^2) \wedge (\|x_5 - x_7\|^2 \leq 10^2) \\ \wedge (\|x_5 - \begin{bmatrix}110 & 20\end{bmatrix}^T\|^2 \leq 5^2)\big)$$

for agent 5 and the noncollaborative formulas

$$\phi_6 := F_{[5,15]}(\|x_6 - \begin{bmatrix}50 & 20\end{bmatrix}^T\|^2 \leq 5^2),$$
$$\phi_7 := F_{[5,15]}(\|x_7 - \begin{bmatrix}110 & 80\end{bmatrix}^T\|^2 \leq 5^2).$$

for the other agents. Under equation (7.12), agents 6 and 7 move to $\begin{bmatrix}50 & 20\end{bmatrix}^T$ and $\begin{bmatrix}110 & 80\end{bmatrix}^T$ by no more than 15 time units, respectively. However, agent 5 is forced to move to $\begin{bmatrix}110 & 20\end{bmatrix}^T$ and be close to agents 6 and 7 by no more than 10 time units. This may lead to critical events where agent 5 violates the funnel in equation (7.9) for some $t \geq 0$. If agents 6 and 7 collaborate, satisfaction of $\phi_6$ and $\phi_7$ can be postponed and $\phi_5$ can be satisfied first, such as by using collaborative control for $\phi_5$.

---

**Hybrid Control Strategy.** Following these two examples, we will design a hybrid control strategy for each agent $i$ to decrease the robustness $r_i$ locally (as in example 7.2) and initiate collaboration (as in example 7.3) when needed. Each agent is described by an augmented state $z_i \in \mathcal{Z}_i \subseteq \mathbb{R}^{n_z}$ with domain $\mathcal{Z}_i$. There are internal and external inputs $u_i \in \mathcal{U}_i$ and $u_i^{\text{ext}} \in \mathcal{U}_i^{\text{ext}}$, respectively, so that we can define the hybrid domain $\mathfrak{H}_i := \mathcal{Z}_i \times \mathcal{U}_i \times \mathcal{U}_i^{\text{ext}}$. A hybrid system is now a tuple $\mathcal{H}_i := (C_i, F_i, D_i, G_i)$, where $C_i \subseteq \mathfrak{H}_i$ and $D_i \subseteq \mathfrak{H}_i$ are the domains where the state can change continuously and discontinuously, respectively. These continuous and discrete changes are described by the set-valued flow and jump maps $F_i : \mathfrak{H}_i \rightrightarrows \mathbb{R}^{n_z}$ and $G_i : \mathfrak{H}_i \rightrightarrows \mathbb{R}^{n_z}$, respectively. The hybrid system dynamics are governed by

$$\begin{cases} \dot{z}_i = F_i(z_i, u_i, u_i^{\text{ext}}) & \text{for } (z_i, u_i, u_i^{\text{ext}}) \in C_i \\ z_i \in G_i(z_i, u_i, u_i^{\text{ext}}) & \text{for } (z_i, u_i, u_i^{\text{ext}}) \in D_i. \end{cases} \tag{7.13}$$

Note that the continuous evolution of $z_i$ in $C_i$ follows an ordinary differential equation (ODE) described by $F_i$, while the discrete change of $z_i$ in $D_i$ follows a set inclusion that is described by $G_i$. Specifically, for each agent $i$, we define the state

$$z_i := \begin{bmatrix} x_i^T & p_i^{\text{f}\,T} & p_i^{\text{r}\,T} \end{bmatrix}^T \in \mathcal{Z}_i$$

with $\mathcal{Z}_i := \mathbb{R}^{n_i} \times \mathbb{R}^6_{\geq 0} \times \mathbb{Z}^2$ and where the parameters

$$p_i^{\mathrm{f}} := \begin{bmatrix} t_i^* & \rho_i^{\max} & r_i & p_i^{\gamma T} \end{bmatrix}^T,$$
$$p_i^{\gamma} := \begin{bmatrix} \gamma_i^0 & \gamma_i^{\infty} & l_i \end{bmatrix}^T,$$

define the funnel in equation (7.9) and

$$p_i^r := \begin{bmatrix} \mathfrak{n}_i & \mathfrak{c}_i \end{bmatrix}^T$$

are states that define the logic of the hybrid control strategy. Here, $\mathfrak{n}_i \in \mathbb{N}$ indicates the number of times that we decrease the robustness $r_i$, while $\mathfrak{c}_i \in \{1, \ldots, N\}$ indicates collaborative control for the formula $\phi_{\mathfrak{c}_i}$. Initially, the state is selected such that $z_i(0) := \begin{bmatrix} x_i(0)^T & p_i^{\mathrm{f}}(0)^T & 0^T \end{bmatrix}^T$ with the time-varying funnel parameters $p_i^{\mathrm{f}}(0) := \begin{bmatrix} t_i^* & \rho_i^{\max} & r_i & \gamma_i^0 & \gamma_i^{\infty} & l_i \end{bmatrix}^T$ are chosen according to equations (7.3)–(7.8). The map $F_i$ is given by $F_i(z_i, u_i, u_i^{\mathrm{ext}}) := \begin{bmatrix} (f_i(x_i) + g_i(x_i)u_i + w_i(x,t))^T & 0^T \end{bmatrix}^T$, where the control law is defined as

$$u_i := \begin{cases} -\epsilon_i(\bar{x}_i, t) g_i(x_i)^T \dfrac{\partial \bar{\rho}^{\psi_i}(\bar{x}_i)}{\partial x_i}^T & \text{if } \mathfrak{c}_i = 0 \qquad (7.14a) \\ -\epsilon_{\mathfrak{c}_i}(\bar{x}_{\mathfrak{c}_i}, t) g_i(x_i)^T \dfrac{\partial \bar{\rho}^{\psi_{\mathfrak{c}_i}}(\bar{x}_{\mathfrak{c}_i})}{\partial x_i}^T & \text{if } \mathfrak{c}_i > 0. \qquad (7.14b) \end{cases}$$

**Critical Event Detection.** We first detect critical events when the funnel in equation (7.9) is violated via the domain

$$D_i' := \{(z_i, u_i, u_i^{\mathrm{ext}}) \in \mathfrak{H}_i | \xi_i(t) \in \{-1, 0\},\ \mathfrak{c}_i = 0\}.$$

We use the notation $\{\hat{z}_i \in \mathcal{Z}_i | \mathit{update}\}$ to denote the set of potential states $\hat{z}_i$ after a discrete change in $z_i$, where its elements change according to *update*. The set $D_i'$ is split into disjoint sets that indicate repairs of three repair stages as discussed next.

**Repair Stage 1.** The first repair stage is detected by

$$D_{i,1}' := D_i' \cap \{(z_i, u_i, u_i^{\mathrm{ext}}) \in \mathfrak{H}_i | \mathfrak{n}_i < N_i\},$$

where $N_i \in \mathbb{N}$ is a design parameter indicating the maximum number of repair attempts in this stage. Let us illustrate this state in an example first.

---

**Example 7.4** Consider $\phi_i := F_{[4,6]}\psi_i$ with $r_i := 0.4$ (initial robustness), which is supposed to be achieved at $t_i^* := 4.5$. The initial funnel with $\rho_i^{\max}$ and $-\gamma_i + \rho_i^{\max}$ is shown in figure 7.2. Without detection of a critical event, it would hold that $\bar{\rho}^{\phi_i}(\bar{x}_i, 0) \geq r_i$ since $\bar{\rho}^{\psi_i}(\bar{x}_i(t_i^*)) \geq r_i$. However, at time

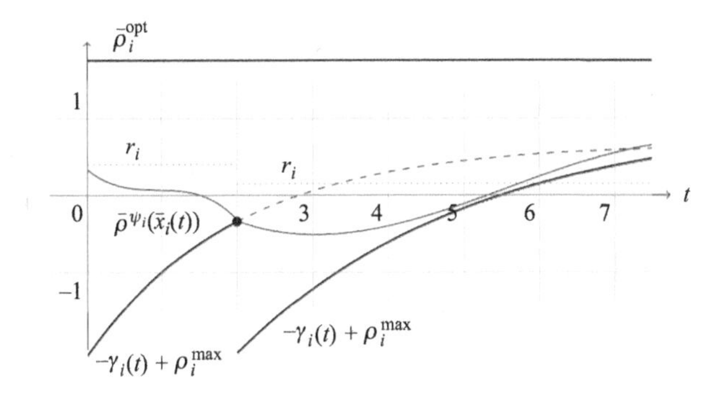


**Figure 7.2**
Funnel repair in the first stage for $\phi_i := F_{[4,6]}\psi_i$. Reprinted with permission from [170, 174].

$t_r := 2$, a critical event is detected when the trajectory $\bar{\rho}^{\psi_i}\big(\bar{x}_i(t)\big)$ touches the lower funnel boundary and repair action is needed. This is done by setting $t_i^* := 6$ (time relaxation), $r_i := 0.15$ (robustness relaxation), and adjusting $\gamma_i$ (lower funnel relaxation). Due to repair action, $\bar{x}_i$ locally satisfies $\phi_i$, as shown in figure 7.2.

---

If $(z_i, u_i, u_i^{\text{ext}}) \in D'_{i,1}$, we relax the parameters $t_i^*$, $r_i$, and $\gamma_i$ as follows:

$$G'_{i,1} := \big\{\hat{z}_i \in \mathcal{Z}_i \,|\, t_i^* = \mathcal{T}_i,\ \hat{r}_i \in \mathcal{R}_i,\ \hat{p}_i^{\gamma} = p_i^{\gamma,\text{new}},\ \hat{\mathfrak{n}}_i = \mathfrak{n}_i + 1\big\},$$

where, to achieve time relaxation, we define

$$\mathcal{T}_i := \begin{cases} \{b_i\} & \text{if } \phi_i = F_{[a_i,b_i]}\psi_i \\ \{a_i\} & \text{if } \phi_i = G_{[a_i,b_i]}\psi_i \\ \{\underline{b}_i + \bar{a}_i\} & \text{if } \phi_i = F_{[\underline{a}_i,\underline{b}_i]}G_{[\bar{a}_i,\bar{b}_i]}\psi_i. \end{cases}$$

The parameter $r_i$ is decreased to $\hat{r}_i \in (0, r_i)$ if $r_i > 0$ and we set $\hat{r}_i := r_i - \delta_i$ otherwise where $\delta_i > 0$ is a design parameter (robustness relaxation); that is, let

$$\mathcal{R}_i := \left\{\hat{r}_i \in \mathbb{R} \,\middle|\, \hat{r}_i \in \begin{cases} (0, r_i) & \text{if } r_i > 0 \\ r_i - \delta_i & \text{if } r_i \leq 0 \end{cases}\right\}.$$

At detection time $t_r$, we set $\gamma_i^r := \rho_i^{\max} - \bar{\rho}^{\psi_i}(\bar{x}_i(t_r)) + \zeta_i^{\text{l}}$, with

$$\zeta_i^{\text{l}} \in \begin{cases} \mathbb{R}_{>0} & \text{if } \hat{t}_i^* > t_r \\ (0, \bar{\rho}^{\psi_i}(\bar{x}_i(t_r)) - \hat{r}_i] & \text{otherwise,} \end{cases}$$

which resembles equation (7.6) (lower funnel relaxation). Let the funnel parameters $p_i^{\gamma,\text{new}} := \begin{bmatrix} \gamma_i^{0,\text{new}} & \gamma_i^{\infty,\text{new}} & l_i^{\text{new}} \end{bmatrix}$ then be updated as

$$\gamma_i^{\infty,\text{new}} \in (\rho_i^{\max} - \bar{\rho}_i^{\text{opt}}, \min(\gamma_i^r, \rho_i^{\max} - \hat{r}_i)]$$

$$l_i^{\text{new}} := \begin{cases} 0 & \text{if } -\gamma_i^r + \rho_i^{\max} \geq \hat{r}_i \\ \frac{-\ln\left(\frac{\hat{r}_i + \gamma_\infty^{i,new} - \rho_i^{\max}}{-(\gamma_i^r - \gamma_\infty^{i,new})}\right)}{\hat{t}_i^* - t_r} & \text{otherwise,} \end{cases}$$

similar to equations (7.7) and (7.8). Finally, set

$$\gamma_i^{0,\text{new}} := (\gamma_i^r - \gamma_i^{\infty,\text{new}}) \exp(l_i^{\text{new}} t_r) + \gamma_i^{\infty,\text{new}}.$$

**Repair Stage 2.** If the first stage is not successful after $N_i$ attempts, the second stage is detected via the set

$$D'_{i,\{2,3\}} := D'_i \cap \{(z_i, u_i, u_i^{\text{ext}}) \in \mathfrak{H}_i | \mathfrak{n}_i \geq N_i\},$$

and collaboration among agents in $\mathcal{V}_i$ is initiated if the timing conditions in the set

$$D'_{i,2} := D'_{i,\{2,3\}} \cap \{(z_i, u_i, u_i^{\text{ext}}) \in \mathfrak{H}_i | \forall j \in \mathcal{V}_i \setminus \{i\}, (\mathfrak{c}_j = -1) \text{ or } (\mathfrak{c}_j = 0, b_i < \mathcal{T}_j)\}$$

are satisfied. The conditions in $D'_{i,2}$ mean that each agent $j \in \mathcal{V}_i \setminus \{i\}$ is either not subject to a task ($\mathfrak{c}_j = -1$) or there is enough time to satisfy $\phi_j$ after $\phi_i$ has been collaboratively satisfied ($\mathfrak{c}_j = 0$, $b_i < \mathcal{T}_j$). In this case, the control law switches from equation (7.14a) to equation (7.14b) via the set

$$G'_{i,2} := \{\hat{z}_i \in \mathcal{Z}_i | \hat{r}_i \in \mathcal{R}_i, \hat{p}_i^\gamma = p_i^{\gamma,\text{new}}, \hat{\mathfrak{c}}_i = i\},$$

where $\hat{\mathfrak{c}}_i := i$ indicates collaborative control for $\phi_{\mathfrak{c}_i}$, while again relaxing the funnel parameters as in the first repair stage. If the conditions in $D'_{i,2}$ hold, agent $i$ requests collaboration from agents $j \in \mathcal{V}_i \setminus \{i\}$ via the set

$$D''_{j,2} := \{(z_j, u_j, u_j^{\text{ext}}) \in \mathfrak{H}_j | \mathfrak{c}_j \in \{-1, 0\}, \exists i \in \{1, \ldots, N\} \setminus \{j\}, j \in \mathcal{V}_i, \mathfrak{c}_i = i\}.$$

In this case, the control law for agent $j \in \mathcal{V}_i \setminus \{i\}$ switches to equation (7.14b) via

$$G''_{j,2} := \{\hat{z}_j \in \mathcal{Z}_j | \hat{p}_j^{\text{f}} = p_i^{\text{f}}, \hat{\mathfrak{c}}_j = \mathfrak{c}_i\},$$

where $\hat{\mathfrak{c}}_j = \mathfrak{c}_i$ and $\hat{p}_j^{\text{f}} = p_i^{\text{f}}$ enforce that all agents in $\mathcal{V}_i$ use the same funnel so the conditions in theorem 7.1 hold.

**Repair Stage 3.** If this second repair stage is not applicable, the third repair stage is detected via the set

$$D'_{i,3} := D'_{i,\{2,3\}} \setminus D'_{i,2},$$

and we successively decrease $r_i$ by $\delta_i > 0$ via the map

$$G'_{i,3} := \{\hat{z}_i \in \mathcal{Z}_i | \hat{r}_i = r_i - \delta_i,\ \hat{p}_i^{\gamma} = p_i^{\gamma,\text{new}}\}.$$

**The Overall System.** It needs to be detected when the task $\phi_i$ is satisfied (i.e., when $r_i \leq \bar{\rho}^{\phi_i}(\bar{x}_i, 0) \leq \rho_i^{\max}$). We detect task satisfaction via the set

$$D_{i,\text{sat}} := \{(z_i, u_i, u_i^{\text{ext}}) \in \mathfrak{H}_i | r_{\nu_i} \leq \rho^{\psi_{\nu_i}}(\bar{x}_{\nu_i}) \leq \rho_{\nu_i}^{\max},\ \mathfrak{c}_i \geq 0, t \in \mathcal{T}_i^{\text{sat}}\} \setminus (D'_i \cup D''_{i,2}),$$

where $\nu_i := \begin{cases} \mathfrak{c}_i & \text{if } \mathfrak{c}_i > 0 \\ i & \text{if } \mathfrak{c}_i = 0 \end{cases}$ denotes the task currently processed and

$$\mathcal{T}_i^{\text{sat}} := \begin{cases} [a_{\nu_i}, b_{\nu_i}] & \text{if } \phi_{\nu_i} = F_{[a_{\nu_i}, b_{\nu_i}]}\psi_{\nu_i} \\ b_{\nu_i} & \text{if } \phi_{\nu_i} = G_{[a_{\nu_i}, b_{\nu_i}]}\psi_{\nu_i} \\ t_{\nu_i}^* + \bar{b}_{\nu_i} & \text{if } \phi_{\nu_i} = F_{[\underline{a}_{\nu_i}, \underline{b}_{\nu_i}]}G_{[\bar{a}_{\nu_i}, \bar{b}_{\nu_i}]}\psi_{\nu_i} \end{cases}$$

defines the time bounds for task satisfaction. Note that the set substraction of $D'_i \cup D''_{i,2}$ exludes the case where $D'_i$ or $D''_{i,2}$ applies simultaneously with $D_{i,\text{sat}}$. If $(z_i, u_i, u_i^{\text{ext}}) \in D_{i,\text{sat}}$, the agents $j \in \mathcal{V}_i \setminus \{i\}$ are then either not subject to a task or need to continue with $\phi_j$. This is initiated by the map

$$G_{i,\text{sat}} := \Big\{\hat{z}_i \in \mathcal{Z}_i | \hat{t}_i^* = \mathcal{T}_i,\ \hat{\rho}_i^{\max} = \rho^{\max},\ \hat{r}_i = r_i \\ \hat{p}_i^{\gamma} = p_i^{\gamma,\text{new}},\ \hat{\mathfrak{c}}_i = \begin{cases} 0 & \text{if } \mathfrak{c}_i > 0 \text{ and } \mathfrak{c}_i \neq i \\ -1 & \text{if } \mathfrak{c}_i = 0 \text{ or } \mathfrak{c}_i = i \end{cases}\Big\}.$$

Finally, the hybrid system $\mathcal{H}_i$ is defined by the sets $D_i := D'_i \cup D''_{i,2} \cup D_{i,\text{sat}}$ and $C_i := \mathcal{Z}_i \setminus D_i$, as well as the map

$$G_i(z_i, u_i, u_i^{\text{ext}}) = \begin{cases} G'_{i,1}(z_i, u_i, u_i^{\text{ext}}) & \text{if } (z_i, u_i, u_i^{\text{ext}}) \in D'_{i,1}, \\ G'_{i,2}(z_i, u_i, u_i^{\text{ext}}) & \text{if } (z_i, u_i, u_i^{\text{ext}}) \in D'_{i,2}, \\ G'_{i,3}(z_i, u_i, u_i^{\text{ext}} & \text{if } (z_i, u_i, u_i^{\text{ext}}) \in D'_{i,3}, \\ G''_{i,2}(z_i, u_i, u_i^{\text{ext}}) & \text{if } (z_i, u_i, u_i^{\text{ext}}) \in D''_{i,2}, \\ G_{i,\text{sat}}(z_i, u_i, u_i^{\text{ext}}) & \text{if } (z_i, u_i, u_i^{\text{ext}}) \in D_{i,\text{sat}}. \end{cases}$$

It is crucial that the behavior of $\mathcal{H}_i$ does not exhibit two or more discrete options at the same time (i.e., control decision within $\mathcal{H}_i$ should be deterministic). Note that $D_i' = D_{i,1}' \cup D_{i,2}' \cup D_{i,3}'$ and $D_{i,1}'$, $D_{i,2}'$, and $D_{i,3}'$ are nonintersecting. Note also that the sets $D_i'$ and $D_{i,\text{sat}}$, as well as $D_{i,2}''$ and $D_{i,\text{sat}}$, are nonintersecting. However, $D_i'$ and $D_{i,2}''$ are intersecting. Therefore, if $(z_i, u_i, u_i^{\text{ext}}) \in D_i' \cap D_{i,2}''$, we only execute the jump induced by $D_{i,2}''$ to account for the logic modeled by the hybrid system. This can be achieved by modifying $D_i'$ to $D_i' \setminus D_{i,2}''$. It can also be shown that the robustness $r_i$ obtained by the hybrid control strategy is lower-bounded due to assumption 7.3.

**Example 7.5** Consider $N := 9$ omnidirectional robots as in example 7.1, with the dynamics

$$\dot{x}_i = f_i^{\text{u}}(x) + \begin{bmatrix} \cos(\theta_i) & -\sin(\theta_i) & 0 \\ \sin(\theta_i) & \cos(\theta_i) & 0 \\ 0 & 0 & 1 \end{bmatrix} \left(B_i^T\right)^{-1} R_i u_i + w_i,$$

where $R_i$ and $B_i$ are defined as before. We use $f_i^{\text{u}}(x) := \begin{bmatrix} f_{i,x}^{\text{u}}(x) & f_{i,y}^{\text{u}}(x) & 0 \end{bmatrix}$ as a means of collision avoidance with the elements

$$f_{i,k}^{\text{u}}(x) := \sum_{j=1, j \neq i}^{9} \kappa_i \frac{p_{k,i} - p_{k,j}}{\|p_i - p_j\| + 0.000001}$$

for $k \in \{x, y\}$, and where $\kappa_i := 10$. The noise $w_i$ is drawn from a truncated normal distribution with mean 0 and variance 100. The simulation example resembles examples 7.2 and 7.3. We also add requirements governing the robot's orientation. For the first set of independent agents (as in example 7.2), we let

$$\begin{aligned}
\phi_1 &:= G_{[0,15]}\big((\|p_1 - p_2\| \leq 25) \wedge (\|p_1 - p_3\| \leq 25) \wedge (\|p_1 - p_4\| \leq 25)\big), \\
\phi_2 &:= G_{[10,15]}\big((\|p_2 - \begin{bmatrix} 10 & 90 \end{bmatrix}^T \| \leq 5) \wedge (|\theta_2 + 45| \leq 7.5)\big), \\
\phi_3 &:= G_{[10,15]}\big((\|p_3 - \begin{bmatrix} 10 & 10 \end{bmatrix}^T \| \leq 5) \wedge (|\theta_3 - 45| \leq 7.5)\big), \\
\phi_4 &:= G_{[10,15]}\big(\|(p_4 - \begin{bmatrix} 45 & 20 \end{bmatrix}^T \| \leq 5) \wedge (|\theta_4 - 135| \leq 7.5)\big).
\end{aligned}$$

We added the requirements that agents 2, 3, and 4 should eventually be oriented with $-45$, 45, and 135 degrees and remain with this orientation from then on. For the second set of independent agents (as in example 7.3), let

$$\begin{aligned}
\phi_5 &:= F_{[5,10]}\big((\|p_5 - p_6\| \leq 10) \wedge (\|p_5 - p_7\| \leq 10) \wedge (\|p_5 - \begin{bmatrix} 110 & 20 \end{bmatrix}^T \| \leq 5)\big), \\
\phi_6 &:= F_{[5,15]}\big((\|p_6 - \begin{bmatrix} 50 & 20 \end{bmatrix}^T \| \leq 5) \wedge (|\theta_6 - 45| \leq 7.5)\big), \\
\phi_7 &:= F_{[5,15]}\big((\|p_7 - \begin{bmatrix} 110 & 80 \end{bmatrix}^T \| \leq 5) \wedge (|\theta_7 + 135| \leq 7.5)\big).
\end{aligned}$$

We here added the requirements that agents 6 and 7 should eventually be oriented with 45 and −135 degrees, respectively. Finally, for agents 8 and 9, we impose the collaborative specification

$$\phi_8 := \phi_9 := F_{[5,15]}\big((p_{x,8} - p_{x,9} \leq 10) \wedge (p_{x,8} - p_{x,9} \geq 5) \\ \wedge (p_{y,8} - p_{y,9} \leq 10) \wedge (p_{y,8} - p_{y,9} \geq 5)\big).$$

We select $\eta := 1$, for which it holds that $\bar{\rho}_1^{\text{opt}} = 23.9$, $\bar{\rho}_2^{\text{opt}} = \bar{\rho}_3^{\text{opt}} = \bar{\rho}_4^{\text{opt}} = 4.92$, $\bar{\rho}_5^{\text{opt}} = 4.97$, $\bar{\rho}_6^{\text{opt}} = \bar{\rho}_7^{\text{opt}} = 4.92$, and $\bar{\rho}_8^{\text{opt}} = \rho_9^{\text{opt}} = 1.11$, such that we initially choose $r_i := 0.5$ for all $i \in \{1, \ldots, N\}$. For the parameters of the hybrid system, we set $\delta_i := 1.5$ and $N_i := 1$.

The resulting trajectories are shown in figure 7.3. It can be seen that agent 1 does not satisfy $\phi_1$, but it does find a least-violating solution by staying as close as possible to agents 2, 3, and 4. These agents independently satisfy their own formulas. For agents 1 and 2, the corresponding funnels are shown in figures 7.4a and 7.4b. It can be seen that agent 1 first tries to repair the parameters in the first repair stage and then successively decreases robustness $r_1$ by $\delta_1$ in the third repair stage. For agents 5, 6, and 7, it can be seen that they all satisfy their formulas. In particular, agent 5 uses collaborative control along with agents 6 and 7 in the second repair stage after an unsuccessful repair attempt in the first stage. This can be seen in figure 7.4d, while figures 7.4e and 7.4f show the behavior of the collaborating agents 6 and 7. Agents 8 and 9 satisfy their formulas collaboratively. The funnel for agent 8 can be seen in figure 7.4c. All the tasks, except $\phi_1$, are satisfied with a robustness of $r_i := 0.5$.

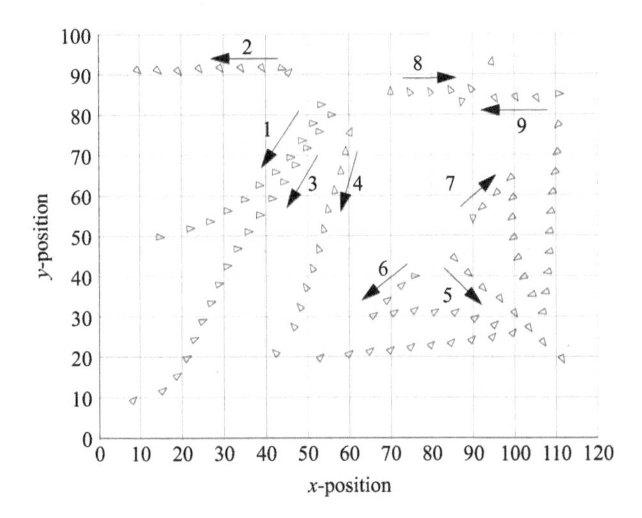


**Figure 7.3**
Agent trajectories: triangles indicate agent orientations. Reprinted with permission from [174].

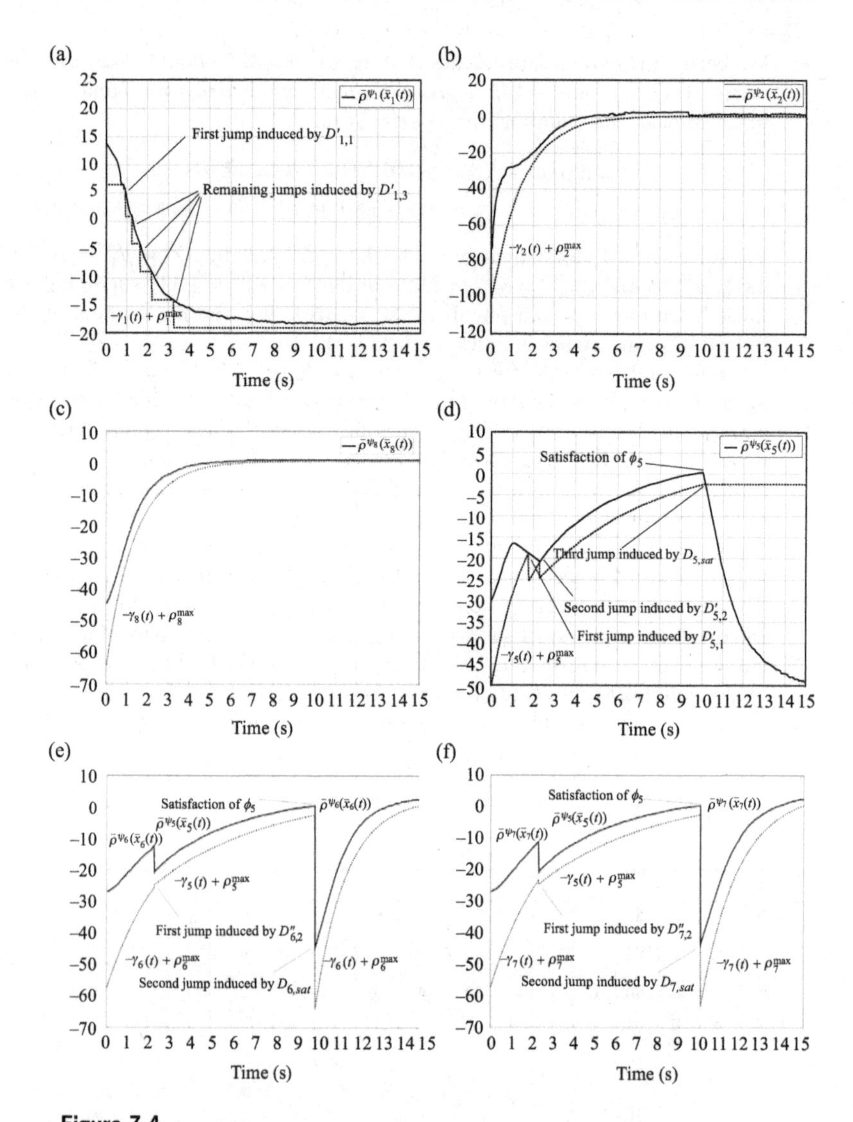


**Figure 7.4**
Funnel repairs for agents 1, 2, 8, 5, 6, and 7. (a) Funnel repairs for agent 1. The first and third repair stages are activated. (b) No funnel repairs for agent 2. The formula $\phi_2$ is satisfied without repairs. (c) No funnel repairs for agent 8. The formula $\phi_8$ is satisfied without repairs. (d) Funnel repairs for agent 5, who requests collaborative help from 6 and 7. (e) Agent 6 starts collaborating with agent 5 at around 2.3 seconds. (f) Agent 7 starts collaborating with agent 5 at around 2.3 seconds. Reprinted with permission from [174].

## 7.3 Experiments and Practical Considerations

Finally, we present a set of experiments with $N$ Nexus 4WD Mecanum Robotic Cars, which are omnidirectional robots as shown in figure 7.5(a). These robots are subject to local STL formulas while operating among $M$ Turtlebots that are executing linear temporal logic (LTL) formulas. Each Nexus Mecanum Robotic Car is described by its position $p_i := [p_{x,i}, p_{y,i}]^T$ and orientation $\theta_i$. The control architecture is shown in figure 7.5(b). The implementation within the Robot Operating System [239] can be found in [4] and also includes the description of the collision avoidance controller.

Algorithm 7.1 summarizes the hybrid control strategy presented in section 7.2. Input limitations, the collision avoidance controller, and the digital (zero-order hold) implementation of the continuous-time funnel controller pose practical challenges and may lead to a violation of the funnel. In this regard,

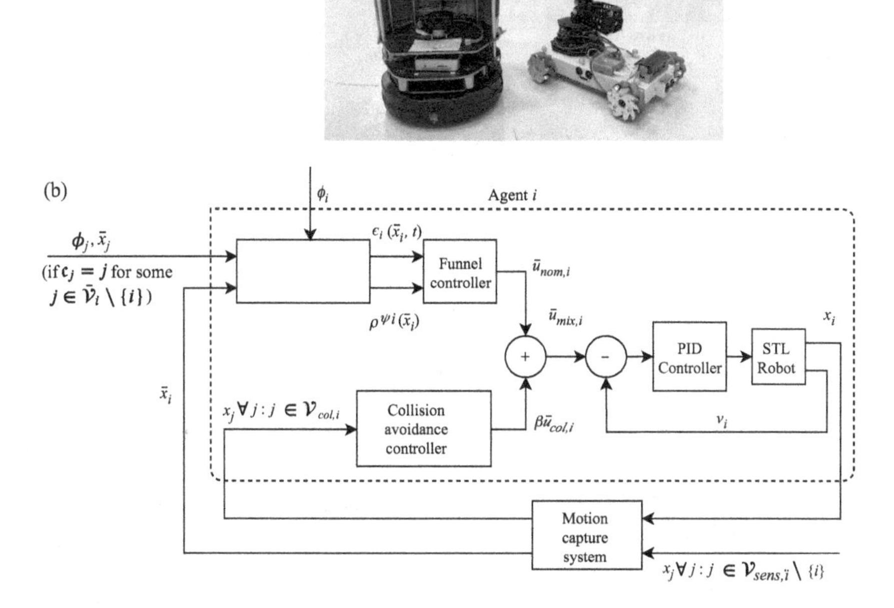


**Figure 7.5**
(a) TurtleBots and Nexus 4WD Mecanum Robotic Cars. (b) Control architecture. Reprinted with permission from [180].

it is important to note that the funnel control law is not defined outside the funnel.[4] The detection and repair scheme, however, recomputes the funnel in these cases, and hence enables us to apply the controller in practice.

**Algorithm 7.1** *1: Calculate initial* $r_i$, $\rho_i^{max}$, $\gamma_i$ *(as instructed in sections 7.1 and 7.2)*

*2: Set* $ce := 0$ ▷ *Counter for critical events*

*3: Set* $\mathfrak{c}_i := 0$ ▷ *Indicates 2nd repair stage*

*4:* ***repeat***

*5:* ***if*** *Critical Event and* $\mathfrak{c}_j := 0$ *for all* $i \in \mathcal{V}_j \setminus \{j\}$ ***then***

*6: Set* $ce := ce + 1$

*7:* ***if*** $ce \leq N_i$ ***then*** ▷ *1st repair stage*

*8: Calculate new* $r_i$, $\rho_i^{max}$, $\gamma_i$ *(section 7.2)*

*9:* ***else*** ▷ *2nd and 3rd repair stages*

*10:* ***if*** *2nd repair stage and* $\mathfrak{c}_i = 0$ ***then***

*11: Calculate new* $r_i$, $\rho_i^{max}$, $\gamma_i$ *(section 7.2)*

*12: Set* $\mathfrak{c}_i := i$*; send* $\rho_i^{max}$, $\gamma_i$, $\psi_i$ *to* $j \in \mathcal{V}_i \setminus \{i\}$

*13:* ***else if*** *2nd repair stage and* $\mathfrak{c}_i = i$ ***then***

*14:* $r_i := r_i - \delta_i$*; Calculate new* $\rho_i^{max}$, $\gamma_i$

*15: Send* $\rho_i^{max}$, $\gamma_i$, $\psi_i$ *to* $j \in \mathcal{V}_i \setminus \{i\}$

*16:* ***else*** *3rd repair stage*

*17:* $r_i := r_i - \delta_i$*; Calculate new* $\rho_i^{max}$, $\gamma_i$

*18:* ***if*** $\mathfrak{c}_j = j$ *and* $i \in \mathcal{V}_j \setminus \{j\}$ ***then*** ▷ *2nd repair stage*

*19: Set* $\rho_i^{max} := \rho_j^{max}$, $\gamma_i := \gamma_j$, $\psi_i := \psi_j$

*20:* ***if*** $\mathfrak{c}_j$ *changed from* $j$ *to* $-1$ *and* $i \in \mathcal{V}_j \setminus \{j\}$ ***then***

*21: Reset* $\psi_i$*, Calculate new* $r_i$, $\rho_i^{max}$, $\gamma_i$

*22: Calculate and output* $\epsilon(\bar{x}_i, t)$ *and* $\bar{\rho}^{\psi_i}(\bar{x}_i)$

*23:* ***until*** $\bar{\rho}^{\phi_i}(\bar{x}_i, 0) \geq \bar{r}_i$ *where* $\bar{r}_i$ *is maximized.*

*24: Set* $\mathfrak{c}_i := -1$

We now present the experimental results for three scenarios that were performed on a rectangular workspace of size $[-3.5, -3.5] \times [3.5, 3.5]$ (measured in meters). Videos of the three scenarios can be found in [5,6, and 7], respectively. The control loops was run at 100 Hz.

[4]This is as opposed to the controller that we obtained from time-varying CBFs that are defined outside the set $\mathcal{C}(t)$.

---

**Example 7.6** We have $M=2$ and $N=3$ robots. For LTL robots 1 and 2, we define the propositions $\alpha_{1,1}$, $\alpha_{1,2}$, $\alpha_{2,1}$, and $\alpha_{2,2}$ as illustrated in figure 7.6. The LTL tasks are

$$\phi_1 := GF(\alpha_{1,1} \wedge \alpha_{1,2}),$$
$$\phi_2 := GF(\alpha_{2,1} \wedge \alpha_{2,2}).$$

Stated in words, robot 1 (robot 2) should periodically visit $\alpha_{1,1}$ and $\alpha_{1,2}$ ($\alpha_{2,1}$ and $\alpha_{2,2}$). For STL robot 3, define the predicates $\mu_{3,1} := (\|p_3 - \begin{bmatrix}0 & -2\end{bmatrix}^T\| \leq 0.1)$, $\mu_{3,2} := (\|p_3 - \begin{bmatrix}1.5 & -1.5\end{bmatrix}^T\| \leq 0.1)$, and $\mu_{3,3} := (\|p_3 - p_4\| \leq 0.7)$. Robot 3 is then subject to the STL specification

$$\phi_3 := \phi_3' \wedge \phi_3'' := G_{[21,30]}\mu_{3,1} \wedge F_{[57,58]}(\mu_{3,2} \wedge \mu_{3,3}).$$

Stated in words, between 21–30 seconds, robot 3 should always be in region $\mu_{3,1}$ and between 57–58 seconds, robot 3 should eventually be in region $\mu_{3,2}$, while being at most 0.7 meters from robot 4. For STL robot 4, define the predicates $\mu_{4,1} := (\|p_4 - \begin{bmatrix}2 & 2\end{bmatrix}^T\| \leq 0.1)$, $\mu_{4,2} := (\|p_4 - p_3\| \leq 1)$, and $\mu_{4,3} := (\|p_4 - p_5\| \leq 1)$. Robot 4 is then subject to

$$\phi_4 := \phi_4' \wedge \phi_4'' := G_{[5,30]}(\mu_{4,2} \wedge \mu_{4,3}) \wedge F_{[83,87]}\mu_{4,1}.$$

Stated in words, always between 5–30 seconds robot 4 should be at most 1 meter from robots 3 and 5 and eventually between 83–87 seconds robot 4 should be in region $\mu_{4,1}$. Robot 5 should always between 21–30 seconds be in region $\mu_{5,1}$ and eventually between 44–47 seconds be in region $\mu_{5,2}$ with predicates $\mu_{5,1} := (\|p_5 - \begin{bmatrix}0 & 2\end{bmatrix}^T\| \leq 0.1)$ and $\mu_{5,2} := (\|p_5 - \begin{bmatrix}-1 & 0.5\end{bmatrix}^T\| \leq 0.1)$. Robot 5 is subject to

$$\phi_5 := \phi_5' \wedge \phi_5'' := G_{[21,30]}\mu_{5,1} \wedge F_{[44,47]}\mu_{5,2}.$$

The trajectories of the robots for 0–30 seconds and 30–90 seconds are shown in figures 7.6a and 7.6b, respectively. The evolution of a robot over time is indicated by increasing color intensity, and it can be seen that collisions are avoided. Note that robots 3 and 4 are coupled to other robots. In figure 7.6a, showing $\phi_3'$, $\phi_4'$, and $\phi_5'$, robot 4 needs to stay close to robots 3 and 5, while robots 3 and 5 are supposed to move to $\mu_{3,1}$ and $\mu_{5,1}$, respectively, such that robot 4 cannot satisfy $\phi_4'$. Robot 4 finds a least-violating solution by successively reducing the robustness $r_4$ (the third repair stage; see figure 7.6c) and consequently staying as close as possible to robots 3 and 5. Robots 3 and 5 satisfy their tasks $\phi_3'$ and $\phi_5'$, respectively, as illustrated for robot 5 in figure 7.6d. More formally, we have $\bar{\rho}^{\phi_3'}(\bar{x}_3, 0) \geq -0.18$, $\bar{\rho}^{\phi_4'}(\bar{x}_4, 0) \geq -1.11$, and $\bar{\rho}^{\phi_5'}(\bar{x}_5, 0) \geq -0.12$. In figure 7.6b, showing $\phi_3''$, $\phi_4''$, and $\phi_5''$, robot 3 needs to move to $\mu_{3,2}$, while staying close to robot 4. Robot 4, however, is supposed to move to $\mu_{4,1}$. Therefore, at some point, robot 3 establishes a collaboration with robot 4 to collaboratively satisfy $\phi_3''$ (the second repair stage; see figure 7.6e). Afterward, robot 4 is continuous with $\phi_4''$ (see figure 7.6f). It holds that $\bar{\rho}^{\phi_3''}(\bar{x}_3, 0) \geq 0.02$, $\bar{\rho}^{\phi_4''}(\bar{x}_4, 0) \geq -0.4$,

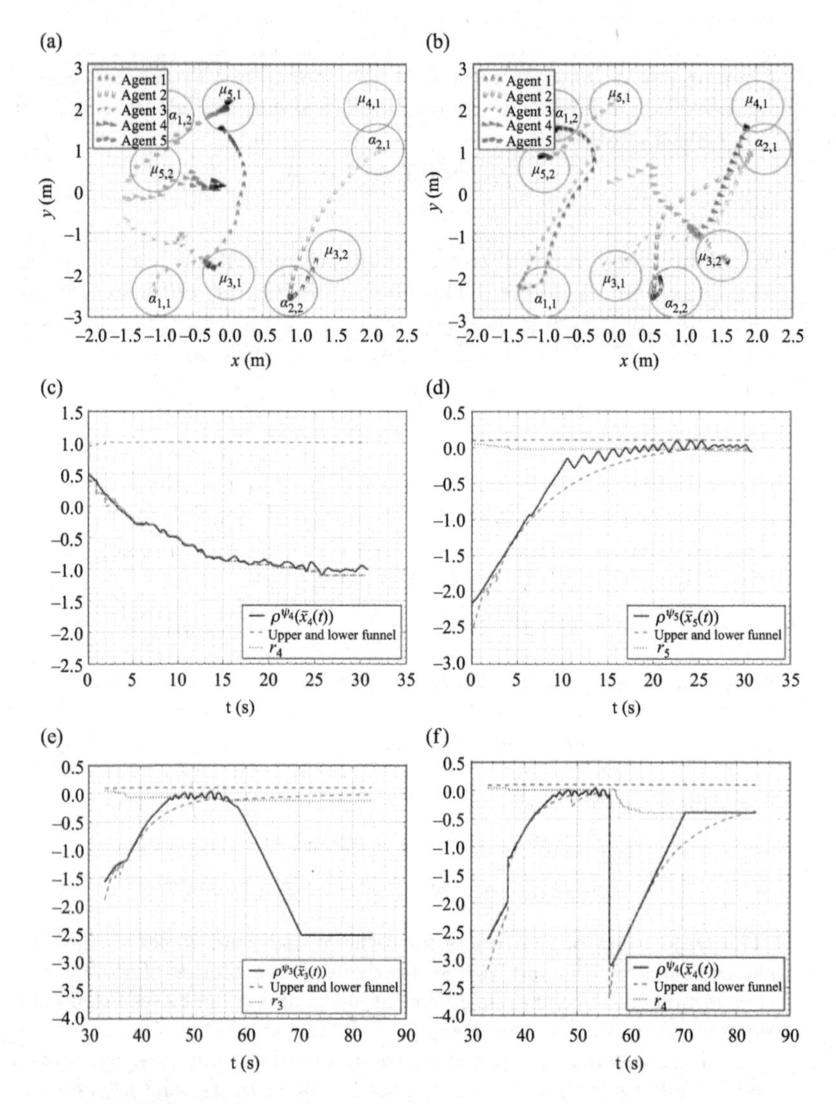


**Figure 7.6**
Robot trajectories and the evolution of the quantitative semantics for selected robots in example 7.6. (a) Robot trajectories from 0–30 seconds. (b) Robot trajectories from 30–90 seconds. (c) Quantitative semantics $\bar{\rho}^{\psi'_4}(\bar{x}_4(t))$. (d) Quantitative semantics $\bar{\rho}^{\psi'_5}(\bar{x}_5(t))$. (e) Quantitative semantics $\bar{\rho}^{\psi''_3}(\bar{x}_3(t))$. (f) Quantitative semantics $\bar{\rho}^{\psi''_4}(\bar{x}_4(t))$. Reprinted with permission from [180].

and $\bar{\rho}^{\phi_5''}(\bar{x}_5, 0) \geq 0.5$. This scenario illustrates how the online detection and repair scheme handles critical events such that, even when obstacles need to be avoided or when local tasks are not satisfiable, $\bar{\rho}^{\phi_i}(\bar{x}_i, 0) \geq \bar{r}_i$ is achieved where $\bar{r}_i$ is maximized. As can be seen in the video in [5], the workspace is densely filled with robots, so collision avoidance is the main reason why the robustness $\bar{r}_i$ is decreased.

---

**Example 7.7** In this scenario, we couple robots under STL tasks to robots under LTL tasks to induce periodic motion. Consider $M = 2$ and $N = 2$ robots. For LTL robots 1 and 2, let

$$\phi_1 := GF(\alpha_{1,1} \wedge \alpha_{1,2}),$$
$$\phi_2 := GF(\alpha_{2,1} \wedge \alpha_{2,2}),$$

again encode the task to periodically visit the regions $\alpha_{1,1}$, $\alpha_{1,2}$ and $\alpha_{2,1}$, $\alpha_{2,2}$, respectively (see figure 7.7a). For STL robots 3 and 4, define the predicates $\mu_3 := (\|p_3 - p_1\| \leq 0.6)$ and $\mu_4 := (\|p_4 - p_2\| \leq 0.6)$ and consider the STL formulas

$$\phi_3 := G_{[0,90]}\mu_3,$$
$$\phi_4 := G_{[0,90]}\mu_4;$$

that is, robots 3 and 4 track robots 1 and 2, respectively. The robots under STL tasks congest the path so that collisions are expected. The trajectories of the robots from 23–45 seconds are shown in figure 7.7a. The LTL tasks $\phi_1$ and $\phi_2$ are satisfied, while the robots under STL tasks closely track the robots under LTL tasks (see figure 7.7b); $\phi_3$ and $\phi_4$ are not satisfied due to collision avoidance and the induced repair stages; however, we obtain $\bar{\rho}^{\phi_3}(\bar{x}_3, 0) \geq -0.45$ and $\bar{\rho}^{\phi_4}(\bar{x}_4, 0) \geq -0.47$.

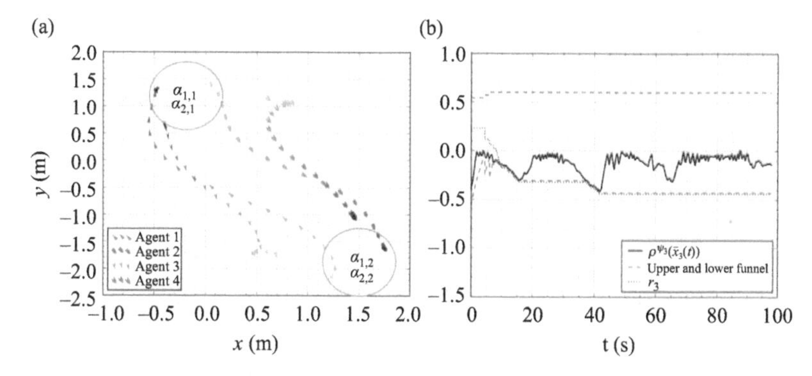


**Figure 7.7**
Robot trajectories and quantitative semantics for example 7.7. (a) Robot trajectories from 23–45 seconds. (b) Quantitative semantics $\bar{\rho}^{\psi_3}(\bar{x}_3(t))$. Reprinted with permission from [180].

**Example 7.8** We again couple robots to other robots, but in a more complex way. Consider $M=1$ and $N=3$ robots. For the LTL robot, we have

$$\phi_1 := F(\alpha_{1,1} \wedge \alpha_{1,2} \wedge \alpha_{1,3} \wedge \alpha_{1,4} \wedge \alpha_{1,5}),$$

where the propositions can be seen in figure 7.8a. The LTL robot finds the shortest path that satisfies $\phi_1$, which is to move to $\alpha_{1,1}$, $\alpha_{1,2}$, $\alpha_{1,3}$, $\alpha_{1,4}$, and $\alpha_{1,5}$ in sequence. The robots under STL tasks are supposed to make a formation with respect to the LTL robot and also track its orientation. Define the predicates $\mu_{2,1} := (\|p_2 - p_1 - \begin{bmatrix}\sin(x_{1,3}) & -\cos(x_{1,3})\end{bmatrix}^T\|) \leq 0.1$ and $\mu_{2,2} := (|x_{2,3} - x_{1,3}|) \leq 0.09$, and let robot 2 be subject to the STL formula

$$\phi_2 := G_{[10,100]}(\mu_{2,1} \wedge \mu_{2,2}),$$

which encodes the task that robot 2 should always be to the right of robot 1 and track its orientation. Further define the predicates $\mu_{3,1} := (\|p_3 - p_1 - \begin{bmatrix}-\sin(x_{1,3}) & \cos(x_{1,3})\end{bmatrix}^T\|) \leq 0.1$ and $\mu_{3,2} := (|x_{3,3} - x_{1,3}|) \leq 0.09$, and let robot 3 be subject to the STL task

$$\phi_3 := G_{[10,100]}(\mu_{3,1} \wedge \mu_{3,2});$$

that is, robot 3 should always be to the left of robot 1 and track its orientation. Similarly, $\phi_4$, omitted here, encodes the task that robot 4 should always be behind robot 1 and track its orientation. The robustness function for robot 3 is shown in figure 7.8b.

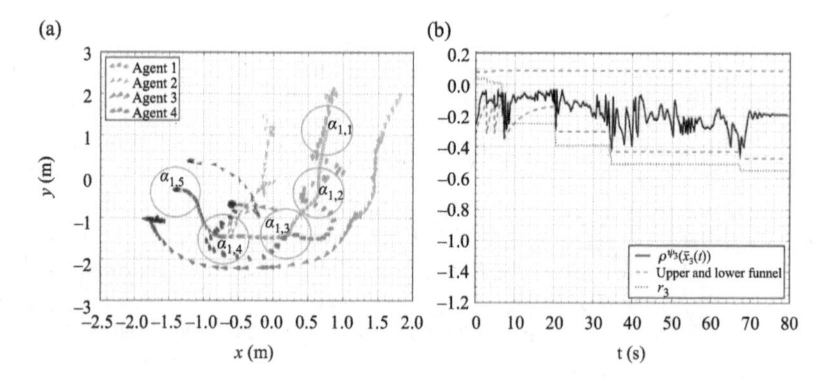


**Figure 7.8**
Robot trajectories and quantitative semantics for example 7.8. (a) Robot trajectories. (b) Quantitative semantics $\bar{\rho}^{\psi_3}(\bar{x}_3(t))$. Reprinted with permission from [180].

## 7.4 Notes and References

The results from sections 7.1 and 7.2 are primarily taken from [173, 174]. Distributed funnel-based control of leader-follower multi-agent systems under STL specifications was considered in [69, 70]. In a different direction, funnel control laws and assume-guarantee contracts were used for compositional design under STL specifications [187]. On the other hand, distributed funnel-based consensus algorithms were designed in [147, 196]. Similarly, the authors in [205] present control laws for distance-based formation control of multi-agent systems, while the authors in [310] design funnel controllers for vehicle platooning.

In section 7.3, we briefly touched on practical challenges when implementing STL control algorithms for multi-agent systems based on [180]. More recently, multi-agent control design frameworks were implemented on a wider range of robotics platforms; for instance, see [270] using an efficient sampling-based multi-agent planner, [269] using prescribed performance control, and [119, 120] using CBFs for reactive specifications.

# IV

# Planning under Spatiotemporal Logic Specifications

# Chapter 8

# Timed Automata-Based Planning

The feedback control laws presented in parts II and III are limited in the sense that they can satisfy specifications from only certain fragments of signal temporal logic (STL). Specifically, we required concavity in the predicate functions to derive globally valid feedback control laws. This is a well-known limitation in feedback control, and it is not only limited to control under STL specifications; for instance, in the context of robot navigation where conflicting convex (obstacle avoidance) and concave (reachability) objectives are considered. While one may be able to relax this assumption in some cases and derive locally valid controllers (e.g., in the case of robot navigation [155]), this always requires a careful analysis of the local optima induced by the particular specification. For the same reason, we also did not consider negation and disjunction operators in the STL fragments of parts II and III. In addition, we did not consider complex nesting of temporal operators, and we preselected the times when an eventually or until operator is satisfied in our CBF and funnel encoding. This comes as no surprise, as the general control problem under STL specifications is NP-hard.

In this chapter, we design timed signal transducers that encode STL specifications without limiting ourselves to specific STL fragments. Using these timed signal transducers, we can decompose an STL specification into a sequence of simpler specifications that can be mapped to the fragments in parts II and III. Consequently, we can apply our previously derived feedback control laws sequentially to satisfy the specification at hand. Importantly, we integrate properties of our feedback control laws into the timed signal transducer to ensure that the sequence of simpler specifications can be satisfied. The consideration of more complex specifications naturally results in an increase in computational complexity. Fortunately, the construction of the timed signal transducers and the synthesis of the sequence of simpler specifications can be performed offline.

In section 8.1, we show how to construct timed signal transducers that encode STL specifications. We then provide an algorithm to check the satisfiability of the STL specification (i.e., to check if there is a continuous-time signal $x:\mathbb{R}\to\mathbb{R}^n$ that satisfies the specification). In section 8.2, we construct a timed abstraction of a dynamical system $\dot{x}=f(x)+g(x)u$. This timed abstraction relies on information of the feedback control laws from parts II and III. We will build this timed abstraction such that it has a similar structure as the timed signal transducer, which enables us to describe their joint behavior by a product automaton that does not contain more states than the timed signal transducer itself. This makes the approach applicable to higher-dimensional multi-agent systems for which existing approaches are known to scale poorly due to the state space explosion. In fact, the size of the timed abstraction will not depend on the number of agents, but rather on the complexity of the specification. Finally, in section 8.3, we show how we can synthesis a plan, which is a sequence of simpler specifications, that we execute by our feedback control laws.

## 8.1 Encoding Signal Interval Temporal Logic

We consider a general class of STL specifications, but we assume that time intervals of temporal operators are rational intervals. The same assumption was previously made when we presented metric interval temporal logic (recall section 3.3). We call this "signal interval temporal logic (SITL)." Formally, let $M$ again denote a set of predicates with $\mu\in M$. Then, the SITL syntax is

$$\phi ::= \top \mid \mu \mid \neg\phi' \mid \phi'\wedge\phi'' \mid \phi' U_I \phi'' \mid \phi' \underline{U}_I \varphi'', \tag{8.1}$$

where all operators are defined as before. We, however, restrict the time interval $I$ to belonging to the nonnegative rationals (i.e., $I\subseteq\mathbb{Q}_{\geq 0}$). In addition, we require that $I$ is not a singleton; that is, $I$ cannot take the form $I:=[a,a]$ for $a\in\mathbb{Q}_{\geq 0}$. As remarked before, note that the former assumption is not restrictive in practice, while the latter assumption excludes punctuality constraints.

In the remainder, the SITL specification $\phi$ is abstracted into a metric interval temporal logic (MITL) specification $\varphi$ that is translated into its language-equivalent timed signal transducer, as instructed in section 3.3. We emphasize that we use the symbols $\varphi$ and $\phi$ for MITL and SITL formulas, respectively. This timed signal transducer will then be modified to account for the error induced by considering propositions in $\varphi$ (MITL) instead of predicates in $\phi$ (SITL). The modified timed signal transducer then describes all real-valued signals that satisfy the SITL specification, and it can hence be checked if the specification is satisfiable.

### 8.1.1 Timed Automata for Signal Interval Temporal Logic

The SITL formula $\phi$ consists of the predicates $\mu_i \in M$ with $i \in \{1, \ldots, |M|\}$. For a moment, let us use the notation $\phi(M)$ instead of $\phi$ to make explicit the dependence of the SITL formula $\phi$ on the predicate $M$. We then abstract the SITL formula $\phi(M)$ into an MITL formula $\varphi := T(\phi(M))$ by the transformation $T(\cdot)$, which essentially replaces predicates $M$ in $\phi(M)$ by a set of atomic propositions $AP$. Formally, we associate a proposition $p_i \in \{\top, \bot\}$ with each predicate $\mu_i \in M$, and we define this set of propositions as $AP := \{p_1, \ldots, p_{|M|}\}$. Informally, we can hence think of $\varphi$ as $\phi(AP)$. For instance, $\phi(M) := F_I(\mu_1 \wedge \mu_2)$ becomes $\varphi := T(\phi(M)) = F_I(p_1 \wedge p_2)$. Let the inverse $T^{-1}(\varphi) = T^{-1}(T(\phi(M))) = \phi(M)$ be obtained by replacing each $p_i \in AP$ in $\varphi$ with the corresponding $\mu_i \in M$.

In the next step, we construct a timed signal transducer

$$TST_\varphi := (S, s_0, \Lambda, \Gamma, c, \iota, \Delta, \lambda, \gamma, \mathcal{A})$$

for the MITL formula $\varphi$ according to section 3.3 with $\Lambda := AP$, where we recall that $AP$ is the set of propositions coming from the $T(\phi)$ abstraction of $\phi$. Recall from theorem 3.4 that $TST_\varphi$ encodes all signals $\bar{\sigma} : \mathbb{R}_{\geq 0} \to \mathbb{B}^{|AP|}$ that satisfy $\varphi$.

As we ultimately aim to satisfy the STL formula $\phi$, not the MITL formula $\varphi$, we have to modify the timed signal transducer $TST_\varphi$ to account for the error that we make by abstracting the predicates in $\phi$ into propositions in $\varphi$. Recall that states $s \in S$ and transitions $\delta \in \Delta$ in the timed signal transducer $TST_\varphi$ are equipped with input labels $\lambda(s)$ and $\lambda(\delta)$, respectively. These labels, as we saw in section 3.3, are Boolean formulas over the propositions in $AP$; that is, any spatial information about the predicates in $M$ is lost. To recover this information, we perform the following operations on the states and transitions in $TST_\varphi$:

[O1] Remove each state $s \in S$ for which there is no point $x \in \mathbb{R}^n$ such that $x \models T^{-1}(\lambda(s))$. Remove the corresponding state $s$ from the acceptance set $\mathcal{A}$. Further, remove the corresponding ingoing transitions $(s', g, r, s) \in \Delta$ for some $s' \in S$ and outgoing transitions $(s, g, r, s') \in \Delta$ for some $s' \in S$.

[O2] Remove each transition $\delta := (s, g, r, s') \in \Delta$ for which there is no point $x \in \mathbb{R}^n$ such that $x \models T^{-1}(\lambda(\delta))$. Remove the corresponding transition $\delta$ from the acceptance set $\mathcal{A}$.

Note that we have to perform a feasibility check for each state $s \in S$ and for each transition $\delta \in \Delta$; that is, we have to check if there is a point $x \in \mathbb{R}^n$ that satisfies the input labels for states $x \models T^{-1}(\lambda(s))$ and transitions $x \models T^{-1}(\lambda(\delta))$. In order to perform this feasibility check, we can set up a feasibility problem as a mixed-integer program; see [44, chapter 2] for details. For linear predicate functions, this indeed reduces to solving a mixed-integer linear program. The timed signal transducer modified by operations [O1] and [O2] is denoted by

$$TST_\phi := (S^\phi, s_0, \Lambda, \Gamma, c, \iota, \Delta^\phi, \lambda, \gamma, \mathcal{A}^\phi). \tag{8.2}$$

Due to its construction, note that it holds that $S^\phi \subseteq S$, $\Delta^\phi \subseteq \Delta$, and $\mathcal{A}^\phi \subseteq \mathcal{A}$.

Knowing that the MITL specification $\varphi$ and the SITL specification $\phi$ differ only in their atomic elements, it is easy to imagine that the modified timed signal transducer $TST_\phi$ now encodes all signals $x : \mathbb{R} \to \mathbb{R}^n$ that satisfy the SITL specification $\phi$. In the next section, we will formally show this result. In fact, we next construct a plan $d_p : \mathbb{R}_{\geq 0} \to BC(M)$ (formally defined next) from $TST_\phi$ that characterizes all signals $x : \mathbb{R}_{\geq 0} \to \mathbb{R}^n$ such that $(x, 0) \models \phi$.

## 8.1.2 Satisfiability of Signal Interval Temporal Logic

The synthesis of a plan $d_p$ is based on the fact that we can translate $TST_\phi$, which is a timed automaton when neglecting the output labels [12], to a region automaton $RA(TST_\phi)$.[1] The region automaton $RA(TST_\phi)$ can then be used to check the emptiness of $TST_\phi$; that is, it can be used to analyze the reachability properties of $TST_\phi$. Since the modified timed signal transducer $TST_\phi$ has invariants for states $\iota(s)$ and guards $g$ for transitions $(s, g, r, s') \in \Delta^\phi$, we modify the standard algorithm to construct a region automaton presented in

[1] Similarly, we could translate $TST_\phi$ into a computationally more efficient zone automaton, but that is avoided here for simplicity. We refer the reader to [261] for more details.

[12]. We start by associating a transition relation $\Rightarrow$ over the extended state space $S^\phi \times \mathbb{R}^O_{\geq 0}$; that is, over the joint space of states $S^\phi$ and clocks (recall that there are $O$ clocks) in $TST_\phi$.

**Definition 8.1 (Equivalent Transition System of $TST_\phi$)** *Let $(S^\phi \times \mathbb{R}^O_{\geq 0}, \Rightarrow)$ be a transition system with transition relation $(s,c) \overset{\delta}{\Rightarrow} (s',c')$ if there is a time $t' \in \mathbb{R}_{\geq 0}$ and a transition $\delta := (s,g,r,s') \in \Delta^\phi$ in $TST_\phi$ such that*

- *for all $\tau \in (0,t')$, it holds that $c+\tau \models \iota(s)$ (invariance condition),*
- *it holds that $c' = r(c+t')$ (reset condition), and*
- *it holds that $c+t' \models g$ (guard condition).*

Note that each transition $(s,c) \overset{\delta}{\Rightarrow} (s',c')$ requires a combination of time and discrete transitions that simulate the transition $\delta := (s,g,r,s') \in \Delta^\phi$ in $TST_\phi$. Each such transition has to satisfy the invariance condition $\iota$, the reset condition $r$, and the guard condition $g$. By construction, it follows that $TST_\phi$ and $(S^\phi \times \mathbb{R}^O_{\geq 0}, \Rightarrow)$ exhibit the same behavior in terms of the states $S^\phi$ that are reachable.

With the definition of the equivalent transition system in hand, we can now analyze the reachability properties of the infinite state modified timed signal transducer $TST_\phi$. What is meant by that is that we can find states in $TST_\phi$ that can be reached from the initial state $s_0$. Consequently, we can check if the acceptance condition $\mathcal{A}^\phi$ can be satisfied. We do so by constructing the finite state region automaton $RA(TST_\phi)$. Formally, this region automaton is the quotient system of the transition system $(S^\phi \times \mathbb{R}^O_{\geq 0}, \Rightarrow)$ under an equivalence relation $\sim \subseteq \mathbb{R}^O_{\geq 0} \times \mathbb{R}^O_{\geq 0}$ (defined in the remainder). More generally, we can think of a quotient system as another transition system that is obtained from a transition system by joining some of the states into equivalence classes. These equivalence classes are defined via the equivalence relation and form the states of the quotient system. In our case, the equivalence relation $\sim$ will define equivalence classes that are clock regions for $c$. Formally, for each clock $c_o$ with $o \in \{1, \ldots, O\}$ of $c$, let $C_o$ be the largest constant occurring in any clock constraint of $TST_\phi$.[2] Denote as $\lfloor c_o \rfloor$ and $\text{fract}(c_o)$ the integral and fractional parts of $c_o$, respectively, so that $c_o = \lfloor c_o \rfloor + \text{fract}(c_o)$.

[2]We remind the reader that clock constraints are Boolean combinations over clock predicates; that is, they follow the syntax $g ::= c_o \leq k \mid c_o \geq k \mid \neg g \mid g_1 \wedge g_2$ for some constant $k \in \mathbb{Q}_{\geq 0}$. We pick the largest such constant $k$ in $TST_\phi$ associated with $c_o$ to be $C_o$.

**Definition 8.2 (Clock Regions)** *Let* $\sim$ *be an equivalence relation* $\sim \subseteq \mathbb{R}^O_{\geq 0} \times \mathbb{R}^O_{\geq 0}$ *with* $c \sim c'$ *if*

- *for each* $o \in \{1, \ldots, O\}$*, either* $\lfloor c_o \rfloor = \lfloor c'_o \rfloor$ *or* $\min(c_o, c'_o) > C_o$*,*
- *for each* $k, l \in \{1, \ldots, O\}$ *with* $c_k \leq C_k$ *and* $c_l \leq C_l$*,* $fract(c_k) \leq fract(c_l)$ *iff* $fract(c'_k) \leq fract(c'_l)$*, and*
- *for each* $o \in \{1, \ldots, O\}$ *with* $c_o \leq C_o$*,* $fract(c_o) = 0$ *iff* $fract(c'_o) = 0$

*A clock region is now an equivalence class that is induced by* $\sim$ *(i.e., a clock region* $[c]$ *is defined as* $[c] := \{c' | c \sim c'\}$*).*

To provide an example, the clock region $\{c' \in \mathbb{R}^2_{\geq 0} | c' \sim [1.5\, 1.2]^T\}$ joins the set of clocks $c' = \begin{bmatrix} c'_1 & c'_2 \end{bmatrix}^T$ such that $1 < c'_2 < c'_1 < 2$ if $\min(C_1, C_2) > 2$. Let $\alpha$ and $\alpha'$ be clock regions and assume that $c \in \alpha$ and $c' \in \alpha'$. If $(s, c) \overset{\delta}{\Rightarrow} (s', c')$ and $c \sim \bar{c}$ for some $\bar{c}$, it then holds that there is a $\bar{c}'$ with $c' \sim \bar{c}'$ such that $(s, \bar{c}) \overset{\delta}{\Rightarrow} (s', \bar{c}')$. It is exactly this property that enables us to reason over clock regions within the region automaton instead of the behavior of the infinite state transition system $(S^\phi \times \mathbb{R}^O_{\geq 0}, \Rightarrow)$. With this definition, we can finally define the region automaton.

**Definition 8.3 (Region Automaton of** $TST_\phi$**)** *The region automaton* $RA(TST_\phi) := (Q, q_0, \Delta_R, \mathcal{A}_R)$ *is the quotient system of the transition system* $(S^\phi \times \mathbb{R}^O_{\geq 0}, \Rightarrow)$ *using clock regions as equivalence classes, and* $RA(TST_\phi)$ *is defined as follows:*

- *The states are pairs of* $(s, \alpha) \in Q := S^\phi \times A$ *with* $s \in S^\phi$ *and* $\alpha \in A$*, where* $A$ *is the set of all clock regions.*
- *The initial states are* $q_0 := (s_0, \alpha_0) \in s_0 \times A$*, where* $\alpha_0$ *is the clock region corresponding to* $c(0)$*.*
- *For* $q := (s, \alpha) \in Q$ *and* $q' := (s', \alpha') \in Q$*, there is a transition* $(q, \delta, q') \in \Delta_R$ *if there is a transition* $(s, c) \overset{\delta}{\Rightarrow} (s', c')$ *for* $c \in \alpha$ *and* $c' \in \alpha'$*.*
- $(s, \alpha) \in \mathcal{A}_R(i)$ *if* $s \in \mathcal{A}^\phi(i)$*.*

**Finding Accepting State and Timing Sequences.** The region automaton $RA(TST_\phi) := (Q, q_0, \Delta_R, \mathcal{A}_R)$ has a finite set of states so we can, in principle, analyze its behavior by just considering all possible state sequences starting from the initial state $q_0$. We can then check if there is a state sequence that satisfies the generalized Büchi acceptance condition $\mathcal{A}_R$. However, there are more efficient solutions. For instance, using standard graph search techniques such as the memory-efficient variant of the nested depth-first search [83], here used to deal with the generalized Büchi acceptance condition as in [294]. As a result, we obtain, if it exists, a state sequence

$$\mathfrak{q} := (q_0, q_1, \ldots)$$

satisfying the generalized Büchi acceptance condition $\mathcal{A}_R$ with states $q_j := (s_j, \alpha_j)$ and transitions $(q_j, \delta_j, q_{j+1}) \in \Delta_R$ for each step $j \in \mathbb{N}$. As we are interested in satisfying the specification $\phi$ (i.e., we would like to know if there is a signal $x : \mathbb{R} \to \mathbb{R}^n$ such that $(x, 0) \models \phi$), we require that the first transition $\delta_0$ is such that $\gamma(\delta_0) = y$ as opposed to $\gamma(\delta_0) = \neg y$, which would indicate the opposite (recall section 3.3). In particular, we note that the sequence $\mathfrak{q}$ will consist of a prefix of length $\mathrm{p}+1$ and a suffix of length s such that

$$\mathfrak{q} := (\mathfrak{q}_{\mathrm{p}}, \mathfrak{q}_{\mathrm{s}}^{\omega}),$$

where we denote the prefix and suffix as $\mathfrak{q}_{\mathrm{p}} := (q_0, \ldots, q_{\mathrm{p}})$ and $\mathfrak{q}_{\mathrm{s}} := (q_{\mathrm{p}+1}, \ldots, q_{\mathrm{p}+\mathrm{s}})$, respectively. Note that the $\omega$ notation in $\mathfrak{q}_{\mathrm{s}}^{\omega}$ is used to indicate that $\mathfrak{q}_{\mathrm{s}}$ is repeated infinitely often.

The state sequence $\mathfrak{q}$ does not contain any timing information (i.e., when a state $q_j$ in $\mathfrak{q}$ should be visited). Naturally, timing information is needed to satisfy an STL specification $\phi$. We add such timing information to the state sequence $\mathfrak{q}$ in the form of the timing sequence

$$\bar{\tau} := (\bar{\tau}_{\mathrm{p}}, \bar{\tau}_{\mathrm{s}}^{\omega}).$$

Similar to $\mathfrak{q}_{\mathrm{p}}$ and $\mathfrak{q}_{\mathrm{s}}$, we have a prefix and a suffix timing sequence $\bar{\tau}_{\mathrm{p}} := (\tau_0 := 0, \ldots, \tau_{\mathrm{p}})$ and $\bar{\tau}_{\mathrm{s}} := (\tau_{\mathrm{p}+1}, \ldots, \tau_{\mathrm{p}+\mathrm{s}})$, where $\tau_j \in \mathbb{R}_{\geq 0}$ for $j \geq 1$ corresponds to the occurrence of $\delta_j$, which happens $\tau_j$ time units after the occurrence of $\delta_{j-1}$. The timing sequence $\bar{\tau}$ can be constructed from the clock regions $\alpha_j$ in $q_j = (s_j, \alpha_j)$, and $\bar{\tau}$ has to be such that the invariance, reset, and guard conditions are satisfied. How can we construct this timing sequence $\bar{\tau}$? The proof of [12, lemma 4.13] shows how to find timings for a simple acceptance condition (i.e., only requiring $\bar{\tau}_{\mathrm{p}}$, while we deal with a generalized Büchi acceptance condition for which a suffix is needed). Algorithmic implementations are provided by tools such as UPPAAL [38]. For now, we assume for simplicity that $\bar{\tau}$ is given to us; later, in section 8.3, we show how to obtain $\bar{\tau}$ in our setting. Importantly, if a state sequence $\mathfrak{q}$ were found that satisfied the generalized Büchi acceptance condition $\mathcal{A}_R$, then there always exists a corresponding timing sequence $\bar{\tau}$.

**Plan Synthesis.** Now, by denoting the total elapsed time at step $j$ as $T_j := \sum_{k=0}^{j} \tau_j$, we can associate a plan $d_p(t)$ with the state and timing sequences $\mathfrak{q}$ with $\bar{\tau}$ as

$$d_p(t) := \begin{cases} \lambda(\delta_j) & \text{if } t = T_j \\ \lambda(s_j) & \text{if } T_j < t < T_{j+1}. \end{cases} \tag{8.3}$$

We now show that there is plan $d_p(t)$ if and only if the SITL formula $\phi$ is satisfiable. Therefore, we first establish a connection between the existence of a plan $d_p(t)$ and the existence of an accepting run of $TST_\phi$.

**Lemma 8.1** ***(Connection between an Accepting Run of*** $TST_\phi$ ***and a Plan*** $d_p$*) Let $\phi$ be an SITL formula from equation (8.1), and let $TST_\phi$ be the modified timed signal transducer in equation (8.2). Let a plan $d_p$ be constructed as in equation (8.3). For any signal $\bar{\sigma} : \mathbb{R}_{\geq 0} \to \mathbb{B}^{|AP|}$, there is an accepting run of $TST_\phi$ over $\bar{\sigma}$ with $(\bar{\sigma}, 0) \models \varphi$ if only if there is a plan $d_p(t)$ so that $\sigma(t) \models d_p(t)$ for all $t \in \mathbb{R}_{\geq 0}$.*

*Proof:* $\Rightarrow$*: Departing from $TST_\phi$, the infinite state transition system $(S^\phi \times \mathbb{R}^O_{\geq 0}, \Rightarrow)$ has by construction the same reachable set as $TST_\phi$; that is, the same reachable configurations $(s_0, c(0)), (s_0, r(c(0))), (s_1, r(c(0)) + \tau_1), \ldots$. By construction of the equivalence relation $\sim$, reachability properties of $TST_\phi$ can then equivalently be analyzed by considering the finite state transition system $RA(TST_\phi)$ [12, lemma 4.13]. If there now is an accepting run of $TST_\phi$ over $\bar{\sigma}$ with $(\bar{\sigma}, 0) \models \varphi$ (i.e., with $\gamma(\delta_0) = y$), the plan $d_p(t)$ can be constructed as described previously by obtaining $\mathfrak{q}$ and $\bar{\tau}$ (from which $d_p$ is built) directly from the accepting run of $TST_\phi$ over $\bar{\sigma}(t)$. It will hold by construction that $\bar{\sigma}(t) \models d_p(t)$ for all $t \in \mathbb{R}_{\geq 0}$.*

$\Leftarrow$*: If there is a plan $d_p(t)$ such that $\bar{\sigma}(t) \models d_p(t)$ for all $t \in \mathbb{R}_{\geq 0}$, then it trivially follows that $TST_\phi$ has an accepting run over $\bar{\sigma}(t)$. This follows simply by the construction of $d_p(t)$, where $\mathfrak{q}$ and $\bar{\tau}$ have been obtained based on $RA(TST_\phi)$, and the equivalence relation $\sim$. Removing states and transitions from $TST_\varphi$ according to operations [O1] and [O2], resulting in $TST_\phi$, only removes behavior from $TST_\varphi$ (not adding other behaviors) so by [102, theorem 6.7], an accepting run of $TST_\phi$ over $\bar{\sigma}$ with $\gamma(\delta_0) = y$ results in $(\bar{\sigma}, 0) \models \varphi$.*

If there is a plan $d_p$ for $TST_\phi$, there has to exist a signal $\bar{\sigma} : \mathbb{R}_{\geq 0} \to \mathbb{B}^{|AP|}$ such that $\bar{\sigma}(t) \models d_p(t)$ for all $t \in \mathbb{R}_{\geq 0}$. Consequently, this result could be framed as "There is an accepting run of $TST_\phi$ over some signal $\bar{\sigma} : \mathbb{R}_{\geq 0} \to \mathbb{B}^{|AP|}$ with $(\bar{\sigma}, 0) \models \varphi$ if and only if there is a plan $d_p(t)$." Note, however, that in general, it cannot hold that there is plan $d_p(t)$ if and only if the MITL formula $\varphi$ is satisfiable. This follows due to the removal of states from $TST_\varphi$ via operations [O1] and [O2] in $TST_\phi$ from which we construct $d_p(t)$. This means that there may be an accepting run of $TST_\varphi$ over $\bar{\sigma}(t)$ such that $(\bar{\sigma}, 0) \models \varphi$, while there is no accepting run of $TST_\phi$ over $\bar{\sigma}(t)$.

From here, we can now present our main result, which states that $TST_\phi$ encodes the SITL specification $\phi$. Note that this result follows particularly from operations [O1] and [O2], which are performed on $TST_\varphi$ to obtain $TST_\phi$.

**Theorem 8.1 (Satisfiability of SITL)** *Let $\phi$ be an SITL formula from equation (8.1), and let $TST_\phi$ be the modified timed signal transducer in equation (8.2). Let a plan $d_p$ be constructed as in equation (8.3), if possible. Then, there is a plan $d_p(t)$ if and only if $\phi$ is satisfiable.*

*Proof:* $\Rightarrow$: *The existence of a plan $d_p(t)$ implies, by lemma 8.1, that any signal $\bar{\sigma} : \mathbb{R}_{\geq 0} \rightarrow \mathbb{B}^{|AP|}$ with $\bar{\sigma}(t) \models d_p(t)$ for all $t \in \mathbb{R}_{\geq 0}$ results in a run of $TST_\varphi$ over $\bar{\sigma}$, and is such that $(\bar{\sigma}, 0) \models \varphi$. Operations [O1] and [O2] remove all states $s$ and transitions $\delta$ from $TST_\varphi$ that are infeasible; that is, for which there is no point $x \in \mathbb{R}^n$ such that $x \models T^{-1}(\lambda(s))$ and $x \models T^{-1}(\lambda(\delta))$, respectively. Note again that the only difference between the semantics of $\phi$ and $\varphi$ is the difference in the semantics of predicates $\mu$ and propositions $p$, respectively. It thus follows that, based on the run of $TST_\varphi$ over some signal $\bar{\sigma}(t)$ (which is guaranteed to exist), we can construct a signal $x : \mathbb{R}_{\geq 0} \rightarrow \mathbb{R}^n$ with $x(t) \models T^{-1}(d_p(t))$ for all $t \in \mathbb{R}_{\geq 0}$ such that $(x, 0) \models \phi$ (i.e., $\phi$ is satisfiable).*

$\Leftarrow$: *If $\phi$ is satisfiable, it means that there is a signal $x : \mathbb{R}_{\geq 0} \rightarrow \mathbb{R}^n$ such that $(x, 0) \models \phi$. Associated with $x(t)$, we define the signal*

$$\bar{\sigma}(t) := \begin{bmatrix} h_1^\top(x(t)) & \dots & h_{|M|}^\top(x(t)), \end{bmatrix}^T$$

*where $h_i^\top(x) := \top$ if $h_i(x) \geq 0$ and $h_i^\top(x) := \bot$ otherwise, with $h_i(x)$ being the predicate function associated with $\mu_i$. The signal $\bar{\sigma}$ is hence such that $(\bar{\sigma}, 0) \models \varphi$. As there is an accepting run of $TST_\varphi$ over $\bar{\sigma}$, we see that there also is an accepting run of $TST_\phi$ over $\bar{\sigma}$ as the corresponding states visited in the run of $TST_\varphi$ over $\bar{\sigma}$ will not be removed in $TST_\phi$ by operations [O1] and [O2]. By lemma 8.1, it follows that there hence is a plan $d_p(t)$.*

We can now slightly redefine the plan $d_p(t)$ into a plan $d_\mu : \mathbb{R}_{\geq 0} \rightarrow BC(M)$ as

$$d_\mu(t) := T^{-1}(d_p(t)). \tag{8.4}$$

The next result is a straightforward consequence of the previous results.

**Corollary 8.1 (Plan Satisfaction Implies Satisfaction of SITL)** *Let $\phi$ be an SITL formula from equation (8.1), and let $TST_\phi$ be the modified timed signal transducer in equation (8.2). Let a plan $d_\mu$ be constructed as in equation (8.4). If a signal $x:\mathbb{R}_{\geq 0}\rightarrow\mathbb{R}^n$ is such $x(t)\models d_\mu(t)$ for all $t\in\mathbb{R}_{\geq 0}$, then it follows that $(x,0)\models\phi$.*

It is now clear why we call $d_p$ (and $d_\mu$) a plan. If we can control a dynamical system $\dot{x}=f(x)+g(x)u$ to follow the plan $d_\mu$, it will result in a satisfaction of the SITL formula $\phi$. Therefore, in section 8.2, we abstract $\dot{x}=f(x)+g(x)u$ into a timed signal transducer $TST_S$ that can be used to check if $d_\mu$ can be executed by a control law $u$. This abstraction is based on the assumption of existing feedback control laws $u$, such as those that were previously designed. In section 8.3, we modify $TST_\phi$ into $TST_\phi^{\mathrm{m}}$ to ensure that the plan $d_\mu:\mathbb{R}_{\geq 0}\rightarrow BC(M)$, now found from $TST_\phi^{\mathrm{m}}$, can be executed by the system $\dot{x}=f(x)+g(x)u$.

## 8.2 Timed Abstraction of Dynamical Control Systems

We will now combine the modified timed signal transducer $TST_\phi$ with the transient feedback control laws $u(x,t)$ designed in parts II and III. The main idea is to define a timed abstraction of the system that tells us which transitions within $TST_\phi$ can be accomplished by such a control law $u(x,t)$. Recall, therefore, that a run of $TST_\phi$ is simply a sequence of discrete and time steps as

$$(s,c(t))\xrightarrow{\tau}(s,c(t)+\tau)\xrightarrow{\delta}(s',r(c(t)+\tau)),$$

where state invariance and guard conditions also need to be satisfied by an input signal $d:\mathbb{R}_{\geq 0}\rightarrow\mathbb{B}^{|AP|}$ over which the run is defined (recall section 3.3). We instead interpret this input signal now to be the real-valued signal $x:\mathbb{R}_{\geq 0}\rightarrow\mathbb{R}^n$ as $TST_\phi$ encodes the STL specification $\phi$. Effectively, for an initial condition $x(0)$ that satisfies the initial state label $\lambda(s)$ (i.e., such that $x(0)\models T^{-1}(\lambda(s))$), we want the feedback control law $u(x,t)$ to result in a trajectory $x(t)$ such that $x(t)$ satisfies the initial state label $\lambda(s)$ for all times $t\in(0,\tau)$; that is, $x(t)\models T^{-1}(\lambda(s))$ for all $t\in(0,\tau)$. After the discrete step $\delta$, we require that the trajectory $x(\tau)$ satisfies the next state label $\lambda(s')$; that is, $x(\tau)\models T^{-1}(\lambda(\tau))$.

**Timed Abstraction.**We will construct a timed abstraction of the system $\dot{x} = f(x) + g(x)u(x,t)$ that aligns with the modified timed signal transducer $TST_\phi$. By that, we mean that each possible input label $\lambda$ within $TST_\phi$ will result in a state of the timed abstraction, and a transition between states in the timed abstraction exists only if the control law $u(x,t)$ can enforce transitions between labels. In this way, the timed abstraction considers transitions between Boolean combinations of predicates as encoded by the label $\gamma$. Therefore, let us define the set of all Boolean combinations of predicates of the states in $TST_\phi$ as

$$BC(TST_\phi) := \{z \in BC(M) | \exists s \in S^\phi, \lambda(s) = T^{-1}(z)\}.$$

As opposed to considering transitions between partitions of the state space directly, as traditionally done and as described in section 3.1.3, we hereby mitigate the state space explosion in the product automaton between the abstraction and the timed signal transducer that is particularly present in multi-agent systems. Such an approach can result in up to $mn$ states of the product automaton, where $m$ and $n$ are the number of states in abstraction and specification automaton, respectively, while the approach pursued here will not need more than $n$ states.

To formalize the timed abstraction, consider the dynamical system given as

$$\dot{x}(t) = f(x(t)) + g(x(t))u(x(t),t),\ x(0) := x_0, \tag{8.5}$$

where $f : \mathbb{R}^n \to \mathbb{R}^n$ and $g : \mathbb{R}^n \to \mathbb{R}^{n \times m}$ are sufficiently continuous and $u : \mathbb{R}^n \times \mathbb{R}_{\geq 0} \to \mathbb{R}^m$ can generally be any feedback control law (e.g., as designed in parts II and III). We abstract the system in equation (8.5) into a timed signal transducer

$$TST_S := (\tilde{S}, \tilde{S}_0, \tilde{\Lambda}, \tilde{c}, \tilde{\Delta}, \tilde{\lambda}),$$

where $\tilde{S}$ and $\tilde{S}_0$ are the set of states and initial states, respectively, while $\tilde{\Lambda}$ are input variables, $\tilde{c}$ are clocks, $\tilde{\Delta}$ is a transition relation, and $\tilde{\lambda}$ are input labels. Note the absence of output variable and labels, invariants, and a Büchi acceptance condition. We define the elements of $TST_S$ subsequently.

For each element in $BC(TST_\phi)$, there is one state $\tilde{s} \in \tilde{S}$. Consequently, we have that the number of states in $\tilde{S}$ is equivalent to the number of Boolean combinations within $TST_\phi$ (i.e., that $|\tilde{S}| = |BC(TST_\phi)|$). The set of input variables $\tilde{\Lambda}$ is simply the set of predicates $M$ (i.e., $\tilde{\Lambda} := M$). The input labels $\tilde{\lambda} : \tilde{S} \cup \tilde{\Delta} \to BC(TST_\phi)$ map each state $\tilde{s} \in \tilde{S}$ to a unique element from the set of Boolean combinations $BC(TST_\phi)$. Thus, for two states $\tilde{s}', \tilde{s}'' \in \tilde{S}$ with $\tilde{s}' \neq \tilde{s}''$, it holds that $\tilde{\lambda}(\tilde{s}') \neq \tilde{\lambda}(\tilde{s}'')$ and each state in $\tilde{S}$ is uniquely labeled by $\tilde{\lambda}$.

Note that $TST_\phi$ and $TST_S$ hence align by means of the set of Boolean combinations $BC(TST_\phi)$. The set of initial states $\tilde{S}_0$ consists of all element $\tilde{s}_0 \in \tilde{S}$ such that $x_0 \models \tilde{\lambda}(\tilde{s}_0)$. In the timed abstraction $TST_S$, we further have only a single clock $\tilde{c}$ (i.e., $\tilde{c}$ is a scalar). Finally, we define the transition relation $\tilde{\Delta}$. Specifically, $\tilde{\Delta}$ is based on the ability of the system to switch in finite time, by means of a feedback control law $u_{\tilde{\delta}}(x,t)$, between states in $\tilde{S}$ while satisfying the input labels $\tilde{\lambda}$. A transition from $\tilde{s}$ to $\tilde{s}'$ is indicated by $\tilde{\delta} := (\tilde{s}, \tilde{g}, 0, \tilde{s}') \in \tilde{\Delta}$ where $\tilde{g}$ is a guard that depends on equation (8.5). In particular, we assume that $\tilde{g}$ encodes intervals of the form $(C', C'')$, $[C', C'')$, $(C', C'']$, $[C', C'']$, or conjunctions of them, where $C', C'' \in \mathbb{Q}_{\geq 0}$ with $C' \leq C''$.

**Definition 8.4 (Transitions in $TST_S$)** *There is a transition $\tilde{\delta} := (\tilde{s}, \tilde{g}, 0, \tilde{s}') \in \tilde{\Delta}$ if, for all $\tau > 0$ with $\tau \models \tilde{g}$ and for all $x_0 \in \mathbb{R}^n$ with $x_0 \models \tilde{\lambda}(\tilde{s})$, there is a control law $u_{\tilde{\delta}}(x,t)$ such that the solution $x$ to equation (8.5) is such that one of the following holds:*

- *$x(t) \models \tilde{\lambda}(\tilde{s})$ for all $t \in [0, \tau)$ and $x(\tau) \models \tilde{\lambda}(\tilde{s}')$, or*
- *$x(t) \models \tilde{\lambda}(\tilde{s})$ for all $t \in [0, \tau]$ and there is a small[3] $\tau' > \tau$ such that $x(\tau') \models \tilde{\lambda}(\tilde{s}')$ for all $t \in (\tau, \tau']$.*

Based on this definition, we define the input label for a transition $\tilde{\delta}$ as $\tilde{\lambda}(\tilde{\delta}) := \tilde{\lambda}(\tilde{s}')$ in the first case and $\tilde{\lambda}(\tilde{\delta}) := \tilde{\lambda}(\tilde{s})$ in the second case. These two types of transitions can be thought of as transitioning from open into closed and closed into open sets, respectively. Note also that the feedback control law $u_{\tilde{\delta}}(x,t)$ that achieves such a transition has to ensure invariance and finite-time reachability properties. In particular, such transitions can be captured by the STL formulas

$$G_{[0,\tau)}\mu_{\text{inv}}(x) \wedge F_\tau \mu_{\text{reach}}(x), \tag{8.6}$$

$$G_{[0,\tau]}\mu_{\text{inv}}(x) \wedge G_{(\tau,\tau']}\mu_{\text{reach}}(x), \tag{8.7}$$

where $\mu_{\text{inv}}(x) := \tilde{\lambda}(\tilde{s})$ and $\mu_{\text{reach}}(x) := \tilde{\lambda}(\tilde{s}')$ are predicates and where $\tau \models \tilde{g}$. The STL formulas in equations (8.6) and (8.7) belong to the STL fragments considered in parts II and III, and we can design feedback control laws $u_{\tilde{\delta}}(x,t)$ for them if the predicate functions associated with $\mu_{\text{inv}}$ and $\mu_{\text{reach}}$ are concave.

**Timed Abstraction Accepting SITL Plans $d_\mu$.** The definition of a plan $d_\mu(t)$ from section 8.1 will now allow the timed abstraction $TST_S$ to be an

[3]We want that $\tau' - \tau$ is close to zero and pick $\tau'$ to be smaller than the smallest clock constant in $TST_\phi$.

acceptor of the plan $d_\mu(t)$; that is, $TST_S$ will indicate if the system in equation (8.5) under a feedback control law $u(x,t)$ can perform the required motion according to $d_\mu(t)$ to satisfy $\phi$. Therefore, we define a run of $TST_S$ slightly different compared to a run of $TST_\phi$. A run of $TST_S$ over the input signal $d_\mu : \mathbb{R}_{\geq 0} \to BC(M)$ again consists of an alternation of time and discrete steps:

$$(\tilde{s}_0, 0) \xrightarrow{\tilde{\delta}_0} (\tilde{s}_1, 0) \xrightarrow{\tau_1} (\tilde{s}_1, \tau_1) \xrightarrow{\tilde{\delta}_1} \ldots.$$

A time step of duration $\tau$ is denoted by

$$(\tilde{s}, 0) \xrightarrow{\tau} (\tilde{s}, \tau),$$

with $d_\mu(t+t') = \tilde{\lambda}(\tilde{s})$ for each $t' \in (0, \tau)$. A discrete step at time $t$ is denoted by

$$(\tilde{s}, \tilde{c}(t)) \xrightarrow{\tilde{\delta}} (\tilde{s}', 0)$$

for some transition $\tilde{\delta} = (\tilde{s}, \tilde{g}, 0, \tilde{s}') \in \tilde{\Delta}$, such that $\tilde{c}(t) \models \tilde{g}$ and for which $x \models \tilde{\lambda}(\tilde{\delta})$ implies that $x \models d_\mu(t)$. If $d_\mu(t)$ does not result in a run of $TST_S$ over $d_\mu(t)$, then it can be concluded that the system in equation (8.5) cannot execute $d_\mu(t)$. Otherwise (i.e., if $d_\mu(t)$ results in a run of $TST_S$ over $d_\mu(t)$), we define the control law $u(x,t)$ based on the plan $d_\mu(t)$ and the run of $TST_S$ over $d_\mu(t)$. Specifically, let

$$u(x,t) := \begin{cases} u_{\tilde{\delta}_1}(x,t) & \text{for all } t \in [0, T_1) \\ u_{\tilde{\delta}_{j+1}}(x, t - T_j) & \text{for all } t \in (T_j, T_{j+1}) \text{ with } j \geq 2, \end{cases} \tag{8.8}$$

where $T_j$ was defined previously, and, for $t = T_j$ with $j \geq 2$, let

$$u(x, T_j) := \begin{cases} u_{\tilde{\delta}_{j+1}}(x, 0) & \text{if } \tilde{\lambda}(\tilde{s}_{j+1}) = d_\mu(T_j) \\ u_{\tilde{\delta}_j}(x, \tau_j) & \text{if } \tilde{\lambda}(\tilde{s}_j) = d_\mu(T_j). \end{cases} \tag{8.9}$$

Note that $u(x, T_j)$ accounts for the two types of transitions in definition 8.4.

**Theorem 8.2** *Let $\phi$ be an SITL formula from equation (8.1), and let $TST_\phi$ be the modified timed signal transducer in equation (8.2). Let a plan $d_\mu$ be constructed as in equation (8.4). Further, let $TST_S$ be a timed abstraction of the system in equation (8.5). If $d_\mu(t)$ results in a run of $TST_S$ over $d_\mu(t)$, then applying $u(x,t)$ in equations (8.8)–(8.9) to the system in equation (8.5) results in $(x,0) \models \phi$.*

*Proof: If $d_\mu(t)$ results in a run of $TST_S$ over $d_\mu(t)$, applying $u(x,t)$ to equation (8.5) results in $x(t) \models d_\mu(t)$ for all $t \in \mathbb{R}_{\geq 0}$ due to the way that transitions $\tilde{\delta}$ in $TST_S$ are defined. According to corollary 8.1, we can infer that $(x,0) \models \phi$.*

## 8.3 Dynamically Feasible Plan Synthesis

So far, we have discussed a way to synthesize a plan $d_\mu(t)$ that can be checked for dynamic feasibility by the timed abstraction $TST_S$. However, it may be the case that the plan $d_\mu(t)$ does not result in a run of $TST_S$; that is, the system in equation (8.5) may be unable to execute the plan $d_\mu(t)$. In this section, we prune the timed signal transducer $TST_\phi$ and instead synthesize, if possible, a dynamically feasible plan $d_\mu(t)$ from the pruned version of $TST_\phi$.

**Pruning of the Timed Signal Transducer $TST_\phi$.** To accomplish this goal, we will remove transition $\delta$ from $TST_\phi$ and further constrain the guard $g$ of transitions within $TST_\phi$. All this is done with the goal to find a sequence of states $\mathfrak{q}$ and timings $\bar{\tau}$ from which we can construct a plan $d_\mu(t)$ that is a run of the timed abstraction $TST_S$. First, we remove transitions by the following two operations:

[O3] Remove each transition $\delta := (s, g, r, s') \in \Delta^\phi$ with $s \neq s'$ from $TST_\phi$ for which there is no transition $\tilde{\delta} := (\tilde{s}, \tilde{g}, 0, \tilde{s}') \in \tilde{\Delta}$ in $TST_S$ for which it holds that $\lambda(s) = T(\tilde{\lambda}(\tilde{s}))$, $\lambda(s') = T(\tilde{\lambda}(\tilde{s}'))$, and for which $x \models \tilde{\lambda}(\tilde{\delta})$ implies $x \models T^{-1}(\lambda(\delta))$. Remove the corresponding transition $\delta$ from $\mathcal{A}^\phi$.

[O4] For each initial transition $\delta_0 := (s_0, g, r, s') \in \Delta$, remove $\delta_0$ if $x_0 \not\models T^{-1}(\lambda(s'))$ or $x_0 \not\models T^{-1}(\lambda(\delta_0))$. Remove the corresponding transition $\delta_0$ from $\mathcal{A}^\phi$.

We note that operation [O3] removes all transitions from $TST_\phi$ that cannot be executed by a control law $u$, while operation [O4] removes all initial transitions that violate the initial condition $x_0$. We denote the pruned sets of transitions and acceptance conditions as $\Delta^{\mathrm{m}}$ and $\mathcal{A}^{\mathrm{m}}$, respectively, and we note that $\Delta^{\mathrm{m}} \subseteq \Delta^\phi$ and $\mathcal{A}^{\mathrm{m}} \subseteq \mathcal{A}^\phi$. Finally, we make sure that the timing of the guard $\tilde{g}$ within the timed abstraction $TST_S$ is taken into account by including an additional clock and additional guards into $TST_\phi$. Therefore, define the extended clock $c^{\mathrm{m}} := \begin{bmatrix} c^T & \tilde{c} \end{bmatrix}^T$ and perform the final operation on $TST_\phi$.

[O5] For each transition $\delta^{\mathrm{m}} := (s, g, r, s') \in \Delta^{\mathrm{m}}$ with $s \neq s'$, define the guard $g^{\mathrm{m}} := g \wedge \tilde{g}$, where the guard $\tilde{g}$ is from the corresponding transition $\tilde{\delta} := (\tilde{s}, \tilde{g}, 0, \tilde{s}') \in \tilde{\Delta}$ in $TST_S$ such that $\lambda(s) = T(\tilde{\lambda}(\tilde{s}))$, $\lambda(s') = T(\tilde{\lambda}(\tilde{s}'))$, and for which $x \models \tilde{\lambda}(\tilde{\delta})$ implies $x \models T^{-1}(\lambda(\delta^{\mathrm{m}}))$. Finally, replace guards $g$ and resets $r$ in $\delta^{\mathrm{m}}$ with $g^{\mathrm{m}}$ and $r^{\mathrm{m}}(c^{\mathrm{m}}) := \begin{bmatrix} r(c) & 0 \end{bmatrix}^T$.[4]

[4]For each self-transition $\delta^{\mathrm{m}} := (s, g, r, s') \in \Delta^{\mathrm{m}}$ with $s = s'$, we simply replace $g$ and $r$ with $g^{\mathrm{m}} := g \wedge [0, \infty)$ and $r^{\mathrm{m}}(c^{\mathrm{m}}) := \begin{bmatrix} r(c) & \tilde{c} \end{bmatrix}^T$.

We emphasize that adding clocks $\tilde{c}$ and guards $\tilde{g}$ from the timed abstraction $TST_S$ is crucial to ensure the feasibility of the plan that we synthesize in the remainder. Let the pruned timed signal transducer now be denoted as

$$TST_\phi^{\mathrm{m}} := (S, s_0, \Lambda, \Gamma, c^{\mathrm{m}}, \iota, \Delta^{\mathrm{m}}, \lambda, \gamma, \mathcal{A}^{\mathrm{m}}). \tag{8.10}$$

The pruned timed signal transducer $TST_\phi^{\mathrm{m}}$ restricts the behavior of $TST_\phi$ to the exact behavior that can be accepted by the timed abstraction $TST_S$. In this way, $TST_\phi^{\mathrm{m}}$ encodes the joint behavior of $TST_\phi$ and $TST_S$ since the input label of each state in $TST_\phi$ corresponds to one state in $TST_S$. As a result, $TST_\phi^{\mathrm{m}}$ has no more states than $TST_\phi$. The usefulness of this fact is illustrated as follows. If $\varphi$ is built from three elementary timed signal transducers (e.g., one until and two eventually operators, as in figure 3.7), $TST_\phi$ will have $4^3 = 64$ states (depending on operations [O1] and [O2]) in the worst case. Assuming that we build an abstraction $DA$ of the system with 100 states based on transitions between partitions of the state space directly, as traditionally done and described in section 3.1.3, the product of $TST_\phi$ and $DA$ may contain up to 6,400 states. The situation gets even worse when forming the region automaton, which induces $\mathcal{O}(|S| \cdot 2^{C_{\max}})$ states where $S$ is the state of the product automaton and $C_{\max}$ is the maximum clock constant contained in $TST_\phi$ [12, theorem 4.16].

**Plan Synthesis.** We can finally derive a dynamically feasible plan $d_p(t)$ from $TST_\phi^{\mathrm{m}}$. Similar to section 8.1, we find an accepting state sequence $\mathfrak{q} := (\mathfrak{q}_{\mathrm{p}}, \mathfrak{q}_{\mathrm{s}}^\omega)$ by a nested depth-first search now performed on the region automaton $RA(TST_\phi^{\mathrm{m}}) := (Q, q_0, \Delta_R, \mathcal{A}_R)$. Recall that the state sequence $\mathfrak{q}$ consists of a prefix $\mathfrak{q}_{\mathrm{p}} := (q_0, \ldots, q_p)$ and a suffix $\mathfrak{q}_{\mathrm{s}} := (q_{p+1}, \ldots, q_{p+s})$ that is repeated infinitely often. For such a sequence $\mathfrak{q}$, we can then find a timing sequence $\bar{\tau} := (\bar{\tau}_{\mathrm{p}}, \bar{\tau}_{\mathrm{s}}^\omega)$ consisting of a prefix $\bar{\tau}_{\mathrm{p}} := (\tau_0 := 0, \ldots, \tau_{\mathrm{p}})$ and a suffix $\bar{\tau}_{\mathrm{s}} := (\tau_{\mathrm{p}+1}, \ldots, \tau_{\mathrm{p+s}})$. As explained before, the timing sequence can be found algorithmically using tools such as UPPAAL [38]. From here, we can construct a dynamically feasible plan $d_\mu$ and a control law $u$, which results in the satisfaction of $\phi$ by the system in equation (8.5).

**Theorem 8.3 (Dynamically Feasible Plan Satisfying $\phi$)** *Let $\phi$ be an SITL formula from equation (8.1) and let $TST_S$ be a timed abstraction of the system in equation (8.5). Let $TST_\phi^m$ be the pruned timed signal transducer in equation (8.10). If the plan $d_\mu(t)$ is defined as in equation (8.4) but based on the accepting sequence of states and timings $\mathfrak{q}$ and $\bar{\tau}$ from $RA(TST_\phi^m)$, then the plan $d_\mu(t)$ results in a run of $TST_S$ over $d_\mu(t)$. Thus, applying*

*$u(x,t)$ in equations (8.8)–(8.9) to the system in equation (8.5) results in $(x,0) \models \phi$.*

*Proof: The plan $d_\mu(t)$ results in a run of $TST_S$ over $d_\mu(t)$ due to operations [O3]–[O5], which lead to the pruned timed signal transducer $TST_\phi^m$ describing the joint behavior of $TST_\phi$ and $TST_S$. Applying $u(x,t)$ to equation (8.5) then results in $(x,0) \models \phi$ according to theorem 8.2. Note that theorem 8.2 holds even when $d_\mu(t)$ is obtained from $TST_\phi^m$ instead of $TST_\phi$ due to the language inclusion $L(TST_\phi^m) \subseteq L(TST_\phi)$.*

---

**Example 8.1** We consider an academic example that is easy to follow, yet rich enough to illustrate the presented theory. Consider a system consisting of the states

$$x := \begin{bmatrix} x_1^T & x_2^T \end{bmatrix}^T \in \mathbb{R}^4;$$

for example, a system consisting of two robots. The SITL formula is

$$\phi := (\mu_1 U_{(0,\infty)} \mu_2) \wedge F_{(0,3)} \mu_3 \wedge F_{(0,3)} \mu_4,$$

with predicate functions

$$\begin{aligned} h_1(x) &:= \epsilon^2 - \|x_1 - x_2 - f_A\|^2, \\ h_2(x) &:= \epsilon^2 - \|x_1 - p_A\|^2, \\ h_3(x) &:= \epsilon^2 - \|x_2 - p_B\|^2, \\ h_4(x) &:= \epsilon^2 - \|x_1 - x_2 - f_B\|^2, \end{aligned}$$

where $\epsilon := 0.25$. Let $f_A := \begin{bmatrix} -0.5 & 0.5 \end{bmatrix}^T$ and $f_B := \begin{bmatrix} -0.5 & 2 \end{bmatrix}^T$ such that $\mu_1$ and $\mu_4$ encode formations between the robots. Further, let $p_A := \begin{bmatrix} 1 & 1 \end{bmatrix}^T$ and $p_B := \begin{bmatrix} -1 & 1 \end{bmatrix}^T$ such that $\mu_2$ and $\mu_3$ encode reachability objectives of robots 1 and 2, respectively. Note that there is no state $x \in \mathbb{R}^4$ such that $x \models \mu_1 \wedge \mu_4$, which is important for operations [O1] and [O2]. We consider an initial state $x_0$ such that $x_0 \models \mu_1$, while $x_0 \not\models \mu_2$, $x_0 \not\models \mu_3$, and $x_0 \not\models \mu_4$, which is important for operation [O4]. The corresponding MITL formula $\varphi = T^{-1}(\phi) = (p_1 U_{(0,\infty)} p_2) \wedge F_{(0,3)} p_3 \wedge F_{(0,1)} p_4$ was translated to $TST_\varphi$, resulting in 65 states.

We assume the dynamics $\dot{x} = f(x) + u$, where $f(x)$ may be unknown and consider centralized CBFs from part II. For simplicity, we assume that each possible transition $\tilde{\delta}$ can be made within 1–4 time units (i.e., $C' := 1$ and $C'' := 4$) such that $\tilde{g} := [C', C'']$, which is used for operation [O5]. The timed signal transducer $TST_\varphi$ was then transformed into the pruned timed signal transducer $TST_\phi^{\mathrm{m}}$ by performing operations [O1]–[O5]. The corresponding region automaton $RA(TST_\phi^{\mathrm{m}})$ has 2,723 states. To illustrate that the spatiotemporal nature of the specification, we compare the nested depth-first search of $RA(TST_\phi^{\mathrm{m}})$ with a nested depth-first search performed on $RA(TST_\varphi)$. For $RA(TST_\varphi)$, a sequence $\mathfrak{q}^\varphi := ((s_0^\varphi, \alpha_0^\varphi), (s_1^\varphi, \alpha_1^\varphi), \ldots)$ is obtained from the following:

$$\lambda(s_0^\varphi) = p_1 \wedge \neg p_2 \wedge \neg p_3 \wedge \neg p_4,$$
$$\lambda(s_1^\varphi) = p_1 \wedge p_2 \wedge p_3 \wedge p_4,$$

for which there is no $x \in \mathbb{R}^n$ such that $x \models T^{-1}(\lambda(s_1^\varphi))$. The sequence $\mathfrak{q} := ((s_0, \alpha_0), (s_1, \alpha_1), \ldots)$ from $RA(TST_\phi^{\mathrm{m}})$, however, results in

$$\lambda(s_0) = p_1 \wedge \neg p_2 \wedge \neg p_3 \wedge \neg p_4,$$
$$\lambda(s_1) = p_1 \wedge p_2 \wedge \neg p_3 \wedge \neg p_4,$$
$$\lambda(s_2) = \neg p_1 \wedge p_3 \wedge p_4,$$

accounting for the spatial properties induced by the predicates; that is, for each $\lambda(s_0)$, $\lambda(s_1)$, and $\lambda(s_1)$, there is $x \in \mathbb{R}^n$ such that $x \models T^{-1}(\lambda(s_0))$, $x \models T^{-1}(\lambda(s_1))$, and $x \models T^{-1}(\lambda(s_2))$, respectively. A timing sequence $\bar{\tau}$ is obtained with $\bar{\tau} := (0, 1, 1, 1, \ldots)$, defining the plan $d_\mu(t)$ that can be implemented as stated in theorem 8.3. The result is $(x, 0) \models \phi$, as shown in figure 8.1.

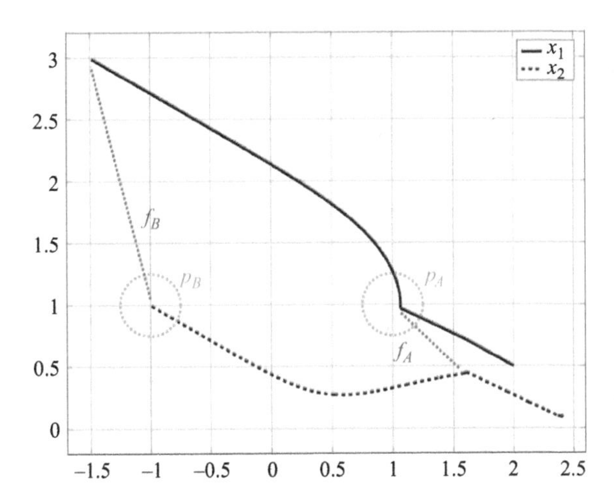


**Figure 8.1**
Execution of plan $d_\mu(t)$ by applying $u(x,t)$. The timing $\bar{\tau}$ is respected since $x(1) \models \mu_2$ and $x(2) \models \mu_3$ (indicated by the red dotted circles $p_A$ and $p_B$). Reprinted with permission from [176].

# 8.4 Notes and References

The results from this chapter are mainly taken from [182, 176]. Automata representations have previously been used to divide complex temporal logic specifications into subspecifications that can be implemented sequentially by feedback control laws. There are numerous approaches that follow this paradigm for control under LTL specifications [97, 157, 39, 141] and under MITL specifications [339, 215, 28, 312]. Timed automata for SITL have recently been used as an open-loop trajectory generator, and the authors in [313] design a learning-based algorithm to track this open-loop trajectory. Timed abstractions, as in section 8.2, were previously presented in [311]. Reactive control synthesis under STL specifications was considered in [119, 120, 182] by using a combination of monitoring algorithms, timed automata-based planning, and CBF–based control. The control problem under SITL specifications is naturally more challenging and, in full generality as presented in this chapter, computationally expensive. As previously discussed, this is no surprise, as LTL and MITL control problems are already NP-hard, which motivated the restrictions on the STL fragments considered in parts II and III of the book. Another line of research considers the encoding of bounded-time STL specifications into mixed-integer linear programs; for instance, see [241, 242, 60, 64, 63, 160, 159]. Solving such programs is computationally more tractable than using timed automata, but is inherently limited to discrete time systems.

# V

# Moving Forward

# Chapter 9

# Outlook and Open Problems

In this book, we presented an introduction to formal methods for multi-agent control systems. The main parts of the book were geared toward mathematically inclined scientists, engineers, and students, and we provided a basic understanding of theory, algorithms, and challenges that may arise. Our starting point were formal languages that provide a convenient way of formalizing system requirements. We stated early on that the inherent complexity of standard, formal methods–based solutions for control design render their application to multi-agent systems computationally intractable due to the infamous state explosion problem. For that reason, we took a different approach from existing textbooks (see, e.g., [43] and [287]), by instead designing scalable feedback control laws that infuse desired transient properties into the closed loop of the system.

In this last chapter, we give the interested reader an outlook of future research directions and point to open problems. Specifically, this book did not touch upon more practical aspects in this problem space, and we only briefly discussed how we can integrate these feedback control laws with multi-agent planning algorithms in part IV. As the original control problem that we are interested in is NP-hard, we had to make several assumptions along the way that one may be able to relax. We hope that the following list inspires the reader to think about, and hopefully address, some of these open problems in the future.

**Feedback Control Laws for Broader Classes of Systems and Specifications.** Throughout this book, we considered limited classes of dynamical systems and fragments of formal specifications. Along the way, we indicated why these limitations were in place, while referring to recent literature that relaxes some of the aforementioned assumptions. In the future, we believe that even broader classes of systems and specifications can be considered by designing novel computational approaches for constructing valid CBFs and funnels that encode spatiotemporal logic specifications. As we aim for principled and general design approaches, one may naturally think of optimization-based solutions. For instance, promising candidates may be based on sum-of-squares or, more generally, semidefinite programming. Success of these techniques has been demonstrated in simpler settings–not considering formal specifications—for computing stability and safety certificates [235, 234, 227]. Similarly, learning-based approaches are promising, and success has been reported in simpler settings for certifying and enforcing stability and safety [178, 248, 53]. As learning approaches would only provide candidate CBFs and candidate funnels, these candidates need to be checked for validity in a formal verification process [144, 9].

**Scalable Planning and Task Decomposition.** In part IV of the book, we briefly explored the possibility of integrating the proposed feedback control laws into a planning framework that can deal with general spatiotemporal logic specifications. Despite being able to compute paths that satisfy these specifications offline, the decidability of timed automata is known to be EXSPACE-complete. In practice, this calls for more efficient planning algorithms, especially for multi-agent planning problems. An interesting starting point in this direction can be the use of large language models. If we keep the NP-hard nature of the control problem in mind, an important task will be to identify meaningful and tractable fragments of specifications that can enable efficient planning. Ultimately, scalable and decentralized planning algorithms are needed, along with an interface for these planners that allows for easy integration of the feedback control laws presented in parts II and III. In a similar direction, task decomposition algorithms should be investigated to increase scalability further. In essence, one would like to decompose a global specification into local (ideally independent) agent specifications. Task decomposition has great potential to decrease the complexity of multi-agent control problems.

**Sensing and State Estimation.** In this book, we assumed to have perfect state knowledge; that is, we assumed that each agent has knowledge of the multi-agent state $x(t)$ within the centralized setting, while each agent may need only partial knowledge of $x(t)$ within the decentralized setting. The assumption of perfect state knowledge is standard in nonlinear and multi-agent control systems. However, with a more holistic view, and with regard to applications in robotics and cyber-physical systems, one has to look at the problem of state estimation for formal methods–based control. The first avenue of future research is the consideration of general sensor modalities. Instead of having knowledge of the state $x(t)$, one will only have access to sensor measurements $y(t)=p(x(t))$, where $p:\mathbb{R}^n\rightarrow\mathbb{R}^p$ models the sensors of the agent. These sensor map $p$ can describe abstract sensor models such as noisy sensors $y(t)=x(t)+v(t)$ or omnidirectional/birectional sensors, but $p$ may also describe more concrete sensor models such as cameras, lidar, and radar. The challenge then is to obtain accurate state estimates $\hat{x}(t)$, and to design robust feedback control laws that satisfy formal specifications despite state estimation errors. State estimates may be obtained using classical solutions using extended Kalman or particle filters or via learning-enabled state estimators. A crucial step will be the error quantification (e.g., using statistical tools such as conformal prediction [22, 183, 169]), and its integration into the temporal logic–based control design.

**Constrained Communication.** We did not touch on limited communication between agents in our exposition, and we thus did not account for communication limitations within the control design. Our motivation was to design scalable multi-agent control laws that result in a satisfaction of formal specifications. Using a graph-theoretic representation of the agents' communication capabilities, the question of how distributed control algorithms can be designed for consensus problems and formation control has been much studied ([209] and [57]). However, it is not clear how distributed control algorithms can be designed for more complex formal specifications. In another direction, constant communication between agents is not resource efficient, so the design of event-triggered control algorithms becomes of interest [286, 87].

**Risk-Aware Feedback Control Design.** This book focused on the study of deterministic multi-agent systems and deterministic control design. However, agent dynamics and their environment may be stochastic (e.g., due to noise and unmodeled dynamics or due to randomness in the environment and the control design). In such settings, the evaluation of a specification $\phi$ becomes subtle as the robust semantics $\rho^{\phi}(x)$ become a random variable [30, 31]. Starting from this observation, there are various ways to think about the satisfaction of the specification $\phi$ (e.g., by computing expected values, probabilities, or more generally tail risk measures [179, 198]). Designing risk-aware control laws for formal specifications is a fairly new research direction that requires more investigation. In a multi-agent system, principled ways to design control algorithms that can handle varying degrees of risk are a great avenue for new research. A specific challenge in the control design is that the distribution of $\rho^\phi(x)$ is unknown due to the composition of the nonlinear robust semantics $\rho^\phi$ with signal $x$.

**Temporal Robustness.** Most of the existing research uses the quantitative semantics $\rho^\phi(x)$ to assess robustness of the system trajectory $x$ with respect to the specification $\phi$. In fact, the control laws designed in this book directly use gradient information about $\rho^\phi(x)$. While spatial robustness captures robustness against spatial perturbations of $x$ (e.g., against model or control errors), this notion does not directly capture robustness against temporal perturbations [184, 251, 250], such as against delays or, more generally, time shifts within subsignals of $x$. Naturally, this is of great interest and importance in multi-agent systems where individual agents may be delayed, such as due to localization errors, not synchronized clocks, or unexpected environmental events that slow the agent. Temporal perturbations can result in violation of the specifications. The design of temporally robust control laws under formal specification is an interesting and largely unexplored area with many unsolved problems.

**Applications and Toolboxes.** The application of formal methods–based control algorithms to robotics and cyber-physical systems is an important area of research. For a long time, formal methods had the reputation of being computationally infeasible and not applicable to these domains. However, with the recent computational advances and the use of scalable control algorithms, we are now at a stage where formal methods–based algorithms can be implemented in real time. In that direction, the importance of well-maintained toolboxes cannot be highlighted enough.

# Bibliography

[1] https://shemesh.larc.nasa.gov/fm/fm-what.html.

[2] https://github.com/Lindemann1989/TCNSrepo.

[3] https://youtu.be/M-kVQLP5xo0.

[4] https://github.com/KTH-SML/CodeRepositoryTCST.

[5] https://youtu.be/pUq4r48p6Sg.

[6] https://youtu.be/M8x9WTgaDU4.

[7] https://youtu.be/4FmDKrHC9rc.

[8] RMIT University. https://sites.rmit.edu.au/cyber-physical-systems/. Accessed July 21, 2020.

[9] Abate, Alessandro, Daniele Ahmed, Mirco Giacobbe, and Andrea Peruffo. Formal synthesis of Lyapunov neural networks. *IEEE Control Systems Letters*, 5(3):773–778, 2020.

[10] Akazaki, Takumi and Ichiro Hasuo. Time robustness in MTL and expressivity in hybrid system falsification. In *Proceedings of the 27th International Conference on Computer Aided Verification*, pages 356–374, San Francisco, July 2015.

[11] Alur, Rajeev. *Principles of cyber-physical systems.* MIT Press, 2015.

[12] Alur, Rajeev and David L Dill. A theory of timed automata. *Theoretical Computer Science*, 126(2):183–235, 1994.

[13] Alur, Rajeev and Thomas A Henzinger. The benefits of relaxing punctuality. *Journal of the ACM*, 43(1):116–146, 1996.

[14] Alur, Rajeev, Thomas A Henzinger, Gerardo Lafferriere, and George J Pappas. Discrete abstractions of hybrid systems. *Proceedings of the IEEE*, 88(7):971–984, 2000.

[15] Ames, Aaron D, Samuel Coogan, Magnus Egerstedt, Gennaro Notomista, Koushil Sreenath, and Paulo Tabuada. Control barrier functions: Theory and applications. In *Proceedings of the 2019 European Control Conference*, pages 3420–3431, Naples, June 2019.

[16] Ames, Aaron D, Jessy W Grizzle, and Paulo Tabuada. Control barrier function based quadratic programs with application to adaptive cruise control. In *Proceedings of the 53rd IEEE Conference on Decision and Control*, pages 6271–6278, Los Angeles, December 2014.

[17] Ames, Aaron D, Gennaro Notomista, Yorai Wardi, and Magnus Egerstedt. Integral control barrier functions for dynamically defined control laws. *IEEE Control Systems Letters*, 5(3):887–892, 2020.

[18] Ames, Aaron D, Xiangru Xu, Jessy W Grizzle, and Paulo Tabuada. Control barrier function based quadratic programs for safety critical systems. *IEEE Transactions on Automatic Control*, 62(8):3861–3876, 2017.

[19] Anand, Mahathi, Abolfazl Lavaei, and Majid Zamani. Compositional construction of control barrier certificates for large-scale interconnected stochastic systems. *IFAC-PapersOnLine*, 53(2):1862–1867, 2020.

[20] Anand, Mahathi, Abolfazl Lavaei, and Majid Zamani. Compositional synthesis of control barrier certificates for networks of stochastic systems against $\omega$-regular specifications. *Nonlinear Analysis: Hybrid Systems*, 51:101427, 2024.

[21] Andersson, Sofie, Alexandros Nikou, and Dimos V Dimarogonas. Control synthesis for multi-agent systems under metric interval temporal logic specifications. In *Proceedings of the 20th World Congress of the International Federation of Automatic Control*, pages 2397–2402, Toulouse, France, July 2017.

[22] Angelopoulos, Anastasios N and Stephen Bates. A gentle introduction to conformal prediction and distribution-free uncertainty quantification. *arXiv preprint arXiv:2107.07511*, 2021.

[23] Antsaklis, Panos J and Anthony N Michel. *Linear systems*, volume 8. Springer, 1997.

[24] Artstein, Zvi. Stabilization with relaxed controls. *Nonlinear Analysis: Theory, Methods & Applications*, 7(11):1163–1173, 1983.

[25] Artstein, Zvi. Stabilization with relaxed controls. *Nonlinear Analysis: Theory, Methods & Applications*, 7(11):1163–1173, 1983.

[26] Aubin, Jean-Pierre and Hélène Frankowska. *Set-valued analysis*. Modern Birkhäuser Classics, 2009.

[27] Baier, Christel and Joost-Pieter Katoen. *Principles of Model Checking*. MIT Press, 2008.

[28] Barbosa, Fernando S, Lars Lindemann, Dimos V Dimarogonas, and Jana Tumova. Integrated motion planning and control under metric interval temporal logic specifications. In *Proceedings of the 2019 European Control Conference*, pages 2042–2049, Naples, June 2019.

[29] Bartocci, Ezio, Luca Bortolussi, Michele Loreti, and Laura Nenzi. Monitoring mobile and spatially distributed cyber-physical systems. In *Proceedings of the 15th ACM-IEEE International Conference on Formal Methods and Models for System Design*, Vienna, pages 146–155, 2017.

[30] Bartocci, Ezio, Luca Bortolussi, Laura Nenzi, and Guido Sanguinetti. On the robustness of temporal properties for stochastic models. *arXiv preprint arXiv:1309.0866*, 2013.

[31] Bartocci, Ezio, Luca Bortolussi, Laura Nenzi, and Guido Sanguinetti. System design of stochastic models using robustness of temporal properties. *Theoretical Computer Science*, 587:3–25, 2015.

[32] Bartocci, Ezio, Jyotirmoy Deshmukh, Alexandre Donzé, Georgios Fainekos, Oded Maler, Dejan Ničković, and Sriram Sankaranarayanan. Specification-based monitoring of cyber-physical systems: A survey on theory, tools and applications. In *Lectures on runtime verification*, pages 135–175. Springer, 2018.

[33] Bechlioulis, Charalampos P. and George A. Rovithakis. Robust adaptive control of feedback linearizable mimo nonlinear systems with prescribed performance. *IEEE Transactions on Automatic Control*, 53(9):2090–2099, 2008.

[34] Bechlioulis, Charalampos P. and George A. Rovithakis. Adaptive control with guaranteed transient and steady state tracking error bounds for strict feedback systems. *Automatica*, 45(2):532–538, 2009.

[35] Bechlioulis, Charalampos P. and George A. Rovithakis. Prescribed performance adaptive control for multi-input multi-output affine in the control nonlinear systems. *IEEE Transactions on Automatic Control*, 55(5):1220–1226, 2010.

[36] Bechlioulis, Charalampos P. and George A. Rovithakis. Robust partial-state feedback prescribed performance control of cascade systems with unknown nonlinearities. *IEEE Transactions on Automatic Control*, 56(9):2224–2230, 2011.

[37] Bechlioulis, Charalampos P. and George A. Rovithakis. A low-complexity global approximation-free control scheme with prescribed performance for unknown pure feedback systems. *Automatica*, 50(4): 1217–1226, 2014. doi:10.1016/j.automatica.2014.02.020.

[38] Behrmann, Gerd, Alexandre David, and Kim G. Larsen. A tutorial on UPPAAL. In *Formal methods for the design of real-time systems*, pages 200–236, Springer, 2004.

[39] Belta, Calin, Antonio Bicchi, Magnus Egerstedt, Emilio Frazzoli, Eric Klavins, and George J. Pappas. Symbolic planning and control of robot motion [grand challenges of robotics]. *IEEE Robotics Automation Magazine*, 14(1):61–70, 2007.

[40] Belta, Calin, Volkan Isler, and George J. Pappas. Discrete abstractions for robot motion planning and control in polygonal environments. *IEEE Transactions on Robotics*, 21(5):864–874, 2005.

[41] Belta, Calin and Vijay Kumar. Abstraction and control for groups of robots. *IEEE Transactions on Robotics*, 20(5):865–875, 2004.

[42] Belta, Calin and Sadra Sadraddini. Formal methods for control synthesis: An optimization perspective. *Annual Review of Control, Robotics, and Autonomous Systems*, 2(1):115–140, 2019.

[43] Belta, Calin, Boyan Yordanov, and Ebru Aydin Gol. *Formal methods for discrete-time dynamical systems*, volume 15. Springer, 2017

[44] Bemporad, Alberto and Manfred Morari. Control of systems integrating logic, dynamics, and constraints. *Automatica*, 35(3):407–427, 1999.

[45] Berger, Thomas, Achim Ilchmann, and Eugene P Ryan. Funnel control of nonlinear systems. *Mathematics of Control, Signals, and Systems*, 33:151–194, 2021.

[46] Berger, Thomas, Huy Hoàng Lê, and Timo Reis. Funnel control for nonlinear systems with known strict relative degree. *Automatica*, 87:345–357, 2018.

[47] Berger, Thomas, Svenja Otto, Timo Reis, and Robert Seifried. Combined open-loop and funnel control for underactuated multibody systems. *Nonlinear Dynamics*, 95:1977–1998, 2019.

[48] Bhat, Sanjay P. and Dennis S. Bernstein. Finite-time stability of continuous autonomous systems. *SIAM Journal Control Optimization*, 38(3):751–766, 2000.

[49] Bianco, Andrea and Luca de Alfaro. Model checking of probabilistic and nondeterministic systems. In *International Conference on Foundations of Software Technology and Theoretical Computer Science*, pages 499–513. Springer, 1995.

[50] Bisoffi, Andrea and Dimos V. Dimarogonas. A hybrid barrier certificate approach to satisfy linear temporal logic specifications. In *Proceedings of the 2018 American Control Conference*, pages 634–639, Milwaukee, June 2018.

[51] Bisoffi, Andrea and Dimos V. Dimarogonas. Satisfaction of linear temporal logic specifications through recurrence tools for hybrid systems. *IEEE Transactions on Automatic Control*, 66(2):818–825, 2020.

[52] Blanchini, Franco. Set invariance in control. *Automatica*, 35(11):1747–1767, 1999.

[53] Boffi, Nicholas, Stephen Tu, Nikolai Matni, Jean-Jacques Slotine, and Vikas Sindhwani. Learning stability certificates from data. In *Conference on Robot Learning*, pages 1341–1350. PMLR, 2021.

[54] Boyd, Stephen and Lieven Vandenberghe. *Convex optimization*. Cambridge University Press, 2004.

[55] Breeden, Joseph and Dimitra Panagou. Compositions of multiple control barrier functions under input constraints. In *2023 American Control Conference (ACC)*, 3688–3695, 2023.

[56] Brim, Lubos, Tomáš Vejpustek, Jana Šafránek, and David Fabriková. Robustness analysis for value-freezing signal temporal logic. In *Proceedings of the Second International Workshop on Hybrid Systems and Biology*, pages 20–36, Taormina, Italy, September 2013.

[57] Bullo, Francesco, Jorge Cortés, and Sonia Martinez. *Distributed control of robotic networks: A mathematical approach to motion coordination algorithms.* Princeton University Press, 2009.

[58] Buyukkocak, Ali Tevfik, Derya Aksaray, and Yasin Yazıcıoğlu. Planning of heterogeneous multi-agent systems under signal temporal logic specifications with integral predicates. *IEEE Robotics and Automation Letters*, 6(2):1375–1382, 2021.

[59] Buyukkocak, Ali Tevfik, Derya Aksaray, and Yasin Yazıcıoğlu. Control barrier functions with actuation constraints under signal temporal logic specifications. In *2022 European Control Conference (ECC)*, pages 162–168. IEEE, 2022.

[60] Buyukkocak, Ali Tevfik, Peter Seiler, Derya Aksaray, and Vijay Gupta. Iterative planner/controller design to satisfy signal temporal logic specifications. In *2023 American Control Conference (ACC)*, pages 3516–3522. IEEE, 2023.

[61] Cai, Mingyu, Makai Mann, Zachary Serlin, Kevin Leahy, and Cristian-Ioan Vasile. Learning minimally-violating continuous control for infeasible linear temporal logic specifications. In *2023 American Control Conference (ACC)*, pages 146–1452, 2023.

[62] Cao, Yongcan, Wenwu Yu, Wei Ren, and Guanrong Chen. An overview of recent progress in the study of distributed multi-agent coordination. *IEEE Transactions on Industrial Informatics*, 9(1):427–438, 2013.

[63] Cardona, Gustavo A., Disha Kamale, and Cristian-Ioan Vasile. Mixed integer linear programming approach for control synthesis with weighted signal temporal logic. In *Proceedings of the 26th ACM International Conference on Hybrid Systems: Computation and Control*, pages 1–12, 2023.

[64] Cardona, Gustavo A. and Cristian-Ioan Vasile. Partial satisfaction of signal temporal logic specifications for coordination of multi-robot systems. In *International Workshop on the Algorithmic Foundations of Robotics*, pages 223–238. Springer, 2022.

[65] Reyes Castro, Luis I., Pratik Chaudhari, Jana Tumova, Sertac Karaman, Emilio Frazzoli, and Daniela Rus. Incremental sampling-based algorithm for minimum-violation motion planning. In *52nd IEEE Conference on Decision and Control*, pages 3217–3224. IEEE, 2013.

[66] Chang, Dong Eui, Shawn C Shadden, Jerrold E. Marsden, and Reza Olfati-Saber. Collision avoidance for multiple agent systems. In *Proceedings of the 42nd IEEE Conference on Decision and Control*, volume 1, pages 539–543, Maui, December 2003.

[67] Charitidou, Maria and Dimos V. Dimarogonas. Control barrier functions for disjunctions of signal temporal logic tasks. In *European Control Conference*, Bucharest, June 13–16, 2023.

[68] Chen, Chi-Tsong. *Linear system theory and design.* Oxford Unviersity Press, 1999.

[69] Chen, Fei and Dimos V. Dimarogonas. Funnel-based cooperative control of leader-follower multi-agent systems under signal temporal logic specifications. In *2022 European Control Conference (ECC)*, pages 906–911. IEEE, 2022.

[70] Chen, Fei and Dimos V. Dimarogonas. Distributed control of coupled leader-follower multi-agent systems under spatiotemporal logic tasks. *IFAC World Congress*, 56(2):10204–10209, 2023.

[71] Chen, Hongkai, Shan Lin, Scott A Smolka, and Nicola Paoletti. An STL-based formulation of resilience in cyber-physical systems. In *International Conference on Formal Modeling and Analysis of Timed Systems*, pages 117–135, 2022.

[72] Chen, Yuxiao, James Anderson, Karanjit Kalsi, Aaron D. Ames, and Steven H. Low. Safety-critical control synthesis for network systems with control barrier functions and assume-guarantee contracts. *IEEE Transactions on Control of Network Systems*, 8(1):487–499, 2020.

[73] Chen, Yuxiao, Andrew Singletary, and Aaron D. Ames. Guaranteed obstacle avoidance for multi-robot operations with limited actuation: A control barrier function approach. *IEEE Control Systems Letters*, 5(1):127–132, 2020.

[74] Cheng, Richard, Mohammad Javad Khojasteh, Aaron D. Ames, and Joel W. Burdick. Safe multi-agent interaction through robust control barrier functions with learned uncertainties. In *2020 59th IEEE Conference on Decision and Control (CDC)*, pages 777–783. IEEE, 2020.

[75] Chowdhury, Dhrubajit and Hassan K. Khalil. Funnel control of higher relative degree systems. In *Proceedings of the 2017 American Control Conference*, pages 598–603, Seattle, May 2017.

[76] Clark, Andrew. Verification and synthesis of control barrier functions. In *2021 60th IEEE Conference on Decision and Control (CDC)*, pages 6105–6112. IEEE, 2021.

[77] Clarke, Edmund M. Model checking. In *International Conference on Foundations of Software Technology and Theoretical Computer Science*, pages 54–56. Springer, 1997.

[78] Clarke, Edmund M. and E. Allen Emerson. Design and synthesis of synchronization skeletons using branching time temporal logic. In *Proceedings of the Workshop on Logic of Programs*, pages 52–71, Yorktown Heights, NY, April 1981.

[79] Clarke, Edmund M. and Jeannette M. Wing. Formal methods: State of the art and future directions. *ACM Computing Surveys (CSUR)*, 28(4): 626–643, 1996.

[80] Cohen, Max and Calin Belta. *Adaptive and learning-based control of safety-critical systems.* Springer Nature, 2023.

[81] Cortes, Jorge. Discontinuous dynamical systems. *IEEE Control Systems Magazine*, 28(3):36–73 2008.

[82] Cortez, Wenceslao Shaw, Xiao Tan, and Dimos V. Dimarogonas. A robust, multiple control barrier function framework for input constrained systems. *IEEE Control Systems Letters*, 6:1742–1747, 2021.

[83] Courcoubetis, Costas, Moshe Vardi, Pierre Wolper, and Mihalis Yannakakis. Memory-efficient algorithms for the verification of temporal properties. *Formal Methods in System Design*, 1(2–3):275–288, 1992.

[84] Dastani, Mehdi, Koen V. Hindriks, and John-Jules Meyer. *Specification and verification of multi-agent systems.* Springer Science & Business Media, 2010.

[85] De Luca, Alessandro, Giuseppe Oriolo, and Marilena Vendittelli. Control of wheeled mobile robots: An experimental overview. In *Lecture notes in control and information sciences*, pages 181–226. Springer, 2001.

[86] De Moura, Leonardo and Nikolaj Bjørner. Satisfiability modulo theories: Introduction and applications. *Communications of the ACM*, 54(9):69–77, 2011.

[87] Dimarogonas, Dimos V., Emilio Frazzoli, and Karl H. Johansson. Distributed event-triggered control for multi-agent systems. *IEEE Transactions on Automatic Control*, 57(5):1291–1297, 2012.

[88] Dimarogonas, Dimos V. and Karl H. Johansson. Stability analysis for multi-agent systems using the incidence matrix: Quantized communication and formation control. *Automatica*, 46(4):695–700, 2010.

[89] Dimarogonas, Dimos V. and Kostas J. Kyriakopoulos. Decentralized navigation functions for multiple robotic agents with limited sensing capabilities. *Journal of Intelligent and Robotic Systems*, 48(3):411–433, 2007.

[90] Dimarogonas, Dimos V., Savvas G. Loizou, Kostas J. Kyriakopoulos, and Michael M. Zavlanos. A feedback stabilization and collision avoidance scheme for multiple independent non-point agents. *Automatica*, 42(2):229–243, 2006.

[91] Dong, Hanlin and Xuebo Yang. Finite-time prescribed performance control for space circumnavigation mission with input constraints and measurement uncertainties. *IEEE Transactions on Aerospace and Electronic Systems*, 58(4):3209–3222, 2022.

[92] Donzé, Alexandre and Oded Maler. Robust satisfaction of temporal logic over real-valued signals. In *Proceedings of the 8th International Conference on Formal Modeling and Analysis of Timed Systems*, pages 92–106, Klosterneuburg, Austria, September 2010.

[93] Donzé, Alexandre, Vasumathi Raman, Goran Frehse, and Matthia Althoff. BluSTL: Controller synthesis from signal temporal logic specifications. In *Proceedings of the 2nd International Workshop on Applied Verification for Continuous and Hybrid Systems*, volume 34, pages 160–168, Seattle, April 2015.

[94] Egerstedt, Magnus. *Robot ecology: constraint-based design for long-duration autonomy.* Princeton University Press, 2021.

[95] Egerstedt, Magnus and Xiaoming Hu. Formation constrained multi-agent control. *IEEE Transactions on Robotics and Automation*, 17(6):947–951, 2001.

[96] Emerson, E. Allen and Joseph Y. Halpern. "Sometimes" and "not never" revisited: On branching versus linear time temporal logic. *Journal of the ACM (JACM)*, 33(1):151–178, 1986.

[97] Fainekos, Georgios E., Antoine Girard, Hadas Kress-Gazit, and George J. Pappas. Temporal logic motion planning for dynamic robots. *Automatica*, 45(2):343–352, 2009.

[98] Fainekos, Georgios E., Hadas Kress-Gazit, and George J. Pappas. Temporal logic motion planning for mobile robots. In *Proceedings of the 2005 IEEE International Conference on Robotics and Automation*, pages 2020–2025, Barcelona, April 2005. doi:10.1109/ROBOT.2005.1570410.

[99] Fainekos, Georgios E. and George J. Pappas. Robustness of temporal logic specifications for continuous-time signals. *Theoretical Computer Science*, 410(42):4262–4291, September 2009. doi:10.1016/j.tcs.2009.06.021.

[100] Farahani, Samira S., Rupak Majumdar, Vinayak S. Prabhu, Sadegh Soudjani, and Esmaeil Zadeh. Shrinking horizon model predictive control with chance-constrained signal temporal logic specifications. In *Proceedings of the 2017 American Control Conference*, pages 1740–1746, Seattle, May 2017.

[101] Farahani, Samira S., Vasumathi Raman, and Richard M. Murray. Robust model predictive control for signal temporal logic synthesis. In *Proceedings of the 5th Conference on Analysis and Design of Hybrid Systems*, pages 323–328, Atlanta, October 2015.

[102] Ferrère, Thomas, Oded Maler, Dejan Ničković, and Amir Pnueli. From real-time logic to timed automata. *Journal of the ACM (JACM)*, 66(3):19, 2019.

[103] Filippidis, Ioannis, Dimos V. Dimarogonas, and Kostas J. Kyriakopoulos. Decentralized multi-agent control from local LTL specifications. In *Proceedings of the 51st IEEE Conference on Decision and Control*, pages 6235–6240, Maui, December 2012.

[104] Filippov, Aleksej Fedorovič. *Differential equations with discontinuous righthand sides: Control systems*, volume 18. Springer Science & Business Media, 2013.

[105] Fisher, Michael. *An introduction to practical formal methods using temporal logic.* John Wiley & Sons, 2011.

[106] Freeman, Randy and Petar V Kokotovic. *Robust nonlinear control design: State-space and Lyapunov techniques.* Springer Science & Business Media, 2008.

[107] Fu, Jie and Ufuk Topcu. Computational methods for stochastic control with metric interval temporal logic specifications. In *Proceedings of the 54th IEEE Conference on Decision and Control*, pages 7440–7447, Osaka, Japan, December 2015.

[108] Garg, Kunal, Ehsan Arabi, and Dimitra Panagou. Fixed-time control under spatiotemporal and input constraints: A quadratic programming based approach. *Automatica*, 141:110314, 2022.

[109] Garg, Kunal and Dimitra Panagou. Control-Lyapunov and control-barrier functions based quadratic program for spatio-temporal specifications. In *Proceedings of the 58th Conference on Decision and Control*, pages 1422–1429, Nice, France, December 2019.

[110] Garg, Kunal and Dimitra Panagou. Robust control barrier and control Lyapunov functions with fixed-time convergence guarantees. In *2021 American Control Conference (ACC)*, pages 2292–2297. IEEE, 2021.

[111] Gastin, Paul and Denis Oddoux. Fast LTL to Büchi automata translation. In *International Conference on Computer Aided Verification*, pages 53–65. Springer, 2001.

[112] Gilpin, Yann, Vince Kurtz, and Hai Lin. A smooth robustness measure of signal temporal logic for symbolic control. *IEEE Control Systems Letters*, 5(1):241–246, 2020.

[113] Girard, Antoine. Reachability of uncertain linear systems using zonotopes. In *Proceedings of the 8th International Workshop on Hybrid Systems: Computation and Control*, pages 291–305, Zurich, March 2005.

[114] Girard, Antoine and Colas Le Guernic. Efficient reachability analysis for linear systems using support functions. In *Proceedings of the 17th World Congress of the International Federation of Automatic Control*, volume 41, pages 8966–8971, Seoul, July 2008.

[115] Girard, Antoine and George J. Pappas. Approximation metrics for discrete and continuous systems. *IEEE Transactions on Automatic Control*, 52(5):782–798, 2007.

[116] Glotfelter, Paul, Ian Buckley, and Magnus Egerstedt. Hybrid nonsmooth barrier functions with applications to provably safe and composable collision avoidance for robotic systems. *IEEE Robotics and Automation Letters*, 4(2):1303–1310, 2019.

[117] Glotfelter, Paul, Jorge Cortés, and Magnus Egerstedt. Nonsmooth barrier functions with applications to multi-robot systems. *IEEE Control Systems Letters*, 1(2):310–315, 2017.

[118] Goebel, Rafal, Ricardo G. Sanfelice, and Andrew R. Teel. *Hybrid dynamical systems: Modeling, stability, and robustness.* Princeton University Press, 2012.

[119] Gundana, David and Hadas Kress-Gazit. Event-based signal temporal logic synthesis for single and multi-robot tasks. *IEEE Robotics and Automation Letters*, 6(2):3687–3694, 2021.

[120] Gundana, David and Hadas Kress-Gazit. Event-based signal temporal logic tasks: Execution and feedback in complex environments. *IEEE Robotics and Automation Letters*, 7(4):10001–10008, 2022.

[121] Guo, Meng and Dimos V. Dimarogonas. Multi-agent plan reconfiguration under local LTL specifications. *International Journal of Robotics Research*, 34(2):218–235, 2015.

[122] Guo, Meng and Dimos V. Dimarogonas. Task and motion coordination for heterogeneous multiagent systems with loosely coupled local tasks. *IEEE Transactions on Automation Science and Engineering*, 14(2):797–808, April 2017. doi:10.1109/TASE.2016.2628389.

[123] Guo, Meng, Charalampos P. Bechlioulis, Kostas J. Kyriakopoulos, and Dimos V. Dimarogonas. Hybrid control of multiagent systems with contingent temporal tasks and prescribed formation constraints. *IEEE Transactions on Control of Network Systems*, 4(4):781–792, 2016.

[124] Guo, Meng and Dimos V. Dimarogonas. Reconfiguration in motion planning of single and multi-agent systems under infeasible local LTL specifications. In *Proceeding of the 52nd IEEE Conference on Decision and Control*, pages 2758–2763, Firenze, Italy, December 2013. IEEE.

[125] Haghighi, Iman, Austin Jones, Zhaodan Kong, Ezio Bartocci, Radu Gros, and Calin Belta. Spatel: A novel spatial-temporal logic and its applications to networked systems. In *Proceedings of the 18th International Conference on Hybrid Systems: Computation and Control*, Seattle, pages 189–198, 2015.

[126] Haghighi, Iman, Noushin Mehdipour, Ezio Bartocci, and Calin Belta. Control from signal temporal logic specifications with smooth cumulative quantitative semantics. In *Proceedings of the 58th IEEE Conference on Decision and Control*, pages 4361–4366, Nice, France, December 2019.

[127] Hansson, Hans and Bengt Jonsson. A logic for reasoning about time and reliability. *Formal Aspects of Computing*, 6(5):512–535, 1994.

[128] Hashemi, Navid, Xin Qin, Jyotirmoy V. Deshmukh, Georgios Fainekos, Bardh Hoxha, Danil Prokhorov, and Tomoya Yamaguchi. Risk-awareness in learning neural controllers for temporal logic objectives. In *2023 American Control Conference (ACC)*, pages 4096–4103. IEEE, 2023.

[129] Hespanha, Joao P. *Linear systems theory.* Princeton University Press, 2018.

[130] Horn, Roger A. and Charles R. Johnson. *Matrix analysis.* Cambridge University Press, 2nd ed., 1990.

[131] Ee263: Introduction to linear dynamical systems. August 2, 2022. http://ee263.stanford.edu/archive/.

[132] Huang, Yiwen, Sze Zheng Yong, and Yan Chen. Guaranteed vehicle safety control using control-dependent barrier functions. In *2019 American Control Conference (ACC)*, pages 983–988. IEEE, 2019.

[133] Ilchmann, Achim. Tracking with prescribed transient behaviour. *ESAIM: Control, Optimisation and Calculus of Variations*, 7:471–493, 2002.

[134] Isidori, Alberto. *Nonlinear control systems: An introduction.* Springer, 1985.

[135] Jagtap, Pushpak and Dimos Dimarogonas. Controller synthesis against omega-regular specifications: a funnel-based control approach. *Authorea Preprints*, 2023.

[136] Jagtap, Pushpak, Sadegh Soudjani, and Majid Zamani. Temporal logic verification of stochastic systems using barrier certificates. In *International Symposium on Automated Technology for Verification and Analysis*, pages 177–193. Springer, 2018.

[137] Jagtap, Pushpak, Abdalla Swikir, and Majid Zamani. Compositional construction of control barrier functions for interconnected control systems. In *Proceedings of the 23rd International Conference on Hybrid Systems: Computation and Control*, Sydney, pages 1–11, 2020.

[138] Jakšić, Stefan, Ezio Bartocci, Radu Grosu, and Dejan Ničković. Quantitative monitoring of STL with edit distance. In *Proceedings of the 16th International Conference on Runtime Verification*, pages 201–218, Madrid, September 2016.

[139] Ji, Meng and Magnus Egerstedt. Distributed coordination control of multiagent systems while preserving connectedness. *IEEE Transactions on Robotics*, 23(4):693–703, 2007.

[140] Jin, Wanxin, Zhaoran Wang, Zhuoran Yang, and Shaoshuai Mou. Neural certificates for safe control policies. *arXiv preprint arXiv:2006.08465*, 2020.

[141] Kamale, Disha, Eleni Karyofylli, and Cristian-Ioan Vasile. Automata-based optimal planning with relaxed specifications. In *2021 IEEE/RSJ International Conference on Intelligent Robots and Systems (IROS)*, pages 6525–6530. IEEE, 2021.

[142] Kantaros, Yiannis and Michael M Zavlanos. Sampling-based optimal control synthesis for multirobot systems under global temporal tasks. *IEEE Transactions on Automatic Control*, 64(5):1916–1931, 2018.

[143] Kantaros, Yiannis and Michael M. Zavlanos. Stylus*: A temporal logic optimal control synthesis algorithm for large-scale multi-robot systems. *International Journal of Robotics Research*, 39(7):812–836, 2020.

[144] Kapinski, James, Jyotirmoy V. Deshmukh, Sriram Sankaranarayanan, and Nikos Arechiga. Simulation-guided Lyapunov analysis for hybrid dynamical systems. In *Proceedings of the 17th International Conference on Hybrid Systems: Computation and Control*, Berlin, pages 133–142, 2014.

[145] Karaman, Sertac and Emilio Frazzoli. Sampling-based algorithms for optimal motion planning with deterministic $\mu$-calculus specifications. In *Proceedings of the 2012 American Control Conference*, pages 735–742, Montrèal, June 2012.

[146] Karaman, Sertac and Emilio Frazzoli. Sampling-based motion planning with deterministic $\mu$-calculus specifications. In *Proceedings of the 48h IEEE Conference on Decision and Control Held Jointly with 2009 28th Chinese Control Conference*, pages 2222–2229, Shanghai, December 2009.

[147] Karayiannidis, Yiannis, Dimos V. Dimarogonas, and Danica Kragic. Multi-agent average consensus control with prescribed performance guarantees. In *Proceedings of the 51st IEEE Conference on Decision and Control*, pages 2219–2225, Maui, December 2012.

[148] Karayiannidis, Yiannis and Zoe Doulgeri. Model-free robot joint position regulation and tracking with prescribed performance guarantees. *Robotics and Autonomous Systems*, 60(2):214–226, 2012.

[149] Karayiannidis, Yiannis, Dimitrios Papageorgiou, and Zoe Doulgeri. A model-free controller for guaranteed prescribed performance tracking of both robot joint positions and velocities. *IEEE Robotics and Automation Letters*, 1(1):267–273, 2016.

[150] Karimadini, Mohammad and Hai Lin. Guaranteed global performance through local coordinations. *Automatica*, 47(5):890–898, 2011.

[151] Khalil, Hassan K. *Nonlinear Systems.* Pearson, 3rd ed., 2013.

[152] Kloetzer, Marius and Calin Belta. A fully automated framework for control of linear systems from temporal logic specifications. *IEEE Transactions on Automatic Control*, 53(1):287–297, February 2008. doi:10.1109/TAC.2007.914952.

[153] Kloetzer, Marius and Calin Belta. Automatic deployment of distributed teams of robots from temporal logic motion specifications. *IEEE Transactions on Robotics*, 26(1):48–61, February 2010. doi:10.1109/TRO.2009.2035776.

[154] Kloetzer, Marius and Calin Belta. Automatic deployment of distributed teams of robots from temporal logic motion specifications. *IEEE Transactions on Robotics*, 26(1):48–61, 2009.

[155] Koditschek, Daniel E. and Elon Rimon. Robot navigation functions on manifolds with boundary. *Advances in Applied Mathematics*, 11(4):412–442, 1990.

[156] Kolathaya, Shishir and Aaron D Ames. Input-to-state safety with control barrier functions. *IEEE Control Systems Letters*, 3(1):108–113, 2018.

[157] Kress-Gazit, Hadas, Georgios E. Fainekos, and George J. Pappas. Temporal-logic-based reactive mission and motion planning. *IEEE Transactions on Robotics*, 25(6):1370–1381, December 2009. doi:10.1109/TRO.2009.2030225.

[158] Kurtz, Vince and Hai Lin. Bayesian optimization for polynomial time probabilistically complete STL trajectory synthesis. *arXiv preprint arXiv:1905.03051*, 2019.

[159] Kurtz, Vince and Hai Lin. A more scalable mixed-integer encoding for metric temporal logic. *IEEE Control Systems Letters*, 6:1718–1723, 2021.

[160] Kurtz, Vincent and Hai Lin. Mixed-integer programming for signal temporal logic with fewer binary variables. *IEEE Control Systems Letters*, 6:2635–2640, 2022.

[161] A Kurzhanskiy, Alex and Pravin Varaiya. Ellipsoidal techniques for reachability analysis of discrete-time linear systems. *IEEE Transactions on Automatic Control*, 52(1):26–38, 2007.

[162] Kwiatkowska, Marta, Gethin Norman, and David Parker. Prism 4.0: Verification of probabilistic real-time systems. In *International Conference on Computer-Aided Verification*, pages 585–591. Springer, 2011.

[163] Lakshmikantham, Vangipuram and Srinivasa Leela. *Differential and integral inequalities: Theory and applications: volume I: ordinary differential equations.* Academic Press, 1969.

[164] Leahy, Kevin, Austin Jones, and Cristian-Ioan Vasile. Fast decomposition of temporal logic specifications for heterogeneous teams. *IEEE Robotics and Automation Letters*, 7(2):2297–2304, 2022.

[165] Ashford Lee, Edward and Sanjit A Seshia. *Introduction to embedded systems: A cyber-physical systems approach.* MIT Press, 2017.

[166] Liberzon, Daniel. *Switching in systems and control.* Springer Science & Business Media, 2003.

[167] Liberzon, Daniel and Stephan Trenn. The bang-bang funnel controller for uncertain nonlinear systems with arbitrary relative degree. *IEEE Transactions on Automatic Control*, 58(12):3126–3141, 2013.

[168] Lin, Zhiyun, Bruce Francis, and Manfredi Maggiore. State agreement for continuous-time coupled nonlinear systems. *SIAM Journal on Control and Optimization*, 46(1):288–307, 2007.

[169] Lindemann, Lars, Matthew Cleaveland, Gihyun Shim, and George J Pappas. Safe planning in dynamic environments using conformal prediction. *IEEE Robotics and Automation Letters*, 2023.

[170] Lindemann, Lars and Dimos V. Dimarogonas. Decentralized robust control of coupled multi-agent systems under local signal temporal logic tasks. In *Proceedings of the 2018 American Control Conference*, pages 1567–1573, Milwaukee, June 2018.

[171] Lindemann, Lars and Dimos V. Dimarogonas. Control barrier functions for multi-agent systems under conflicting local signal temporal logic tasks. *IEEE Control Systems Letters*, 3(3):757–762, 2019.

[172] Lindemann, Lars and Dimos V. Dimarogonas. Control barrier functions for signal temporal logic tasks. *IEEE Control Systems Letters*, 3(1):96–101, 2019.

[173] Lindemann, Lars and Dimos V. Dimarogonas. Decentralized control barrier functions for coupled multi-agent systems under signal temporal logic tasks. In *Proceedings of the 2019 European Control Conference*, pages 89–94, Naples, June 2019.

[174] Lindemann, Lars and Dimos V. Dimarogonas. Feedback control strategies for multi-agent systems under a fragment of signal temporal logic tasks. *Automatica*, 106:284–293, 2019.

[175] Lindemann, Lars and Dimos V. Dimarogonas. Barrier function-based collaborative control of multiple robots under signal temporal logic task. *IEEE Transactions on Control of Network Systems*, 7(4):1916–1928, 2020.

[176] Lindemann, Lars and Dimos V. Dimarogonas. Efficient automata-based planning and control under spatio-temporal logic specifications. In *2020 American Control Conference (ACC)*, pages 4707–4714. IEEE, 2020.

[177] Lindemann, Lars and Dimos V. Dimarogonas. Funnel control for fully actuated systems under a fragment of signal temporal logic specifications. *Nonlinear Analysis: Hybrid Systems*, 39:100973, 2021.

[178] Lindemann, Lars, Haimin Hu, Alexander Robey, Hanwen Zhang, Dimos Dimarogonas, Stephen Tu, and Nikolai Matni. Learning hybrid control barrier functions from data. In *Conference on Robot Learning*, pages 1351–1370. PMLR, 2021.

[179] Lindemann, Lars, Lejun Jiang, Nikolai Matni, and George J. Pappas. Risk of stochastic systems for temporal logic specifications. *ACM Transactions on Embedded Computing Systems*, 22(3):1–31, 2023.

[180] Lindemann, Lars, Jakub Nowak, Lukas Schönbächler, Meng Guo, Jana Tumova, and Dimos V. Dimarogonas. Coupled multi-robot systems under linear temporal logic and signal temporal logic tasks. *IEEE Transactions on Control Systems Technology*, 29(2):858–865, 2019.

[181] Lindemann, Lars, George J. Pappas, and Dimos V. Dimarogonas. Control barrier functions for nonholonomic systems under risk signal temporal logic specifications. In *Proceedings of the 59th Conference on Decision and Control*, Jeju Island, Republic of Korea, December 2020.

[182] Lindemann, Lars, George J. Pappas, and Dimos V. Dimarogonas. Reactive and risk-aware control for signal temporal logic. *IEEE Transactions on Automatic Control*, 67(10):5262–5277, 2021.

[183] Lindemann, Lars, Xin Qin, Jyotirmoy V. Deshmukh, and George J. Pappas. Conformal prediction for stl runtime verification. In *Proceedings of the ACM/IEEE 14th International Conference on Cyber-Physical Systems (with CPS-IoT Week 2023)*, San Antonio, pages 142–153, 2023.

[184] Lindemann, Lars, Alena Rodionova, and George Pappas. Temporal robustness of stochastic signals. In *25th ACM International Conference on Hybrid Systems: Computation and Control*, pages 1–11, 2022.

[185] Lindemann, Lars, Christos K. Verginis, and Dimos V. Dimarogonas. Prescribed performance control for signal temporal logic specifications. In *Proceedings of the 56th Conference on Decision and Control*, pages 2997–3002, Melbourne, December 2017.

[186] Liu, Jun and Pavithra Prabhakar. Switching control of dynamical systems from metric temporal logic specifications. In *Proceedings of the 2014 IEEE International Conference on Robotics and Automation*, pages 5333–5338, Hong Kong, May 2014.

[187] Liu, Siyuan, Adnane Saoud, Pushpak Jagtap, Dimos V. Dimarogonas, and Majid Zamani. Compositional synthesis of signal temporal logic tasks via assume-guarantee contracts. In *2022 IEEE 61st Conference on Decision and Control (CDC)*, pages 2184–2189. IEEE, 2022.

[188] Liu, Yong, Jim J Zhu, Robert L. Williams, and Jianhua Wu. Omnidirectional mobile robot controller based on trajectory linearization. *Robotics and Autonomous Systems*, 56(5):461–479, 2008.

[189] Liu, Zhiyu, Jin Dai, Bo Wu, and Hai Lin. Communication-aware motion planning for multi-agent systems from signal temporal logic specifications. In *Proceedings of the 2017 American Control Conference*, pages 2516–2521, Seattle, May 2017.

[190] Liu, Zhiyu, Bo Wu, Jin Dai, and Hai Lin. Distributed communication-aware motion planning for multi-agent systems from STL and SpaTeL specifications. In *Proceedings of the 56th IEEE Conference on Decision and Control*, pages 4452–4457, Melbourne, December 2017.

[191] Johan Löfberg. YALMIP: A toolbox for modeling and optimization in MATLAB. In *Proceeding of the 2004 IEEE International Symposium on Computer Aided Control Systems Design*, pages 284–289, Taipei, April 2004.

[192] Loizou, Savvas G. and Kostas J. Kyriakopoulos. Automatic synthesis of multi-agent motion tasks based on LTL specifications. In *Proceeding of the 43rd IEEE Conference on Decision and Control*, pages 153–158, Nassau, December 2004.

[193] Loreti, Michele, Luca Bortolussi, Ezio Bartocci, and Laura Nenzi. A logic for monitoring dynamic networks of spatially-distributed cyber-physical systems. *Logical Methods in Computer Science*, 18, 2022.

[194] Lyapunov, Aleksandr Mikhailovich. The general problem of the stability of motion. *International Journal of Control*, 55(3):531–534, 1992.

[195] Ma, Meiyi, Ezio Bartocci, Eli Lifland, John Stankovic, and Lu Feng. Sastl: Spatial aggregation signal temporal logic for runtime monitoring in smart cities. In *2020 ACM/IEEE 11th International Conference on Cyber-Physical Systems (ICCPS)*, pages 51–62. IEEE, 2020.

[196] Macellari, Luca, Yiannis Karayiannidis, and Dimos V. Dimarogonas. Multi-agent second order average consensus with prescribed transient behavior. *IEEE Transactions on Automatic Control*, 62(10):5282–5288, 2017.

[197] Maity, Dipankar and John S. Baras. Event-triggered controller synthesis for dynamical systems with temporal logic constraints. In *Proceedings of the 2018 American Control Conference*, pages 1184–1189, Milwaukee, June 2018.

[198] Majumdar, Anirudha and Marco Pavone. How should a robot assess risk? Towards an axiomatic theory of risk in robotics. In *Robotics Research*, pages 75–84. Springer, 2020.

[199] Maler, Oded and Dejan Nickovic. Monitoring temporal properties of continuous signals. In *Proceedings of the Joint International Conferences on Formal Modelling and Analysis of Timed Systems and Formal Techniques in Real-Time and Fault Tolerant System*, pages 152–166, Grenoble, France, September 2004.

[200] Maler, Oded, Dejan Nickovic, and Amir Pnueli. From MITL to timed automata. In *Proceedings of the 4th International Conference on Formal Modeling and Analysis of Timed Systems*, pages 274–289, Paris, September 2006.

[201] Manna, Zohar and Amir Pnueli. *The temporal logic of reactive and concurrent systems.* Springer-Verlag New York, 1992.

[202] Mastellone, Silvia, Dušan M. Stipanović, Christopher R. Graunke, Koji A. Intlekofer, and Mark W. Spong. Formation control and collision avoidance for multi-agent non-holonomic systems: Theory and experiments. *International Journal of Robotics Research*, 27(1):107–126, 2008.

[203] Mattingley, Jacob and Stephen Boyd. CVXGEN: A code generator for embedded convex optimization. *Optimization and Engineering*, 13(1):1–27, 2012.

[204] Mehdifar, Farhad, Charalampos P. Bechlioulis, and Dimos V. Dimarogonas. Funnel control under hard and soft output constraints. In *2022 IEEE 61st Conference on Decision and Control (CDC)*, pages 4473–4478. IEEE, 2022.

[205] Mehdifar, Farhad, Charalampos P. Bechlioulis, Farzad Hashemzadeh, and Mahdi Baradarannia. Prescribed performance distance-based formation control of multi-agent systems. *Automatica*, 119:109086, 2020.

[206] Mehdifar, Farhad, Lars Lindemann, Charalampos P. Bechlioulis, and Dimos V. Dimarogonas. Control of nonlinear systems under multiple time-varying output constraints: A single funnel approach. In *2023 62nd IEEE Conference on Decision and Control (CDC)*, 6743–6748, 2023.

[207] Mehdipour, Noushin, Cristian-Ioan Vasile, and Calin Belta. Arithmetic-geometric mean robustness for control from signal temporal logic specifications. In *Proceedings of the 2019 American Control Conference*, pages 1690–1695, Philadelphia, June 2019.

[208] Mehdipour, Noushin, Cristian-Ioan Vasile, and Calin Belta. Average-based robustness for continuous-time signal temporal logic. In *Proceedings of the 58th IEEE Conference on Decision and Control*, pages 5312–5317, Nice, France, December 2019.

[209] Mesbahi, Mehran and Magnus Egerstedt. *Graph theoretic methods in multiagent networks*. Princeton University Press, 2010.

[210] Mishra, Pankaj K. and Pushpak Jagtap. Approximation-free prescribed performance control with prescribed input constraints. *IEEE Control Systems Letters*, 7:1261–1266, 2023.

[211] Mitra, Sayan. *Verifying cyber-physical systems: A path to safe autonomy*. MIT Press, 2021.

[212] Morris, Benjamin J., Matthew J. Powell, and Aaron D. Ames. Continuity and smoothness properties of nonlinear optimization-based feedback controllers. In *Proceedings of the 54th IEEE Conference on Decision and Control*, pages 151–158, Osaka, Japan, December 2015.

[213] Na, Jing, Qiang Chen, Xuemei Ren, and Yu Guo. Adaptive prescribed performance motion control of servo mechanisms with friction compensation. *IEEE Transactions on Industrial Electronics*, 61(1):486–494, 2014.

[214] Nejati, Ameneh, Sadegh Soudjani, and Majid Zamani. Compositional construction of control barrier certificates for large-scale stochastic switched systems. *IEEE Control Systems Letters*, 4(4):845–850, 2020.

[215] Nikou, Alexandros, Jana Tumova, and Dimos V. Dimarogonas. Cooperative task planning of multi-agent systems under timed temporal specifications. In *Proceedings of the 2016 American Control Conference*, pages 7104–7109, Boston, July 2016.

[216] Nikou, Alexandros, Christos K. Verginis, and Dimos V. Dimarogonas. Robust distance-based formation control of multiple rigid bodies with orientation alignment. In *Proceedings of the 20th World Congress of the International Federation of Automatic Control*, pages 15458–15463, Toulouse, France, July 2017.

[217] Nilsson, Petter and Necmiye Ozay. Control synthesis for large collections of systems with mode-counting constraints. In *Proceedings of the 19th International Conference on Hybrid Systems: Computation and Control*, Montreal, pages 205–214, 2016.

[218] Olfati-Saber, Reza. Near-identity diffeomorphisms and exponential e-tracking and 6-stabilization of first-order nonholonomic SE (2) vehicles. In *Proceedings of the 2002 American Control Conference*, pages 4690–4695, Anchorage, May 2002.

[219] Olfati-Saber, Reza. Flocking for multi-agent dynamic systems: Algorithms and theory. *IEEE Transactions on Automatic Control*, 51(3):401–420, 2006.

[220] Olfati-Saber, Reza, James A. Fax, and Richard M. Murray. Consensus and cooperation in networked multi-agent systems. *Proceedings of the IEEE*, 95(1):215–233, 2007.

[221] Olfati-Saber, Reza and Richard M. Murray. Consensus problems in networks of agents with switching topology and time-delays. *IEEE Transactions on Automatic Control*, 49(9):1520–1533, 2004.

[222] Paden, Brad and Shankar Sastry. A calculus for computing Filippov's differential inclusion with application to the variable structure control of robot manipulators. *IEEE Transactions on Circuits and Systems*, 34(1):73–82, 1987.

[223] Panagou, Dimitra, Dušan M. Stipanovič, and Petros G. Voulgaris. Multi-objective control for multi-agent systems using Lyapunov-like barrier functions. In *52nd IEEE Conference on Decision and Control*, pages 1478–1483. IEEE, 2013.

[224] Panagou, Dimitra, Dušan M. Stipanović, and Petros G. Voulgaris. Distributed coordination control for multi-robot networks using Lyapunov-like barrier functions. *IEEE Transactions on Automatic Control*, 61(3):617–632, 2015.

[225] Pant, Yash V., Houssam Abbas, Rhudii A. Quaye, and Rahul Mangharam. Fly-by-logic: Control of multi-drone fleets with temporal logic objectives. In *Proceedings of the 11th IEEE International Conference on Cyber-Physical Systems*, pages 186–197, Porto, Portugal, April 2018.

[226] Pant, Yash V., Houssam Abbas, and Rahul Mangharam. Smooth operator: Control using the smooth robustness of temporal logic. In *Proceedings of the 2017 Conference on Control Technology and Applications*, pages 1235–1240, Kohala Coast, HI, August 2017.

[227] Papachristodoulou, Antonis and Stephen Prajna. On the construction of Lyapunov functions using the sum of squares decomposition. In *Proceedings of the 41st IEEE Conference on Decision and Control, 2002.*, volume 3, pages 3482–3487. IEEE, 2002.

[228] Pappas, George J. Bisimilar linear systems. *Automatica*, 39(12):2035–2047, 2003.

[229] Parwana, Hardik, Aquib Mustafa, and Dimitra Panagou. Trust-based rate-tunable control barrier functions for non-cooperative multi-agent systems. In *2022 IEEE 61st Conference on Decision and Control (CDC)*, pages 2222–2229. IEEE, 2022.

[230] Pepy, Romain, Alain Lambert, and Hugues Mounier. Path planning using a dynamic vehicle model. In *2006 2nd International Conference on Information & Communication Technologies*, volume 1, pages 781–786. IEEE, 2006.

[231] Piterman, Nir, Amir Pnueli, and Yaniv Sa'ar. Synthesis of reactive (1) designs. In *Proceedings of the International Workshop on Verification, Model Checking, and Abstract Interpretation*, pages 364–380, Charleston, SC, 2006.

[232] Pnueli, Amir. The temporal logic of programs. In *Proceedings of the 18th Annual Symposium on Foundations of Computer Science*, pages 46–57, Washington, DC, October 1977.

[233] Prajna, Stephen. *Optimization-based methods for nonlinear and hybrid systems verification*. PhD thesis, California Institute of Technology, 2005.

[234] Prajna, Stephen and Ali Jadbabaie. Safety verification of hybrid systems using barrier certificates. In *International Workshop on Hybrid Systems: Computation and Control*, pages 477–492. Springer, 2004.

[235] Prajna, Stephen, Ali Jadbabaie, and George J. Pappas. A framework for worst-case and stochastic safety verification using barrier certificates. *IEEE Transactions on Automatic Control*, 52(8):1415–1428, 2007.

[236] Prajna, Stephen and Anders Rantzer. Convex programs for temporal verification of nonlinear dynamical systems. *SIAM Journal on Control and Optimization*, 46(3):999–1021, 2007.

[237] Psomopoulou, Efi, Achilles Theodorakopoulos, Zoe Doulgeri, and George A. Rovithakis. Prescribed performance tracking of a variable stiffness actuated robot. *IEEE Transactions on Control Systems Technology*, 23(5):1914–1926, 2015.

[238] Qin, Zengyi, Kaiqing Zhang, Yuxiao Chen, Jingkai Chen, and Chuchu Fan. Learning safe multi-agent control with decentralized neural barrier certificates. *arXiv preprint arXiv:2101.05436*, 2021.

[239] Quigley, Morgan, Ken Conley, et al. Ros: An open-source robot operating system. In *Proceedings of the ICRA Workshop on Open Source Software*, volume 3, page 5. Kobe, Japan, May 2009.

[240] Rajkumar, Ragunathan, Insup Lee, Lui Sha, and John Stankovic. Cyber-physical systems: The next computing revolution. In *Proceedings of the 47th Design Automation Conference*, pages 731–736, Anaheim, CA, June 2010.

[241] Raman, Vasumathi, Antoine Donzé, Mehdi Maasoumy, Richard M. Murray, Alberto Sangiovanni-Vincentelli, and Sanjit A. Seshia. Model predictive control with signal temporal logic specifications. In *Proceedings of the 53rd IEEE Conference on Decision and Control*, pages 81–87, Los Angeles, December 2014.

[242] Raman, Vasumathi, Antoine Donzé, Dorsa Sadigh, Richard M Murray, and Sanjit A Seshia. Reactive synthesis from signal temporal logic specifications. In *Proceedings of the 18th International Conference on Hybrid Systems: Computation and Control*, pages 239–248, Seattle, April 2015.

[243] Raman, Vasumathi and Eric M. Wolff. Mixed-integer linear programming for planning with temporal logic tasks [position paper]. In *Proceedings of the Twenty-Ninth AAAI Conference on Artificial Intelligence*, Austin, TX, January 2015.

[244] Ren, Wei and Randal W. Beard. Consensus seeking in multiagent systems under dynamically changing interaction topologies. *IEEE Transactions on Automatic Control*, 50(5):655–661, 2005.

[245] Reynolds, Craig W. Flocks, herds and schools: A distributed behavioral model. In *Proceedings of the 14th Annual Conference on Computer Graphics and Interactive Techniques*, volume 21, pages 25–34, New York, November 1987.

[246] Rimon, Elon and Daniel E. Koditschek. Exact robot navigation using artificial potential functions. *IEEE Transactions on Robotics and Automation*, 8(5):501–518, 1992.

[247] Rizk, Aurélien, Grégory Batt, François Fages, and Sylvain Soliman. On a continuous degree of satisfaction of temporal logic formulae with applications to systems biology. In *Proceedings of the 6th International Conference on Computational Methods in Systems Biology*, pages 251–268, Rostock, Germany, October 2008.

[248] Robey, Alexander, Haimin Hu, Lars Lindemann, Hanwen Zhang, Dimos V. Dimarogonas, Stephen Tu, and Nikolai Matni. Learning control barrier functions from expert demonstrations. In *Proceedings of the 59th Conference on Decision and Control*, Jeju Island, Republic of Korea, December 2020.

[249] Rodionova, AlLena, Ezio Bartocci, Dejan Nickovic, and Radu Grosu. Temporal logic as filtering. In *Proceedings of the 19th International Conference on Hybrid Systems: Computation and Control*, pages 11–20, Vienna, April 2016.

[250] Rodionova, Alëna, Lars Lindemann, Manfred Morari, and George J. Pappas. Temporal robustness of temporal logic specifications: Analysis and control design. *ACM Transactions on Embedded Computing Systems*, 22(1):1–44, 2022.

[251] Rodionova, Alëna, Lars Lindemann, Manfred Morari, and George J. Pappas. Time-robust control for STL specifications. In *2021 60th IEEE Conference on Decision and Control (CDC)*, pages 572–579. IEEE, 2021.

[252] Rodionova, Alëna, Lars Lindemann, Manfred Morari, and George J. Pappas. Combined left and right temporal robustness for control under STL specifications. *IEEE Control Systems Letters*, 7:619–624, 2022.

[253] Romdlony, Muhammad Z. and Bayu Jayawardhana. On the new notion of input-to-state safety. In *Proceedings of the 55th IEEE Conference on Decision and Control*, pages 6403–6409, Las Vegas, December 2016.

[254] Romdlony, Muhammad Z. and Bayu Jayawardhana. Robustness analysis of systems' safety through a new notion of input-to-state safety. *International Journal of Robust and Nonlinear Control*, 29(7):2125–2136, 2019.

[255] Sadraddini, Sadra and Calin Belta. Robust temporal logic model predictive control. In *Proceedings of the 53rd Conference on Communication, Control, and Computing*, pages 772–779, Monticello, IL, Sept 2015.

[256] Sadraddini, Sadra and Calin Belta. Formal synthesis of control strategies for positive monotone systems. *IEEE Transactions on Automatic Control*, 64(2):480–495, 2018.

[257] Sahin, Yunus E., Petter Nilsson, and Necmiye Ozay. Provably-correct coordination of large collections of agents with counting temporal logic constraints. In *Proceedings of the 8th International Conference on Cyber-Physical Systems*, pages 249–258, Pittsburgh, April 2017.

[258] Sahin, Yunus E., Petter Nilsson, and Necmiye Ozay. Synchronous and asynchronous multi-agent coordination with cLTL+ constraints. In *Proceedings of the 56th IEEE Conference on Decision and Control*, pages 335–342, Melbourne, IEEE, 2017.

[259] Sahin, Yunus E., Rien Quirynen, and Stefano Di Cairano. Autonomous vehicle decision-making and monitoring based on signal temporal logic and mixed-integer programming. In *Proceedings of the 2020 American Control Conference*, pages 454–459, Denver, July 2020.

[260] Saif, Osamah, Isabelle Fantoni, and Arturo Zavala-Río. Real-time flocking of multiple-quadrotor system of systems. In *Proceedings of the 10th Conference on System of Systems Engineering Conference*, pages 286–291, San Antonio, May 2015.

[261] Sankur, Ocan and B Srivathsan. Zone-based verification of timed automata: Extrapolations, simulations and what next? In *Formal Modeling and Analysis of Timed Systems: 20th International Conference, FORMATS 2022, Warsaw, Poland, September 13–15, 2022, Proceedings*, volume 13465, page 16. Springer Nature, 2022.

[262] Santoyo, Cesar, Maxence Dutreix, and Samuel Coogan. A barrier function approach to finite-time stochastic system verification and control. *Automatica*, 125:109439, 2021.

[263] Sarsilmaz, Selahattin B. and Tansel Yucelen. Control of multiagent systems with local and global objectives. In *Proceedings of the 57th IEEE Conference on Decision and Control*, pages 5096–5101, Miami, December 2018.

[264] Sastry, Shankar. *Nonlinear systems: analysis, stability, and control*, volume 10. Springer Science & Business Media, 2013.

[265] Saxena, Naman, Gorantla Sandeep, and Pushpak Jagtap. Funnel-based reward shaping for signal temporal logic tasks in reinforcement learning. *IEEE Robotics and Automation Letters*, 2023.

[266] Schillinger, Philipp, Mathias Bürger, and Dimos V. Dimarogonas. Decomposition of finite LTL specifications for efficient multi-agent planning. In *Proceedings of the 13th International Symposium on Distributed Autonomous Robotic Systems*, pages 253–267, London, November 2016.

[267] Schillinger, Philipp, Mathias Bürger, and Dimos V. Dimarogonas. Multi-objective search for optimal multi-robot planning with finite LTL specifications and resource constraints. In *Proceedings of the 2017 IEEE International Conference on Robotics and Automation*, pages 768–774, Singapore, May 2017.

[268] Schneider, Klaus, Jimmy Shabolt, and John G Taylor. *Verification of reactive systems: Formal methods and algorithms*. Springer, 2004.

[269] Sewlia, Mayank, Christos K. Verginis, and Dimos V. Dimarogonas. Cooperative object manipulation under signal temporal logic tasks and uncertain dynamics. *IEEE Robotics and Automation Letters*, 7(4):11561–11568, 2022.

[270] Sewlia, Mayank, Christos K. Verginis, and Dimos V. Dimarogonas. Maps$^2$: Multi-robot anytime motion planning under signal temporal logic specifications. *arXiv preprint arXiv:2309.05632*, 2023.

[271] Seyboth, Georg S., Dimos V. Dimarogonas, and Karl H. Johansson. Event-based broadcasting for multi-agent average consensus. *Automatica*, 49(1):245–252, 2013.

[272] Sheeran, Mary, Satnam Singh, and Gunnar Stålmarck. Checking safety properties using induction and a sat-solver. In *International Conference on Formal Methods in Computer-Aided Design*, pages 127–144. Springer, 2000.

[273] Shevitz, Daniel and Brad Paden. Lyapunov stability theory of nonsmooth systems. *IEEE Transactions on Automatic Control*, 39(9):1910–1914, 1994.

[274] Shoukry, Yasser, Pierluigi Nuzzo, Indranil Saha, Alberto L. Sangiovanni-Vincentelli, Sanjit A. Seshia, George J. Pappas, and Paulo Tabuada.

Scalable lazy SMT-based motion planning. In *Proceedings of the 55th IEEE Conference on Decision and Control*, pages 6683–6688, Las Vegas, December 2016.

[275] Shoukry, Yasser, Pierluigi Nuzzo, Alberto L. Sangiovanni-Vincentelli, Sanjit A. Seshia, George J. Pappas, and Paulo Tabuada. SMC: Satisfiability modulo convex optimization. In *Proceedings of the 20th International Conference on Hybrid Systems: Computation and Control*, pages 19–28, Pittsburgh, April 2017.

[276] Shvartsman, Rina, Andrew R. Teel, Denny Oetomo, and Dragan Nešić. System of funnels framework for robust global non-linear control. In *Proceedings of the 55th IEEE Conference on Decision and Control*, pages 3018–3023, Las Vegas, December 2016.

[277] Sistla, Aravinda P. and Edmund M. Clarke. The complexity of propositional linear temporal logics. *Journal of the ACM (JACM)*, 32(3):733–749, 1985.

[278] Slotine, Jean-Jacques E., Weiping Li, et al. *Applied nonlinear control*, volume 199. Prentice Hall, 1991.

[279] Sontag, Eduardo and Héctor J. Sussmann. Nonsmooth control-Lyapunov functions. In *Proceedings of 34th IEEE Conference on Decision and Control*, volume 3, pages 2799–2805, San Antonio, December 1995.

[280] Sontag, Eduardo D. A "universal" construction of artstein's theorem on nonlinear stabilization. *Systems & Control Letters*, 13(2):117–123, 1989.

[281] Sontag, Eduardo D. *Mathematical control theory: Deterministic finite dimensional systems.* Springer Science & Business Media, Berlin, 2nd ed., 2013.

[282] Srinivasan, Mohit and Samuel Coogan. Control of mobile robots using barrier functions under temporal logic specifications. *IEEE Transactions on Robotics*, 37(2):363–374, 2020.

[283] Srinivasan, Mohit, Samuel Coogan, and Magnus Egerstedt. Control of multi-agent systems with finite time control barrier certificates and temporal logic. In *Proceedings of the 57th IEEE Conference on Decision and Control*, pages 1991–1996, Miami, December 2018.

[284] Srinivasan, Mohit, Park H. Nak-seung, and Samuel Coogan. Weighted polar finite time control barrier functions with applications to multi-robot systems. In *Proceedings of the 58th IEEE Conference on Decision and Control*, pages 7031–7036, Nice, France, December 2019.

[285] Tabuada, Paulo and Daniel Neider. Robust linear temporal logic. In *Proceedings of the 25th Conference on Computer Science Logic*, pages 10:1–10:21, Marseille, France, August 2016.

[286] Tabuada, Paulo. Event-triggered real-time scheduling of stabilizing control tasks. *IEEE Transactions on Automatic Control*, 52(9):1680–1685, 2007.

[287] Tabuada, Paulo. *Verification and control of hybrid systems: a symbolic approach.* Springer Science & Business Media, 2009.

[288] Tan, Xiao, Wenceslao Shaw Cortez, and Dimos V. Dimarogonas. High-order barrier functions: Robustness, safety, and performance-critical control. *IEEE Transactions on Automatic Control*, 67(6):3021–3028, 2021.

[289] Tan, Xiao and Dimos V. Dimarogonas. Distributed implementation of control barrier functions for multi-agent systems. *IEEE Control Systems Letters*, 6:1879–1884, 2021.

[290] Tanner, Herbert G., Ali Jadbabaie, and George J. Pappas. Stable flocking of mobile agents, part I: fixed topology. In *Proccedings of the 42nd IEEE Conference on Decision and Control*, pages 2010–2015, Maui, December 2003.

[291] Tanner, Herbert G., Ali Jadbabaie, and George J. Pappas. Stable flocking of mobile agents part II: dynamic topology. In *Proccedings of the 42nd IEEE Conference on Decision and Control*, pages 2016–2021, Maui, December 2003.

[292] Tanner, Herbert G., George J. Pappas, and Vijay Kumar. Leader-to-formation stability. *IEEE Transactions on Robotics and Automation*, 20(3):443–455, 2004.

[293] Tarbouriech, Sophie, Antoine Girard, and Laurentiu Hetel. *Control subject to computational and communication constraints.* Springer, 2018.

[294] Tauriainen, Heikki. Nested emptiness search for generalized Büchi automata. *Fundamenta Informaticae*, 70(1–2):127–154, 2006.

[295] Tee, Keng P., Shuzhi S. Ge, and Eng H. Tay. Barrier Lyapunov functions for the control of output-constrained nonlinear systems. *Automatica*, 45(4):918–927, 2009.

[296] Trakas, Panagiotis S. and Charalampos P. Bechlioulis. Approximation-free adaptive prescribed performance control for unknown siso nonlinear systems with input saturation. In *2022 IEEE 61st Conference on Decision and Control (CDC)*, pages 4351–4356. IEEE, 2022.

[297] Trakas, Panagiotis S. and Charalampos P. Bechlioulis. Robust adaptive prescribed performance control for unknown nonlinear systems with input amplitude and rate constraints. *IEEE Control Systems Letters*, 7:1801–1806, 2023.

[298] Tumová, Jana, Luis I. Reyes Castro, Sertac Karaman, Emilio Frazzoli, and Daniela Rus. Minimum-violation ltl planning with conflicting specifications. In *2013 American Control Conference*, pages 200–205. IEEE, 2013.

[299] Tumova, Jana and Dimos V. Dimarogonas. Decomposition of multi-agent planning under distributed motion and task ltl specifications. In *2015 54th IEEE Conference on Decision and Control (CDC)*, pages 7448–7453. IEEE, 2015.

[300] Tumova, Jana and Dimos V. Dimarogonas. Multi-agent planning under local LTL specifications and event-based synchronization. *Automatica*, 70:239–248, August 2016. doi:10.1016/j.automatica.2016.04.006.

[301] Tumova, Jana, Gavin C. Hall, Sertac Karaman, Emilio Frazzoli, and Daniela Rus. Least-violating control strategy synthesis with safety rules. In *Proceedings of the 16th International Conference on Hybrid Systems: Computation and Control*, pages 1–10, 2013.

[302] Ulusoy, Alphan, Stephen L. Smith, Xu Chu Ding, Calin Belta, and Daniela Rus. Optimality and robustness in multi-robot path planning with temporal logic constraints. *International Journal of Robotics Research*, 32(8):889–911, 2013.

[303] Varnai, Peter and Dimos V. Dimarogonas. Prescribed performance control guided policy improvement for satisfying signal temporal logic tasks. In *Proceedings of the 2019 American Control Conference*, pages 286–291, Philadelphia, July 2019.

[304] Varnai, Peter and Dimos V. Dimarogonas. Guided policy improvement for satisfying STL tasks using funnel adaptation. *arXiv preprint arXiv:2004.05653*, 2020.

[305] Varnai, Peter and Dimos V. Dimarogonas. On robustness metrics for learning STL tasks. In *Proceedings of the 2020 American Control Conference*, pages 5394–5399, Denver, June 2020.

[306] Vasile, Cristian-Ioan and Calin Belta. Sampling-based temporal logic path planning. In *Proceedings of the 2013 International Conference on Intelligent Robots and Systems*, pages 4817–4822, Tokyo, November 2013.

[307] Vasile, Cristian-Ioan, Vasumathi Raman, and Sertac Karaman. Sampling-based synthesis of maximally-satisfying controllers for temporal logic specifications. In *2017 IEEE/RSJ International Conference on Intelligent Robots and Systems (IROS)*, pages 3840–3847. IEEE, 2017.

[308] Vasile, Cristian-Ioan, Jana Tumova, Sertac Karaman, Calin Belta, and Daniela Rus. Minimum-violation scLTL motion planning for mobility-on-demand. In *2017 IEEE International Conference on Robotics and Automation (ICRA)*, pages 1481–1488. IEEE, 2017.

[309] Verginis, Christos K. Funnel control for uncertain nonlinear systems via zeroing control barrier functions. *IEEE Control Systems Letters*, 7:853–858, 2022.

[310] Verginis, Christos K., Charalampos P. Bechlioulis, Dimos V. Dimarogonas, and Kostas J. Kyriakopoulos. Robust distributed control protocols for large vehicular platoons with prescribed transient and steady-state performance. *IEEE Transactions on Control Systems Technology*, 26(1):299–304, 2017.

[311] Verginis, Christos K. and Dimos V. Dimarogonas. Timed abstractions for distributed cooperative manipulation. *Autonomous Robots*, 42(4):781–799, 2018.

[312] Verginis, Christos K., Constantinos Vrohidis, Charalampos P. Bechlioulis, Kostas J. Kyriakopoulos, and Dimos V. Dimarogonas. Reconfigurable motion planning and control in obstacle cluttered environments under timed temporal tasks. In *Proceedings of the 2019 IEEE International Conference on Robotics and Automation*, pages 951–957, Montreal, May 2019.

[313] Verginis, Christos K., Zhe Xu, and Ufuk Topcu. Non-parametric neuro-adaptive control subject to task specifications. *arXiv preprint arXiv: 2106.13498*, 2021.

[314] Wang, Li, Aaron Ames, and Magnus Egerstedt. Safety barrier certificates for heterogeneous multi-robot systems. In *Proceedings of the 2016 American Control Conference*, pages 5213–5218, Boston, July 2016.

[315] Wang, Li, Aaron D. Ames, and Magnus Egerstedt. Multi-objective compositions for collision-free connectivity maintenance in teams of mobile robots. In *Proceedings of the 55th IEEE Conference on Decision and Control*, pages 2659–2664, Las Vegas, December 2016.

[316] Wang, Li, Aaron D. Ames, and Magnus Egerstedt. Safety barrier certificates for collisions-free multirobot systems. *IEEE Transactions on Robotics*, 33(3):661–674, 2017.

[317] Wang, Li, Dongkun Han, and Magnus Egerstedt. Permissive barrier certificates for safe stabilization using sum-of-squares. In *Proceedings of the 2018 Annual American Control Conference*, pages 585–590, Milwaukee, June 2018.

[318] Wei, Caisheng, Qifeng Chen, Jun Liu, Zeyang Yin, and Jianjun Luo. An overview of prescribed performance control and its application to spacecraft attitude system. *Proceedings of the Institution of Mechanical Engineers, Part I: Journal of Systems and Control Engineering*, 235(4):435–447, 2021.

[319] Wieland, Peter and Frank Allgöwer. Constructive safety using control barrier functions. In *Proceedings of the 7th IFAC Symposium on Nonlinear Control Systems*, pages 462–467, Pretoria, South Africa, August 2007.

[320] Wiltz, Adrian and Dimos V. Dimarogonas. Handling disjunctions in signal temporal logic based control through nonsmooth barrier functions. In *2022 IEEE 61st Conference on Decision and Control (CDC)*, pages 3237–3242. IEEE, 2022.

[321] Wolper, Pierre. Constructing automata from temporal logic formulas: A tutorial? In *Proceedings of Lectures on Formal Methods and Performance Analysis*, pages 261–277, Berg en Dal, the Netherlands, July 2000.

[322] Wongpiromsarn, Tichakorn, Ufuk Topcu, and Richard M. Murray. Receding horizon control for temporal logic specifications. In *Proceedings of the 13th Conference on Hybrid Systems: Computation and Control*, pages 101–110, New York, April 2010.

[323] Woodcock, Jim, Peter G. Larsen, Juan Bicarregui, and John Fitzgerald. Formal methods: practice and experience. *ACM Computing Surveys (CSUR)*, 41(4):1–36, 2009.

[324] Wooldridge, Michael. *An introduction to multiagent systems*. John Wiley & Sons, 2009.

[325] Xiao, Wei and Calin Belta. Control barrier functions for systems with high relative degree. In *2019 IEEE 58th Conference on Decision and Control (CDC)*, pages 474–479. IEEE, 2019.

[326] Xiao, Wei, Christos G. Cassandras, and Calin Belta. *Safe autonomy with control barrier functions: theory and applications*. Springer Nature, 2023.

[327] Xu, Xiangru. Control sharing barrier functions with application to constrained control. In *Proceedings of the 55th IEEE Conference on Decision and Control*, pages 4880–4885, Las Vegas, December 2016.

[328] Xu, Xiangru. Constrained control of input–output linearizable systems using control sharing barrier functions. *Automatica*, 87:195–201, 2018.

[329] Xu, Xiangru, Jessy W Grizzle, Paulo Tabuada, and Aaron D. Ames. Correctness guarantees for the composition of lane keeping and adaptive cruise control. *IEEE Transactions on Automation Science and Engineering*, 15(3):1216–1229, 2017.

[330] Xu, Xiangru, Paulo Tabuada, Jessy W. Grizzle, and Aaron D. Ames. Robustness of control barrier functions for safety critical control. In *Proceedings of the 5th Conference on Analysis and Design of Hybrid Systems*, pages 54–61, Atlanta, October 2015.

[331] Xu, Zhe and Agung Julius. Census signal temporal logic inference for multiagent group behavior analysis. *IEEE Transactions on Automation Science and Engineering*, 15(1):264–277, 2016.

[332] Yordanov, Boyan, Jana Tumova, Ivana Cerna, Jiří Barnat, and Calin Belta. Temporal logic control of discrete-time piecewise affine systems. *IEEE Transactions on Automatic Control*, 57(6):1491–1504, 2011.

[333] Zavlanos, Michael M., Magnus B. Egerstedt, and George J. Pappas. Graph-theoretic connectivity control of mobile robot networks. *Proceedings of the IEEE*, 99(9):1525–1540, 2011.

[334] Zavlanos, Michael M. and George J. Pappas. Distributed connectivity control of mobile networks. *IEEE Transactions on Robotics*, 24(6):1416–1428, 2008.

[335] Zehfroosh, Ashkan and Herbert G. Tanner. Non-smooth control barrier navigation functions for STL motion planning. *Frontiers in Robotics and AI*, 9:782783, 2022.

[336] Zhao, Hengjun, Xia Zeng, Taolue Chen, Zhiming Liu, and Jim Woodcock. Learning safe neural network controllers with barrier certificates. In *Proceedings of the Symposium on Dependable Software Engineering: Theories, Tools, and Applications*, pages 177–185, Guangzhou, China, November 2020.

[337] Zhao, Kai, Yongduan Song, Tiedong Ma, and Liu He. Prescribed performance control of uncertain euler–lagrange systems subject to full-state constraints. *IEEE Transactions on Neural Networks and Learning Systems*, 29(8):3478–3489, 2017.

[338] Zhou, Yuchen, Dipankar Maity, and John S. Baras. Optimal mission planner with timed temporal logic constraints. In *2015 European Control Conference (ECC)*, pages 759–764. IEEE, 2015.

[339] Zhou, Yuchen, Dipankar Maity, and John S Baras. Timed automata approach for motion planning using metric interval temporal logic. In *Proceedings of the 2016 European Control Conference*, pages 690–695, Ålborg, Denmark, October 2016.

# Index

## Cyber-Physical Systems Series

*Calin Belta, Editor*

*Model Checking,* second edition, Edmund M Clarke Jr., Orna Grumberg, Daniel Kroening, Doron Peled, and Helmut Veith, 2018

*Verifying Cyber-Physical Systems: Path to Safe Autonomy,* Sayan Mitra, 2021

*Information-Driven Planning and Control,* Silvia Ferrari and Thomas A. Wettergren, 2021

*Formal Methods for Multi-Agent Feedback Control Systems,* Lars Lindemann and Dimos V. Dimarogonas, 2025